Techniques and Principles in Three–Dimensional Imaging:

An Introductory Approach

Martin Richardson
De Montfort University, UK

T0320583

A volume in the Advances in Multimedia
and Interactive Technologies (AMIT) Book
Series

Information Science
REFERENCE
An Imprint of IGI Global

Managing Director:	Lindsay Johnston
Production Editor:	Jennifer Yoder
Development Editor:	Vince D'Imperio
Acquisitions Editor:	Kayla Wolfe
Typesetter:	John Crodian
Cover Design:	Jason Mull

Published in the United States of America by
Information Science Reference (an imprint of IGI Global)
701 E. Chocolate Avenue
Hershey PA 17033
Tel: 717-533-8845
Fax: 717-533-8661
E-mail: cust@igi-global.com
Web site: http://www.igi-global.com

Library of Congress Cataloging-in-Publication Data

Techniques and principles in three-dimensional imaging : an introductory approach / [compiled] by Martin Richardson.
    pages cm
  Includes bibliographical references and index.
    ISBN 978-1-4666-4932-3 (hardcover) -- ISBN 978-1-4666-4934-7 (print & perpetual access) -- ISBN 978-1-4666-4933-0 (ebook) 1. Three-dimensional imaging. I. Richardson, Martin, 1958- editor of compilation.
  TA1560.T43 2014
  006.6'93--dc23
                              2013032898

This book is published in the IGI Global book series Advances in Multimedia and Interactive Technologies (AMIT) (ISSN: 2327-929X; eISSN: 2327-9303)

Advances in Multimedia and Interactive Technologies (AMIT) Book Series

ISSN: 2327-929X
EISSN: 2327-9303

MISSION

Traditional forms of media communications are continuously being challenged. The emergence of user-friendly web-based applications such as social media and Web 2.0 has expanded into everyday society, providing an interactive structure to media content such as images, audio, video, and text.

The **Advances in Multimedia and Interactive Technologies (AMIT) Book Series** investigates the relationship between multimedia technology and the usability of web applications. This series aims to highlight evolving research on interactive communication systems, tools, applications, and techniques to provide researchers, practitioners, and students of information technology, communication science, media studies, and many more with a comprehensive examination of these multimedia technology trends.

COVERAGE

- Audio Signals
- Digital Games
- Digital Technology
- Digital Watermarking
- Gaming Media
- Internet Technologies
- Mobile Learning
- Multimedia Services
- Social Networking
- Web Technologies

IGI Global is currently accepting manuscripts for publication within this series. To submit a proposal for a volume in this series, please contact our Acquisition Editors at Acquisitions@igi-global.com or visit: http://www.igi-global.com/publish/.

Titles in this Series

For a list of additional titles in this series, please visit: www.igi-global.com

Video Surveillance Techniques and Technologies
Vesna Zeljkovic (New York Institute of Technology, Nanjing Campus, China)
Information Science Reference • copyright 2014 • 369pp • H/C (ISBN: 9781466648968) • US $215.00 (our price)

Techniques and Principles in Three-Dimensional Imaging An Introductory Approach
Martin Richardson (De Montfort University, UK)
Information Science Reference • copyright 2014 • 300pp • H/C (ISBN: 9781466649323) • US $180.00 (our price)

Computational Solutions for Knowledge, Art, and Entertainment Information Exchange Beyond Text
Anna Ursyn (University of Northern Colorado, USA)
Information Science Reference • copyright 2014 • 511pp • H/C (ISBN: 9781466646278) • US $180.00 (our price)

Perceptions of Knowledge Visualization Explaining Concepts through Meaningful Images
Anna Ursyn (University of Northern Colorado, USA)
Information Science Reference • copyright 2014 • 418pp • H/C (ISBN: 9781466647039) • US $180.00 (our price)

Exploring Multimodal Composition and Digital Writing
Richard E. Ferdig (Research Center for Educational Technology - Kent State University, USA) and Kristine E. Pytash (Kent State University, USA)
Information Science Reference • copyright 2014 • 352pp • H/C (ISBN: 9781466643451) • US $175.00 (our price)

Multimedia Information Hiding Technologies and Methodologies for Controlling Data
Kazuhiro Kondo (Yamagata University, Japan)
Information Science Reference • copyright 2013 • 497pp • H/C (ISBN: 9781466622173) • US $190.00 (our price)

Media in the Ubiquitous Era Ambient, Social and Gaming Media
Artur Lugmayr (Tampere University of Technology, Finland) Helja Franssila (University of Tampere, Finland) Pertti Näränen (TAMK University of Applied Sciences, Finland) Olli Sotamaa (University of Tampere, Finland) Jukka Vanhala (Tampere University of Technology, Finland) and Zhiwen Yu (Northwestern Polytechnical University, China)
Information Science Reference • copyright 2012 • 312pp • H/C (ISBN: 9781609607746) • US $195.00 (our price)

Multimedia Services and Streaming for Mobile Devices Challenges and Innovations
Alvaro Suarez Sarmiento (Universidad de las Palmas de Gran Canaria, Spain) and Elsa Macias Lopez (Universidad de las Palmas de Gran Canaria, Spain)
Information Science Reference • copyright 2012 • 350pp • H/C (ISBN: 9781613501443) • US $190.00 (our price)

www.igi-global.com

701 E. Chocolate Ave., Hershey, PA 17033
Order online at www.igi-global.com or call 717-533-8845 x100
To place a standing order for titles released in this series, contact: cust@igi-global.com
Mon-Fri 8:00 am - 5:00 pm (est) or fax 24 hours a day 717-533-8661

Table of Contents

Detailed Table of Contents

Roger Taylor, De Montfort University, UK

The birth of three-dimensional photography was highly controversial with much heated debate and rivalry from its proponents who competed to be first. This chapter highlights the historical background of 3-D by quoting The Great Exhibition of 1850 and the birth of commercial stereoscopic photography. For the first time audiences were introduced to Brewster's stereoscope, allied to Daguerre's photographic images and so successful was public reaction that by 1852 the system had become a commercial triumph with much to be gained. During the following decade the new steam railway network rapidly engulfed much of the British Isles and Europe, making the distribution of mass replicated stereo-views within easy reach for many. Indeed entrepreneurs, such as John Nottage, commissioned sets of stereo-views and built a catalogue of more than 10,000 stereo image pairs that are still highly sought after in todays auction houses. Public interest peaked around 1870 and thereafter began to decline, due in part to the stereo photograph, or pair, becoming ubiquitous. Roger Taylor offers a glimpse into that period of history when enthusiasm for the stereo photograph has never been surpassed.

Graham Saxby, 3 Honor Avenue, Wolverhampton, UK
John Emmett, Broadcast Project Research, UK

In this chapter, the authors discuss models for the behaviour of light and explain the modern units of light measurement and the types of lighting used in photography. The theoretical models of Huygens, Abbe, Young, Maxwell, and Fresnel are outlined, emphasising the effects of diffraction and polarisation. They describe the structure and physiology of the human eye and the stereoscopic principle. The development of 3-D cinema and television is discussed with a summary of viewing parameters.

Chapter 3

Geoff Ogram, Stereoscopic Society, UK

The relation between visual perception and the recorded image is discussed in this chapter, emphasising the historical growth of the understanding of depth perception and its visual cues. The stereoscopic principle is explained in detail, and figures are given for comfortable viewing of stereoscopic images.

Chapter 4

Geoff Ogram, Stereoscopic Society, UK

The origins of stereoscopic imagery are discussed briefly, and a practical method of producing stereoscopic pairs of images with a single-lens camera is explained in this chapter. The history of stereoscopic cameras is summarised, and the models and formats listed. The various formats for 35 mm formats are discussed and the fundamental geometry reviewed. Instructions for the correct mounting of stereo pairs of images are given. Equipment for digital stereoscopic photography is discussed, with descriptions of available models and their control devices and associated software listed for each model. Specifically designed programs for stereo image processing are available.

Chapter 5

Philip Surman, De Montfort University, UK

This chapter covers the work carried out on head tracked 3-D displays in the past ten years that has been funded by the European Union. These displays are glasses-free (auto-stereoscopic) and serve several viewers who are able to move freely over a large viewing region. The amount of information that is displayed is kept to a minimum with the use of head position tracking, which allows images to be placed in the viewing field only where the viewers are situated so that redundant information is not directed to unused viewing regions. In order to put the work into perspective, a historical background and a brief description of other display types are given first.

Chapter 6

Hans I. Bjelkhagen, Centre for Modern Optics, UK

In 1891 the optical physicist Gabriel Lippmann developed a method of reproducing colour in photography without dyes, instead using pure light from the solar spectrum. Later study took his interest into the research of three-dimensional imaging via a method of integral photography in which a fly's eye lens array is used to record images in complete three-dimensional fidelity. Other noteworthy workers in the field such as Ives, Burckhart & Doherty, Bonnet and Montabello followed up the principle, but today Lippmann is acknowledged as being a founding father of the micro-lens technique for three-dimensional imaging. Advances in micro-lens production has led to the easy availability of lenticular print and consumer electronic companies are eager to develop 3-D TV system that incorporates much Lippmann theory. This chapter offers a brief history of Gabriel Lippmann and his subsequent legacy.

Chapter 7

The discovery of diffraction and interference led eventually to the holographic principle, the recording and reconstruction of the shape of a wavefront. Transmission and reflection holograms are detailed in this chapter, along with the principles of rainbow holograms and holographic stereograms and their applications. Digital holography is described in the form of multiplexed images. The psychological and philosophical implications of the holographic image are discussed with some examples from the field of creative art.

Chapter 8

In this chapter, the authors examine the concept of 3-D displays with sampled spatial spectra in discrete hoxels, reducing the redundancy present in digital holography, which prints each individual hogel. The underlying theory deals with the principles of 3-D imaging using the quantum model and 4-D (x,y,z,t) Fourier transforms. The image information is treated as a single spatial spectrum and its 3-D information recovered by digital vector treatment through hoxels. This enable the achievement of real-time communication in 3-D.

This chapter explores the cultural contexts in which three-dimensional imaging has been developed, disseminated, and employed. It surveys the diverse technologies and intellectual domains that have contributed to spatial imaging and argues that it is an important example of an interdisciplinary field. Over the past century-and-a-half, specialists from distinct fields have devised explanations and systems for the experience of 3-D imagery. Successive audiences have found these visual experiences compelling, adapting quickly to new technical possibilities and seeking new ones. These complementary interests, and their distinct perspectives, have co-evolved in lock-step. A driver for this evolution is visual culture, which has grown to value and demand the spectacular. As a result, professional and popular engagements with 3-D have had periods of both popularity and indifference, and cultural consensus has proven to be ephemeral.

Chapter 10

Martin Richardson, De Montfort University, UK

This final chapter focuses on the potential of three-dimensional imaging. In particular the medium's ability to record three-dimensional objects, as with the holograms made of John Harrison's famous fourth timekeeper "H4" for the Royal Observatory, National Maritime Museum in Greenwich, London, and the strange case of Professor Günter von Hagen and his "BODY WORLDS: The Anatomical Exhibition of Real Human Bodies," who seriously explored its potential but relinquished its further exploration due to negative public opinion of his exhibitions at that time. Holographic stereograms are also discussed, in particular their ability to capture animation, as detailed in Case Study Three: Holograms of David Bowie. The text also explores some future applications of wavefront reconstruction.

Foreword

In 1981, my school friend Ian Frankland brought a curiously thick book for us to see. It was called *Der Kampf im Westen*. Inside the covers were compartments that held 100 black-and-white photographs of German soldiers from World War II. I flipped through these pictures with the dismissive eye common to any 13-year-old looking at old stuff. The only thing I wondered about was why the pictures were printed in pairs. My friend handed me a folding metal picture holder that came with the book, and showed me how to use it. I had never seen a stereoscope before and was not prepared for the transportation I experienced when I saw the pair of images stereoscopically for the first time. I was no longer looking at "old stuff," but was faced with a spatial reality of living people, like myself (from another time and place), but now as if I was one of them standing right there in the trenches.

As a visual artist, this experience stayed with me even though I did not fully understand the process that made the pictures. Ten years later, I found myself in a holography class run by Professor Martin Richardson at the Royal College of Art in London. As he explained stereoscopy, the German soldier pictures from *Der Kampf im Westen* jumped from my memory. I finally understood what I had seen, and even more exciting was that now I understood it, I could make my very own pictures in three-dimensional pictures. Thank you, Martin.

During my time at the Royal College of Art, I made several holograms. A favourite one depicts the pouring of 1000 needles over my motorcycle helmet clad head. At the same time, I was also experimenting with stereographic transparencies using a custom-made stereo camera rig that I had cobbled together. I graduated from the Royal College with a Master's degree in furniture design, but it was that one day in the holographic laboratory that turned my interest in stereoscopic imaging into a lifelong vocation.

Somehow, through luck and my homegrown imaging skills, three-dimensional imaging became centre stage in my life during 2005 when I worked on the 3-D version of Disney's animated classic, *Chicken Little*. In fact, this was the first digitally projected stereoscopic movie, and it opened the doors for the flood of stereoscopic imaging that surrounds us today; the rest is history.

I hope that Martin Richardson's book inspires you to start making great 3-D because it definitely inspires me!

Phil Captain 3-D McNally
Disney Digital 3-D, USA

Phil Captain 3-D McNally *graduated from the Royal College of Art with an MA, RCA, and later moved to Los Angeles in the United States to join Industrial Light and Magic. In true Hollywood style, he legally changed his name to Captain 3-D and went on to work with Steven Spielberg's company DreamWorks. He is currently Senior Stereographic Supervisor for Disney Digital 3-D.*

Preface

OVERVIEW

Today's world is littered by complex visual information. Two-dimensional images fill every instant of our lives from portable communications and text messaging to marketing displays interlaced with critical news items concerning our changing world. Our culture thrives on this proliferation of data and society depends on it. However, what if this two-dimensional fog obscures other areas that require inspection. For example, when medical diagnosis necessitates accurate tissue positioning inside our three-dimensional bodies if essential surgery be required, or air traffic control, where staff requires better understanding of the three-dimensional airspace in flight path logistics. These critical problems may be overcome by the introduction of 3-D parallax and stereoscopic imaging. The ability to visualize three-dimensional data is becoming increasingly important for engineers and scientists seeking insight into structural forms and improved design, but just as important for communicating the results of such studies to those with a non-technical background.

Three-dimensional imaging contributes to our understanding of computational dynamics, data visualization, and the production of hyper-realistic 3-D displays. It is a developing science and a remarkable concept. Communication with light—pure information—is our ultimate means of communiqué and may be found with the development of real-time digital 3-D. It can be a tool to visualize complex engineering data in 3-D using digital holographic technology or numerical data that depict complex features of airflow around objects such as cars or buildings combined with rendered photo-realistic digital models captured three-dimensionally, perhaps as a digital hologram.

For the last few years, I have taught the principles of three-dimensional imaging and holography to successive students both at the undergraduate and postgraduate levels in the Imaging and Displays Research Group at De Montfort University, Leicester. While preparing modules, course materials, and lectures, I became impressed by the abundance of recent research on specific aspects of 3-D and its applications. I also noticed the lack of a concise, in depth, handbook that could integrate the results of this research. A broad scope was needed to consider the large amount of journals and literature flooding the market at a time when 3-D was experiencing resurgence with televisions stations such as SKY-3D and BBC-3D bringing all to the masses. It was as if someone had opened a Pandora's box where every week something new could be found on a newsstand that was relevant to the great 3-D bonanza. In detailing much of the most recent information it became clear that wavefront reconstruction obtained from finer structures hitherto not been research in any great depth. This presents a number of unique and fundamental research questions regarding optical science and 3-D. If true 3-D is to reflect our world as we see it then a means of wavelength-selective inspection system will need to be developed that measures the angular performance of diffracted light. In holography, Optical Variable Devices (OVDs) are commonly employed to authenticate identity documents such as passports, ID cards, credit cards, travel cards, etc., or to protect valuable products/brands such as bank notes, software discs, and designer. An OVD is an iridescent image that exhibits various optical effects such as movement or colour changes. They cannot be photocopied or scanned, nor can they be accurately replicated or reproduced. OVDs also find use in tamper-evident seals e.g. on pharmaceutical products or dutiable goods. These products use state of the art security devices because they are key targets for criminal targeting. Widely employed optical security devices include watermarking, intaglio printing, micro-printing, security threads, magnetic inks, florescent dyes, and embossed holograms. The most secure OVD incorporates a hologram because holographic images have the ability to create an easily recognizable three-dimensional image, which appears to change or move depending on the viewing angle. Whilst holographic OVDs currently define the state of the art, the ready availability of low-cost holographic embossing technology and the level of scrutiny required by these devices increasingly compromise the current generation of holographic security devices. The security industry has been concerned for some years by counterfeit holograms, in particular on the € 500 bank note (although there has also been pressure to remove this denomination due to its ease of use in money laundering).

Medical applications for 3-D include three-dimensional scans from MIR and computational diagnoses of early brain traumas for example. More advanced research suggests that patient interaction with three-dimensional imaging could assist with learning difficulties. The system will incorporate speech analysis software with 3-D to model the correct articulation as a visual template for the patient to match and will also provide the accompanying speech sound. The patient must then produce the speech sound based on both their interpretation of the 3-D visual image of the tongue and mouth contours and the speech sound signal. By superimposing a three-dimensional electronic image of the patient's articulation on a hologram, for example, the patient will see exactly where they need to make adjustments by how closely their articulation visually and auditorily matches the template. The system will also provide a matching auditory speech sound together with a voice sign-wave display to aid the targeted articulation. The system will thus be able to assess individual attempts and save a profile of progress. The patient will benefit from having tangible evidence of success, and the speech and language therapist can use the profiles to assess and measure outcomes and to manage therapy. This augments and reinforces existing therapy techniques and provides real-time information in an accessible non-intrusive form. By stimulating the speech motor programme, it facilitates the transition to perceptually normal speech. The administration and implementation of a regular non-intrusive and simple instrumental analysis advances current technology. The portable device has two fundamental applications. Firstly, it is intended as a tool to help the patient by providing detailed visual and auditory information to facilitate guided practice in private. Secondly, the speech and Language Therapist can use it in the clinical management of patients with MSDs and as an audit tool, identifying outcomes measuring the efficacy and progress of treatment through evidence-based practice. The object of this proposal, which is clinically endorsed, is to investigate the potential of developing a visual and auditory biofeedback system for use in the therapy and clinical management of patients with MSDs. This objective, if met, will be of enormous benefit to both patients and healthcare professionals working within this disparate population by improving clinical audit, therapy techniques, and facilitating perceptually normal speech through self monitoring.

This book declares a new milieu in the field of 3-D, suggesting new areas the student may wish to follow. The research herein is cutting-edge and cross disciplinary, involving an unusual combination of physics, chemistry, graphics, and control electronics expertise, all of which have been brought together specifically for this project. The book employs standard mechanisms for the wider dissemination of scientific outputs. Its authors join the long list of eminent optical scientists that are members of peer-reviewed journals such as *Optics Express, Journal of the Optical Society of America A*, and international publications such as *Reconnaissance* and *Product and Security Magazine*. In addition, each of the authors is methodically published in the proceedings of refereed international and UK conferences.

This book is intended to provide the minimum necessary information about making good 3-D with comprehensive references and further reading suggestions seen as vital for further examination. *Techniques and Principles in Three-Dimensional Imaging: An Introductory Approach* is a lyceum full of rich information about three-dimensional imaging, a technological tool currently reflected in modern culture through its captivating visual effects popular within the entertainment industry, especially with cinema going audience of today. New spectators, in fresh settings, repeatedly make 3-D its own and invest new meanings into it. This publication introduces the reader to the various techniques available for three-dimensional image capture including the use of holography, lenticular technology, and stereo photography. Specifically, it examines digital advances providing the most up to date research available set out in an easily accessible style featuring a unique exploration of three-dimensional technology through an understanding based on first-hand experience of advocates of the field. It also counsels an engaging historical study that positions the beginnings of stereo photography, with its role in popular culture and relevance to contemporary understanding of three-dimensional imaging, with a vibrancy and passion for technological progress like no other.

TARGET AUDIENCE

Techniques and Principles in Three-Dimensional Imaging: An Introductory Approach consists of ten chapters written by the foremost authorities on the subject of 3-D image capture. Together, they provide the student with information required for grasping the principles and practice of 3-D imaging and are aimed at two broad varieties of reader: first, technicians concerned with the image capture of three-dimensional space, and second, an extensive spectrum of intelligent popular reader interested in the wider implications of three-dimensional technology that defines the character of the twentieth century. Its intended readership will include those concerned with communications and visual studies, fine art and museology, and practitioners of STEM subjects (Science, Technology, Engineering, and Mathematics) such as scientists and students of physics and optical engineering.

THE CHAPTERS

Gathering of material for this volume began in 2011, when I was fortunate to read Geoff Ograms *Magical Images*. This was a self published manual written for members of the Stereoscopic Society who hold regular meetings in London. This volume has been carefully selected from a litany of previously unpublished works, and I would especially like to thank Graham Saxby, Sean Johnston, and Geoff Ogram, without whose help this publication would not be as high quality as it is. Its progress burned through three editors before reaching the publisher. Much of the book's structure derives from hard-learnt lessions from the industry where perfect 3-D is a constant demand.

Chapter one has been written by Roger Taylor, Emeritus Professor of Historical Photography. "The Optical Wonder of the Age" reflects on a time when jarring controversy took hold of early stereographic photography in a battle between Dr. Charles Wheatstone and Sir David Brewster, charting beginnings of the stereo photograph from the 1850 to its mass replication and its decline toward the end of the 18th century.

Dr. John Emmett and Graham Saxby have jointly written chapter two, "Models for the Behaviour of Light." Together, they explore light as energy in a waveform and unfold the theory of light miraculously, involving such great minds as Isaac Newton, Thomas Young, Christian Huygens, Max Planck, Neil's Bohr, and of course, Albert Einstein. The dual characteristic of light remains one of the many puzzles of nature. The particle wave problem to which they refer was clarified somewhat in the year 1900 when Max Planck proposed that all electromagnetic energy is radiated in discrete packages which he called quanta, or singular quantum. Einstein later confirmed Plank's theory via the photoelectric effect and used the word photon to refer to these energy packages. Scientists today refer to light sometimes as particles (photons or quanta) and other times as continuous waves depending on the situation or experiment. The problem is not with nature but with our models or concepts of nature. Together, Saxby and Emmett explore a number of models used to understand the physics of light in an easy-to-follow development designed for undergraduate students.

Dr. Geoff Orgam provides quintessential chapters three and four, "Principles of Binocular Stereoscopic Imaging" and "Analogue and Digital Stereo-Photography," to provide essential reading and establish a technical cornerstone through firsthand experience. His practical approach is aimed at both amateur and professional photographers, suggesting the complicated problem of capturing our three-dimensional world through stereo photography. Two-dimensional photography provides graphic evidence of objects from our surrounding world in a far-from-perfect way, but the information covered by Ogram, and his understanding of stereo imaging, elevates simple photography beyond all expectations.

Chapter five, "Head Tracked Auto-Stereoscopic Displays," focuses on three-dimensional television research – the Holy Grail! Dr. Philip Surman provides a clear overview of current three-dimensional displays for televised interaction and the development of integral video. The pre-requisite of "multi-viewer" capability is an important one, and Surman suggests that the 3-D effect generated must be consistent, regardless of the viewers' seating or head position. These viewing constraints are met by "spectacle-based" systems employing either polarising, anaglyph, or active shuttering glasses. However, Surman takes this further as his accredited research involves non-eyewear auto-stereoscopic television. He addresses the realization of 3-D television systems and tackles the major problems associated with integrated electronic 3-D.

Professor Hans Bjelkhagen returns to the historical theme by contributing chapter six, "Intergral 3-D Imaging Techniques." In it, he discusses to a major figure in the chronicle of three-dimensional imaging in homage to Nobel Prize-winner Gabriel Lippmann and his experiments with integral 3-D imaging. Bjelkhagen traces his development and provides a number of up-to-date advances that are attributed to Lipmann in 1908. For example, his original proposal for integral photography comprised a "fly's-eye" lens sheet and photographic plate. Interestingly, the first practical realization of integral photography actually used a pinhole sheet due to difficulties in producing a lens. Integral photography preserves both horizontal and vertical parallax. It was argued that since our eyes are horizontally disposed, it should be possible to dispense with the vertical parallax information; this led to the development of lenticular photography and its popular use today.

In chapter seven, "Principles of Holography: Wavefront Reconstruction and Holographic Theory," I describe how the diffraction of light can form a three-dimensional image and be re-constructed with holographic theory. I also describe a number of practical applications of this awe-inspiring imaging medium, including its special security qualities, its potential of holographic data storage, and the capture of three-dimensional pulsed holographic portraits. I argue that holography is not simply a way of providing an optical security device but far more diverse and, as the "new" kid on the block, still finding new applications for its ability to manipulate light. Significantly, it explores the technology as "Art." Those who take an interest in the science of modern holography will equally find interest in the explanation and theory of coherent light and the methods employed to capture it exploding the limitations of its forbearers and launching its new development as "hyper-media." Its ultimate conclusion, however, suggests audiences, and some who use it, aren't intrigued by technology; they are more concerned with fantasy.

Chapter eight reaches us from China, where Frank Fan et al. contribute pioneering research into a method of transmitting the holographic presence in "The Reinvention of Holography." Fan describes a system of holographic video presence for potential mass use. The escalating cost of long-distance travel and the increasing availability of high-bandwidth fiber optic transmission lines make the use of HOLO-conferencing an attractive medium for person-to-person communication. Videophone equipment available today is potentially suitable for HOLO-conferencing, so consider the benefit of conveying a rapid visual impression of complex three-dimensional objects via HOLO-conferencing. Fan offers his research in this field combining developments of Holographic Optical Elements (HOEs) and micro-projectors to offer a groundbreaking prototype.

Techniques for developing stereographic images were developed very early on in the history of photography, and stereoscopy became an immensely popular medium in Victorian popular culture. Yet stereoscopy has never enjoyed the same curatorial and academic research interest as photography. While there are extensive collections of stereoscopic images in public museums galleries and archives, they tend not to be on public display. In recent years, my research group has been actively perusing and developing a working relationship with the three major National Museums in London, namely The National Science Museum, The National Maritime Museum, and The Natural History Museum. It has been important to develop a trust with the curators of the museums, as many companies offering different 3-D methods of recording the museum artifacts and constantly approaching them. It has been a very slow and often intense process of negotiation but it seems the promise of 3-D has finally found its application – this is because museums in the UK have been instructed by the government to begin the process of returning many of its most valuable and scientifically important antiquities.

We are presented with an interesting oxymoron in that the museums that house the most scientifically relevant artifacts of stereoscopic history now using 3-D to present accurate records of collections for public engagement. In an article published in the November issue of the science magazine *Nature*, titled "Aboriginal Remains Head For Home," Jim Giles details this process as it heralds the end of museums as we have known them. Subsequently, work has now begun on the three-dimension image capture of collections using "Digital Technology" as measurement and through Professor Chris Stringer, Head of the Department of Paleontology at The Natural History Museum and globally respected scientist. Discussions are currently underway with both The Natural History Museum and The Victoria and Albert Museum regarding the future three-dimensional recordings.

Public consciousness of early stereoscopy is low, and the definitive history of the origins of stereoscopy has yet to be written. Our introductory approach asks questions, such as what effect of the early history of stereoscopy have as a medium on popular culture, or why and how did stereoscopy tap into

Victorian sensibilities so successfully? There are parallels between the history of stereoscopy in the early 19th century and digital technologies in the early 21st. In both cases, the new technology provided a window on the world that broadened horizons, understanding, and interests through the provision of high-quality images of faraway places and events. Stereoscopic images were used inter-alia to provide information and education to the masses, to support sales and marketing activities, document important events, to provide titillation, and to satisfy a growing market for artworks.

Techniques and Principles in Three-Dimensional Imaging: An Introductory Approach's chapter nine is written by Sean Johnston, Professor of Science, Technology, and Society at the University of Glasgow. Johnston explores the changing stance of popular cultural and its engagement with 3-D. For many, three-dimensional imaging is little more than an obscure research interest whose fruits have been of dubious practical or commercial value when considered historically. However, with recent technological advances, and the demands of more sophisticated methods of interaction, contemporary culture has engendered an unprecedented interest in 3-D imagery. The chapter identifies the cultural benefits of 3-D and discusses an optimistic future. The technology progressed rapidly from simple monochrome to colour and animation. Similarly, the Internet has progressed rapidly from static, monochromatic text to full-colour, audio, video, and animation. Current research in holography and lenticular imaging points the way to three- and four-dimensional imaging on the Web in the foreseeable future. Like the Internet, stereoscopy enjoyed massive popular market success. In the 1850s, the marketing slogan of the London stereoscopic company was "no home should be without one."

My final chapter focuses on the potential of three-dimensional imaging. In particular its ability to record three-dimensional objects, as with the holograms made of John Harrison's famous fourth timekeeper "H4" for the Royal Observatory, National Maritime Museum in Greenwich, London, and the strange case of Professor Günter von Hagen and his "BODY WORLDS: The Anatomical Exhibition of Real Human Bodies," who seriously explored its potential but relinquished its further exploration due to negative public opinion of his exhibitions at that time. Holographic stereograms are also discussed, in particular their ability to capture animation, as was the case with David Bowie, and their potential for capturing landscape imagery. The text also explores some future applications of holographic imaging.

CONCLUSION

The rich knowledge included in this publication will eventually filter into design practice where timeliness of response, flexibility, and access to information and fully rendered 3-D on demand become a reality. Advances in display and image capture using electro-optic technology will facilitate this. These characteristics imply a more geographically distributed and collaborative kind of information dissemination than is commonplace today as a complex network of digital 3-D CAD designers primed for holographic exploitation. It is a startling concept and a radical departure from traditional representational imagery required in the 21st century. To identify these research issues and clarify the exchange of collaborative of ideas remains a primary academic focus and will take place within the current decade of digital convergence. The increasing speed at which technology is advancing has resulted in the appearance of new cameras such as the Lytro TM, which has the ability to capture the entire Z-axis depth of field and allows the user to change what is in focus, post capture, and thus move through its space using light-field capture of its Plenoptic lens. There are many interesting parallels in history, partially in the development of painting where artists such as Albrecht Durer made paintings in accordance with the invention of perspective to see whether an object is situated in the foreground or in the background, and to locate the point where all the lines describing the depth of field converge to a vanishing point. Still, the invention of perspective is a false one, a man made two-dimensional calculation that only served to constrain the observer's view rather than enlighten it.

My hope is that this book demonstrates that three-dimensional imaging is more than a recording medium. It is a tool that can substantiate both fantasy and fact, commenting subtly on the real world, blurring our grasp of that reality to the point where objectivity is closer to nature. Those who take an interest in the science of three-dimensional imaging will find interest in the explanation and theory of light and the methods employed by those clever enough to capture its ability to record "real" space via digital media. As 3-D continues its speedy march into the future, discovering new human needs and new potentials for old ideas, you might by now share a sense that its further integration into every aspect of life is not just a technological speculation; it is a 21st century inevitability.

Martin Richardson
De Montfort University, UK
January 2014

Introduction

In a universe with (so far) three spatial dimensions, it is perhaps surprising that we treat two-dimensional images in films, photographs, and paintings as realistic. Most animals do not appear to do so: a rabbit that gets excited over the sight of a cabbage will ignore a painted one and the tales of bees and birds flocking to the painted flowers and bunches of grapes in seventeenth century Dutch still-lifes are probably apocryphal. However, when stereoscopic photography burst into the consciousness of the Victorian middle classes, they could not get enough of it, though interest faded in time, as Roger Taylor points out. Since then, though, there have been a number of resurrections of the technology, usually coinciding with some novel presentation technique, and today we are in the midst of yet another upsurge.

What is the reason for these rises and falls in popularity? It may be in part the fickleness of the general public, embracing one craze until another comes along; it may also be that so many inexperienced practitioners join in, much of their output exceeding the bounds of physiological tolerance; as Geoff Ogram explains: there are limitations built into the geometry of stereo pairs, and these are easy to exceed. However, there are also factors in the technology itself. One that may not be obvious at first is that focusing the eyes on a real object involves two separate forms of kinaesthetic feedback, one from the amount of eye convergence, the other from accommodation. In stereoscopic viewing, the two may be in conflict, as convergence is controlled by the nature of the images, while focusing remains fixed on the plane of the photographs – a point also discussed by Ogram. In prolonged viewing sessions such as full-length films, this conflicting feedback effect can result in anything between unease and actual nausea, as pointed out by Jeff Hecht in a recent *New Scientist* article ('The Barfogenic Zone', 18 Dec 2010). Another factor limiting the realism of binocular stereo images is that the perspective does not change when the viewing position changes; there is no dynamic parallax. This means the loss of one of the most important factors in depth perception in viewing real objects and landscapes. In this respect, a projected stereo image is no better than a mono-image. This is equally true whatever the means of viewing. Less important in general, though critical for a large minority of viewers, is the fact that a sizeable part of the population (some estimates suggest as many as one in five) has poorly developed stereopsis (binocular vision), and obtains little added depth information from viewing a stereo pair. About one person in 20 has no stereopsis at all, usually the result of an uncorrected muscular weakness in one or other eye at the critical time of development of binocular vision (typically 3 to 5 years). Of course this does not prevent these people from perceiving depth: there are many other clues: after all, there are plenty of people with only one functioning eye who can still play a decent game of tennis.

So what are the other clues to depth? Well, there is the effect of light and shade, which painters have always used to give the effect of solidity; there is the clue that nearer objects appear larger in proportion to farther ones and may partially obscure them; and there is the change in contrast (and hue) in very distant objects such as mountains. However, the most important is one that is most noticeably absent from both single images and stereo pairs: dynamic parallax. If you move your head, objects that are close to you appear to move in the opposite direction to objects that are farther away. The ability to perceive this effect is so powerful that it works even if your head is still and only your eyes move. It is present when you look at a sculpture or a multi-image stereogram, although the originals are stationary, and it is paramount when you view a movie. The early cinema often involved scenes of onrushing trains and the like, and although there was no depth in the image, the audience still ducked. Today's more sophisticated audiences are less moved, but when stereo is added, the realism still startles, because now all the clues to perception are involved, including the dynamic parallax generated by the relative movement of the image components.

It would be foolish to suggest that binocular stereoscopy is certain to return to limbo as it has done before; for now, we have exciting computer games viewed by advanced binocular optical techniques, with images under the control of the viewer; the added realism of stereo is sufficient to ensure that before long every computer game will have it. New techniques of projection based on the barrier principle, such as lenticular techniques, and head tracking systems for individual viewers, seem set to reduce if not to eliminate the necessity for image isolating eyewear. I would suggest, though, that the future of three-dimensional imaging lies more in multi-viewpoint stereography and the full-parallax image that can be obtained from the holographic approach, even though the latter requires a good deal of further development. Another promising approach, discussed by Hans Bjelkhagen, involves a process known as integral photography, the principle of which was first outlined by none other than Gabriel Lippmann in the early 1900s as a method of obtaining fully three-dimensional images by pure photography; but techniques and materials have only recently advanced to the point where it can be realised.

There are still a great many places where research is continuing on both improved binocular imaging and full parallax, and this publication cannot claim to have followed every possible avenue of approach to the subject. Many are under wraps for commercial reasons; many, no doubt, though theoretically valid, are destined to remain in obscurity at least until some viable method of production is realised. However, some of these things are undoubtedly going to happen, and thanks to modern publishing methods, this book can be updated as they do.

Graham Saxby
3 Honor Avenue, Wolverhampton, UK
June 2012

Chapter 1
The Optical Wonder of the Age

Roger Taylor
De Montfort University, UK

ABSTRACT

The birth of three-dimensional photography was highly controversial with much heated debate and rivalry from its proponents who competed to be first. This chapter highlights the historical background of 3-D by quoting The Great Exhibition of 1850 and the birth of commercial stereoscopic photography. For the first time audiences were introduced to Brewster's stereoscope, allied to Daguerre's photographic images and so successful was public reaction that by 1852 the system had become a commercial triumph with much to be gained. During the following decade the new steam railway network rapidly engulfed much of the British Isles and Europe, making the distribution of mass replicated stereo-views within easy reach for many. Indeed entrepreneurs, such as John Nottage, commissioned sets of stereo-views and built a catalogue of more than 10,000 stereo image pairs that are still highly sought after in todays auction houses. Public interest peaked around 1870 and thereafter began to decline, due in part to the stereo photograph, or pair, becoming ubiquitous. Roger Taylor offers a glimpse into that period of history when enthusiasm for the stereo photograph has never been surpassed.

THE DAWN OF STEREOSCOPY

After weeks of waiting, months of building, and years of detailed planning, the morning of May Day 1851 dawned bright and clear – a perfect day for the state opening of the Great Exhibition in Hyde Park by Queen Victoria and Prince Albert. From the earliest hours it seemed that the whole of London was on the move, with countless carriages, cabs, gigs and landaus 'pouring in from all parts of the metropolis and the surrounding districts... whole masses of pedestrians marched in mighty phalanx towards the scene of action', some having travelled great distances to be present at this historic moment (*Ill. Lond.*

DOI: 10.4018/978-1-4666-4932-3.ch001

News, 1851). Huge crowds gathered in the park, and in the surrounding streets every window was filled with onlookers hopeful of catching a glimpse of the royal party. With the pervading air of excitement and anticipation it was if the entire nation was aware of the day's significance.

The idea of holding an international exhibition in London had been suggested by Prince Albert during the summer of 1849 and, during the months that followed, his initial proposal gradually took shape, growing both in scale and significance. In January 1850 the scheme was formally assigned to a Royal Commission under the presidency of the Prince himself; this took charge of every aspect of its management, from invitations for submissions to the licensing of refreshments and plans for the restoration of lost property. Invitations to submit exhibits were sent to all parts of the world, and to house everything the Commission planned a building of an unprecedented scale, to cover more than twenty acres.

Joseph Paxton's radical design for a structure made entirely of glass and steel to impart lightness and elegance helped to capture the imagination of the public, who promptly named the building the 'Crystal Palace'. The *Times* pronounced it 'the largest building ever made by human hands without mortar, brick, or stone', and marvelled at the speed of its prefabricated construction (*Times,* 1850).

From the outset the Royal Commissioners imposed a classification system that provided a clear rationale and a coherent structure to the organisation of the exhibits throughout the exhibition. Thirty classes of object were grouped under four broad headings, namely *raw materials, machinery, manufactures* and *fine arts,* which it was believed would allow the diligent observer to appreciate the link between the raw materials and the

finished article. The early decision to make the exhibition competitive proved influential, as this would be an important incentive for manufacturers to enter exhibits of the highest quality. Juries made up of international specialists representing each of the classes were appointed. With more than 7000 exhibitors and a huge number of individual exhibits, the task facing them was Herculean. Starting in May, they worked diligently until the close of the exhibition in October, when they announced more than 5000 awards, ranging from the Council Medal awarded for work of outstanding excellence, to Honourable Mention for those less deserving (Clowes *et al,* 1863). To commemorate the exhibition, every contributor was given a bronze medal bearing a portrait of Prince Albert.

As the exhibition building was so vast, and the exhibits it housed so numerous, there was too much for the visitor to see in a day. Indeed, it took Queen Victoria herself thirty-five visits to work her way through everything. For most people this leisurely approach was impracticable. Instead, visitors with only a day to spare needed to navigate their way through the exhibition using one of the published guides, which drew attention to the most significant objects on chosen routes through the building. Even then it was no easy task, as there was a bewildering amount of detail to absorb, and even the most dedicated visitor could do little more than skim the surface. A wiser option was to pay attention to objects of personal interest, whether the glittering attraction of the Koh-I-Noor diamond or something more practical such as Ferrabee's Patent Grass Cutter for Mowing Lawns, or Culverwell's Portable Domestic Vapour Bath.

Although most of the exhibition was devoted to championing the cause of Britain's manufacturing industries, a significant section was dedicated to the ingenuity and skill of the nation's scientific instrument makers, whose displays were said to be worth £64 000 [about £5M today], making them the third most valuable group on display (*1st Rep.Comm*, 1852). It was here among instruments used in astronomy, surveying, weighing and measuring, telegraphy, chemistry and meteorology that photography and stereoscopy could be found on modest display under the heading of *Class 10, Philosophical Instruments, and Processes Depending Upon Their Use*. The jury for this class was made up of fifteen leading figures from Britain, mainland Europe and the United States, under the chairmanship of Sir David Brewster, whose improvements in perfecting a device for viewing stereoscopic images is central to this essay.

The possibility of recreating stereoscopic vision by means of a viewer was already well known by 1851, having been demonstrated by Charles Wheatstone at a meeting of the British Association for the Advancement of Science in 1838. His thesis was based on the principle that a pair of drawings of a single object made from slightly different (left and right) viewpoints would by the mechanism of visual perception be fused into a single three-dimensional image when seen through a specially designed optical viewer (for which he had coined the term *stereoscope*). Despite its being large and cumbersome, and owing more to science than to domestic entertainment, Wheatstone's viewer and the optical theory underpinning it sparked a lively debate about the principles of binocular vision for much of the 1840s.

One of Wheatstone's chief critics was Brewster himself. His numerous scientific papers on the nature of light, optics, and the physiology of human vision had placed him at the forefront of his field. During the course of these investigations he had decided that Wheatstone's viewer would be 'of little service, and ill-fitted, not only for popular use, but for the application of the instrument to various useful purposes.' (Brewster, 1866,1971) After experimenting with various designs he emerged with a simpler viewer easily held in the hand, which he called a *lenticular stereoscope*. Disheartened by its indifferent reception in Britain, Brewster offered his design to the eminent Parisian instrument maker Jules Duboscq in 1850. In the words of one commentator, Duboscq created a viewer of such beauty that it attracted 'a spontaneous and unanimous cry of admiration' from all who saw it (Moigno, 1856).

Brewster's timing could not have been better, as the following year Duboscq included his stereoscope among his displays in the French Courts of the Great Exhibition, where the exceptional quality of his scientific instruments won high praise from the jury – who rewarded him with a Council Medal (Clowes, 1852). According to Brewster's account of events, at the Great Exhibition his viewer caught 'the particular attention of the Queen, and before the closing of the Crystal Palace M. Duboscq executed a beautiful stereoscope, which I presented to Her Majesty in his name. Brewster's tale of royal approval has been widely quoted as the principal reason for stereoscopy's finding such immediate and universal favour in the years immediately following the Great Exhibition. Queen Victoria herself, however. makes no mention of seeing the stereoscope during her visits to the French displays, despite reporting in some detail on the embroidery, clocks and candelabra. She does refer to photographs, but only in passing. The gift of the stereoscope goes unrecorded in the Royal Archives. (For Queen Victoria's exhibition journal see C.R.

Fay, Palace of Industry, Cambridge.) But the history of how stereoscopy transformed itself from the realms of science into one of the most notable marketing successes of the mid-nineteenth century is far more complicated than Brewster suggests. To understand what made the stereoscope such a success we need to return to January 1839, and trace its evolution up to 1851.

The idea of recording images made by light within a camera obscura had been around since Renaissance times. It was also understood within scientific circles that light itself was a source of energy capable of physically altering the nature of substances, sometimes causing them to lighten and sometimes to darken. But attempts at permanently fixing such changes had always failed. However, during the 1830s two men, Louis Jacques Mandé Daguerre in Paris, and William Henry Fox Talbot in Lacock, independently overcame this difficulty, and announced their separate discoveries in January 1839 (Schaaf, 1992).

A daguerreotype recorded the optical image on the surface of a highly polished silvered metal plate made sensitive to light by fumes of iodine. It offered sharp highly-detailed grainless images. However, the process had two major drawbacks. As the image was on a highly reflective surface it was difficult to view, and as each exposure created a unique image – a little like Polaroid photographs of recent years – making further copies involved a complex and troublesome procedure (Wood, 1989; *Illustrated London News*, 1852). In addition, the image was laterally reversed.

Talbot's process, which he called 'photogenic drawing', used ordinary writing paper, made sensitive to light by coating it first with a solution of common salt, then with a solution of silver nitrate. When exposed to strong light, the image slowly darkened to a deep rich brown. As this image had its light and dark tones reversed, Talbot placed this 'negative' in contact with a second sheet of sensitized paper and exposed both to light once more. The resulting positive bore an image that was tonally correct, and this process could be repeated any number of times. Talbot's negative/positive approach created the physical and conceptual framework that was to underpin photography for the next 150 years (Taylor, 2007).

Throughout the 1840s these two competing technologies attracted much research and investment in the quest to enable exposure times to be reduced to commercially practicable levels. Once this had been achieved it was the fine detail and subtle tones of the daguerreotype that found favour with the public, who flocked in great numbers to newly established photographic portrait studios, which became a feature of towns throughout Britain. With each portrait costing over a pound – 25% more if hand coloured – there were huge profits to be made, some studios accommodating dozens of sitters a week during the summer months when the light was at its strongest. The daguerreotype quickly established itself as the dominant commercial process, its success dependent upon its novelty and on the ostentatious presentation by its practitioners. It was the first photographic process to become a fashionable craze among the middle classes, whose parlours and marble mantelpieces offered ideal settings for the latest consumer goods.

Talbot's process presented a different aspect. Unlike the highly detailed image provided by the daguerreotype, prints from his paper negatives offered a quieter, contemplative view of the world. Subtle tones, warm tints, soft detail and a pleasing balance of light and shadow tones created images that resembled lithographs or mezzotints. As a result the process was adopted by a different class of photographer: folk who, like Talbot, belonged to a leisured class with enough time and disposable income to invest in this new pastime. Taking advantage of an expanding railway network, these gentlemen amateurs traversed the countryside in search of picturesque landscapes, Gothic ruins, ivy-clad castles and scenes of rustic village life. Crossing and re-crossing Europe with their cameras and paraphernalia, they followed in the footsteps of their forbears on their Grand Tours, capturing as they went the grandeur of the Alps and the ancient splendours of Italy. Their photographs became a perfect substitute for sketchpads. By the end of the decade the distinction between the daguerreotype and 'calotype' (as Talbot's improved process had become known) had been drawn, with the daguerreotype the domain of trade and the calotype that of Society. The 1840s were fallow years for stereoscopy, with few advances beyond the scientific investigation of the nature of binocular vision and stereopsis. Writing about the topic some years later, Lady Elizabeth Eastlake noted:

The simple principle of the stereoscope... could never have been illustrated, far less multiplied as it now is, without photography. A few diagrams, of sufficient identity and difference to prove the truth of the principle, might have been constructed by hand for the gratification of a few sages, but no artist, it is to be hoped, could have been found possessing the requisite ability and stupidity to execute two portraits, or two groups, or two interiors, or two landscapes identical in every minutia...by which the principle is brought to any level of capacity. Here, therefore, the accuracy and the insensibility of the machine could alone avail; and if in the order of things the cheap popular toy which the stereoscope now represents was necessary for the use of man, the photograph was first necessary for the service of the stereoscope. (Eastlake, 1850)

Although Queen Victoria was alleged to have given her royal stamp of approval at the Great Exhibition, it was the marriage between the daguerreotype and Brewster's stereoscope that revealed the true potential of binocular stereoscopy. When seen through the two lenses the two diminutive daguerreotypes presented a three-dimensional spatial realism rendered almost as large as life. And, to a society familiar only with the traditional forms of illustration, the optical magic of the stereoscope was a revelation that transformed their appreciation of the world as profoundly as the arrival of television was to do a century later. For the first time the public became spectators of remote scenes and events (see Figure 1).

Duboscq's decision to exhibit a Brewster viewer at the Great Exhibition, using daguerreotypes specially made for the occasion, may have been the first public demonstration of stereoscopic photography, but it did not take long for other instrument makers and photographers to recognise its commercial and social potentials. Leading the way was the London based Anglo-Frenchman Antoine Claudet, whose approach to photography was a shrewd mix of flair and pragmatism. As one of the first workers to adopt Daguerre's process for portraiture, he applied his considerable technical abilities to its improvement. In June 1841, he disclosed a method of reducing exposure times from several minutes to a few seconds; he gave his discovery freely to the world. The shorter exposure times offered the novelty of group portraits with 'three to six persons... engaged at Tea, Cards, Chess, or in

Figure 1. Brewster stereoscope (private collection)

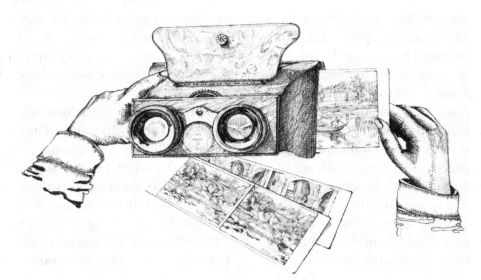

Conversation.' (*Atheneum*, 1841). This proved singularly attractive to his clientele. For the next decade, Claudet's name was linked to advances in photography, with advertisements for his photographic studio a regular feature of London newspapers and periodicals. His contributions to photography were rewarded with a Council Medal at the Great Exhibition for the superior quality of his daguerreotypes, and for his cameras and other photographic devices.

By the summer of 1851, Claudet had established himself in fashionable Regent Street, a thoroughfare favoured by the aristocracy and described as 'the most handsome street in the metropolis' (Cunningham, 1850). Here, potential customers had the opportunity to examine his latest improvements in the luxuriously furnished specimen room, before ascending to his rooftop portrait studio (*Atheneum*, 1851). In October Claudet announced that his sitters could now be photographed stereoscopically, an innovation that marked the first commercial development of stereoscopy. This enhanced mode of taking portraits meant that the pho-

tographs were, in his own words, 'no longer pictures, but real statues, and, when coloured... life itself (Cunningham, 1850). He used the specimen room to exhibit his 'Stereoscopic Daguerreotype Views of the Great Exhibition [which had] excited the interest of her Majesty and Prince Albert.' According to the editor of the *Morning Chronicle* these views of the building offered 'the perfect illusion of reality' (*Atheneum*, 1851).

By the closing months of 1851, everything was now in place for stereoscopy to take centre stage, though initially its entrance was hesitant. One of the factors limiting the widespread adoption of the stereoscope was the relatively high cost of the viewer and slides, at this early date predominantly daguerreotypes, which were difficult to make and complicated to replicate. The calotype process was easy to replicate, but was less than ideal as its softer definition did not lend itself to being magnified in a stereoscope. For several years photographers had been experimenting with glass as a more satisfactory support for the photographic emulsion when making a negative. During the late 1840s a way of coating the glass with albumen (egg white) was adopted, but the results were disappointing except in adept hands.

A better alternative turned out to be *wet collodion*, a solution of pyroxylin (a close relative of gun-cotton) in ether, a transparent, viscous solution ideal for coating glass and, as long as the solvent was not fully evaporated, readily taking up the light-sensitive silver solutions. Its use had been first suggested by Frederick Scott Archer and used experimentally by his photographic colleague the lawyer Peter Wickens Fry, who showed rudimentary examples of the process for the first time at the Great Exhibition (*Morning Chronicle*, 1852). As Claudet had done, he gave his invention to the world unencumbered by the sort of patent restrictions imposed by Daguerre and Talbot on their processes. Now for the first time a photographer had a process he could use without having to buy a licence or surrender a percentage of his profits. The collodion process was to introduce the second epoch of photography, in which commercial enterprise and mass marketing went hand in hand to create an unprecedented demand for photographs.

The first popular and comprehensive article on stereoscopy appeared in January 1852 in the *Illustrated London News* (1852) the first broadsheet to have been fully illustrated with large detailed woodcuts. With weekly sales of around 100,000 copies and a readership reaching four or five times that number, its target readership was the middle classes, who since the 1850s had become increasingly influential in shaping fashions and consumer demand. They largely comprised the very audience who had made the Crystal Palace so successful, with some 43,000 paying visitors each day (*Ill. London News,* 1852).

Throughout 1852, further articles appeared sporadically in the press, with a handful of photographers both in London and the provinces advertising stereoscopic portraiture as an extension to their regular business. Generally such initiatives were tentative, as were advertisements for photography (apart from portraiture), so the public tended to remain largely unaware of its existence. The situation changed in December 1852 when a dedicated group of amateurs mounted the first full scale photographic exhibition in London. With nearly 800 prints drawn from leading British and European photographers, the exhibition attracted the widespread attention of press and public (*J. Phot. Soc.*, 1853). The strategy of these enthusiasts was to raise the profile of photography as part of their campaign to form a society whose aims were to advance the science and art of photography. On 20 January 1853, the *Photographic Society* was formally established. Through regular meetings, exhibitions and a monthly journal the society soon attracted a membership of influential practitioners, the majority being amateurs dedicated to advancing their skill and knowledge through the free interchange of ideas and in published articles. Attracted by these high ideals and noble aspirations, gentlemen amateurs from the professional classes of society were attracted to the society and its meetings, one of which, in May 1853, was dedicated exclusively to the subject of stereoscopy (*J. Phot. Soc.*, 1853).

Stereoscopy had by then languished for almost fifteen years before being taken up by the general public; like all newly introduced technologies it was initially expensive. In that year, for example, Horne, Thornthwaite and Wood, a leading London manufacturer, offered a range of Brewster viewers ranging in price from 3 shillings (in japanned tinplate) to 30 shillings (in walnut with ivory eyepieces). But that was just the beginning of the expense, as the price of stereoscopic slides ranged in price from six to twelve shillings each (Horne *et al*, 1853). Calotypes were cheaper at two shillings each, presumably because the quality was poor in comparison with daguerreotypes. The total cost of buying such a viewer with six of the most expensive daguerreotype slides would have meant an outlay of over five pounds. At that time the total weekly outgoings of a household with four children and a servant in 1852 came to considerably less than this, with little to spare for luxuries (*Family Economist*, 1852).

Within a year, though, an increasing number of instrument makers, opticians, and specialised retailers had entered the field of stereoscopy and introduced a degree of healthy rivalry into the market, and prices were lowered. By 1854 Horne & Thornthwaite had reduced the prices of their stereoscopic pictures by around 40% compared with 1853 prices. In part this was due to the competition, but also to the increased efficiency in printing that came with the widespread adoption of the collodion process. As the market for stereoscopy grew, both amateur and professional photographers began to capitalise on the opportunity. Apart from those who followed Claudet's example and added stereoscopy to their portrait businesses, others recognised that stereoscopy could transform even the most mundane bit of countryside and ivy-covered ruin into something extraordinary. In small market towns and coastal resorts throughout the country, photographers augmented their portrait businesses by establishing themselves as photographic publishers, providing views of their locality for residents and visitors. Most remained small-scale and regional, but others like George Washington Wilson,

James Valentine and Francis Bedford developed significant publishing enterprises who found a ready market with the tourists who flocked into the countryside each summer in search of rural tranquillity. This was a new market for photography and one that relied on stereoscopy for its initial success. A number of factors underpinned these changes. Having demonstrated her industrial supremacy at the Great Exhibition of 1851, Britain set aside the social and economic difficulties of the 1840s to become a supremely self-confident and prosperous nation. The middle classes found they had more disposable income to spend on travel, literature, furnishings and other small luxuries.

The expanding influence of the railways was central to these improvements: by the late 1850s there were over six thousand miles of track, almost double that of the previous decade (Simmonds & Biddle, 1997). This intricate network modified the way in which society thought about travel. Long tedious journeys by stagecoach were consigned to memory; most destinations could be reached within a day in unaccustomed comfort. Queen Victoria's enthusiasm for train journeys gave rail travel a fashionable cachet, and her annual summer excursions to Balmoral established Scotland as a prime tourist destination. Today we associate railways chiefly with passenger travel, but originally they were designed for the bulk transport of coal, iron, fish, milk, newspapers and the Royal Mail. This efficient distribution network opened up new markets for manufacturers, and became the means by which photographers, photographic publishers and instrument makers could expand their horizons and operate nationally. For example, the London based company Horne, Thornthwaite and Wood was no longer reliant on metropolitan custom, but could now respond to postal enquiries and orders from retailers based elsewhere in the country. Equally, a photographer based in the north of Scotland, such as George Washington Wilson in Aberdeen, could utilise this efficient distribution network to supply his stereoscopic views to wholesalers and individual retail outlets anywhere in Britain (Taylor, 1981).

Given the speed at which stereoscopy was taken up by the public, and the profits to be made from their enthusiasm, it was inevitable that some entrepreneur would seize the opportunity to become agent, publisher, distributor, wholesaler, retailer and promoter for this lucrative branch of photography. In early 1855, the Stereoscopic Company established itself in a prime location at 313 Oxford Street, a few doors west of Regent Street, from where it offered an improved stereoscope with 'six charming Photographic Subjects, by the first artists, together with 6 diagrams' for the remarkably low price of 'only one guinea' (*Atheneum*, 1855). The company was established by George Swan Nottage, probably with financial support from his socially well-connected uncles, both of whom had extensive business interests (*Times*, 1885). Within six months the name of the company was changed to the more authoritative sounding 'London Stereoscopic Company', whose metropolitan cachet would attract out-of-town customers (*Daily News*, 1855). From the start Nottage understood that the path to success lay in good publicity and extensive advertising, and in September 1855 he launched a vigorous and sustained campaign to raise public awareness of the latest developments in stereoscopy and of the increasing range of subjects available through the company. Such was the consequent demand for stereoscopy that from its modest start in February 1855 the business progressed to stocking 'a most remarkable collection [of] 10,000 exquisite groups and views' within a year (*Daily News*, 1855). To achieve such remarkable growth Nottage needed to acquire the rights to the works of a wide range of photographers, both British and European, and to act as their agent – though from the outset, Nottage never credited his photographic artists, retaining their identity as a company secret. But anyone familiar with individual stereo photographers would know that leading practitioners enjoyed some kind of non-exclusive distribution arrangement, as they still continued selling under their own names (*Atheneum*, 1856) (see Figure 2).

Throughout 1856 the company continued to expand, taking additional premises in Cheapside, appointing agents throughout Britain, and publishing a catalogue of their ever-increasing inventory. By Christmas 1856 they had adopted the slogan 'No Home without a Stereoscope' and claimed to have laid in 'an immense stock of Subjects from all parts of the world.' This amounted to some 100,000 slides. With so extensive a stock every subject imaginable came before their cameras, ranging from the usual antiquities and topographic views to 'Drawing-Room Elegancies' and scenes of 'Single Life – and its Discomforts, Married Life – and its Pleasures.' There were also scenes of high and low life, love scenes, and humorous studies of 'toothache and other woes' (*Atheneum*, 1856). Every taste was catered for, from intellectual pastimes to popular amusement. This was photography in the dual guise of entertainment and instruction, and it achieved widespread appeal when viewed through the lenses of a stereoscopic viewer (see Figure 3).

Figure 2. Cartoon in Punch, 17 May 1856

Figure 3. Stereoscopes in the Home (attributed title) Anonymous, c. 1856, private collection

As with many fashionable crazes, public interest in stereoscopy peaked during the 1860s and began a slow decline during the 1870s, though never completely disappearing from view during the closing decades of the nineteenth century. In common with novel technologies whose popularities ebb and flow, stereoscopy kept being 'rediscovered'. In the early twentieth century, for example, manufacturers exploited the latest optical and manufacturing advances to create ever more compact cameras and viewers without sacrificing anything of their visual magic. This cycle of rediscovery, reinvention and repackaging of stereoscopy became a characteristic of its progress throughout the twentieth century – remember the Viewmaster and the Weetabix promotions of the 1950s – with its reintroduction to the cinema and its appearance on broadcast television being its latest manifestation. Whether the enthusiasm will be sustained this time round is debatable, but we can be sure that its symbiosis with human vision suggests that it is a technology that will continue to attract innovation.

REFERENCES

Alexis Gaudin & Brother. (1856, January 12). *Athenaeum*, p. 54. London Stereoscopic Company. (1856, December 20). *Athenaeum*, p. 1588.

A Portable Stereoscope. (1855, August 22). *Daily News*.

Barger, M. S., & White, W. B. (1991). *The daguerreotype: Nineteenth-century technology and modern science*. Washington, DC: Smithsonian Institution Press.

Brewster, D. (1866). *The stereoscope, its history, theory and construction*. London: John Murray.

Claudet's Stereoscopic Daguerrotype Portraits and Pictures. (1852, February 23). *Morning Chronicle*, p. 6.

Clowes, W. et al. (1852). *Reports by the juries*. London: William Clowes & Sons.

Cunningham, P. (1850). *Hand-book of London*. London: John Murray.

Death of Lord Mayor. (1885, April 13).. . *Times (London, England)*, 7.

Eastlake, L. E. (1850). Photography. *The Quarterly Review*, *101*, 4.

Editorial. (1850, October 26). *Times*, p. 4.

(1852). *Family Economist*. London: Groombridge & Sons.

Horne, Thornthwaite, & Wood. (1853). *Guide to photography*. London.

Illustrated London News (1851). May 3, p.348.

Journal of the Photographic Society (1853, May 31). pp. 57-66.

Moigno, A. (1856). *Brewster* (p. 30). La Presse.

Mr. Claudet, Daguerreotype Portrait Establishment. (1851a, June 28). *Athenaeum*, p. 673.

Mr. Claudet, Stereoscopic Daguerreotype Portraits. (1851b, October 4). *Athenaeum*, p. 1034.

Mr. Claudet's Gallery of Practical Science. (1841, June 26). *Athenaeum*, p. 491.

Schaaf, L. J. (1992). *Out of the shadows*. London: Yale University Press.

Simmons & Biddle. (1997). *The Oxford companion to British railway history*. Oxford, UK: Oxford University Press.

Stereoscopic Company. (1855, February 3). *Athenaeum*, p. 157.

Stereoscopic Pseudoscopic and Solid Daguerreotypes. (1852, January 25). *Illustrated London News*, p. 77.

Taylor, R. (1981). *George Washington Wilson: Artist and photographer 1923-1893*. Aberdeen, UK: Aberdeen University Press.

Taylor, R. (2007). *Impressed by light*. London: Yale University Press.

Wade, J. J. (1852). *Receipts* (p. 85). First Report of the Commissioners.

Whittingham, C. (1852). *A catalogue of an exhibition of recent specimens of photography*. London: Society of Arts. Retrieved 28 January 2012 from http://peib.dmu.ac.uk

Winter Evenings' Enjoyment. (1855, December 27). *Daily News*.

Wood, J. (Ed.). (1989). *The daguerreotype*. London: Duckworth.

Chapter 2
Models for the Behaviour of Light

Graham Saxby
3 Honor Avenue, Wolverhampton, UK

John Emmett
Broadcast Project Research, UK

ABSTRACT

In this chapter, the authors discuss models for the behaviour of light and explain the modern units of light measurement and the types of lighting used in photography. The theoretical models of Huygens, Abbe, Young, Maxwell, and Fresnel are outlined, emphasising the effects of diffraction and polarisation. They describe the structure and physiology of the human eye and the stereoscopic principle. The development of 3-D cinema and television is discussed with a summary of viewing parameters.

INTRODUCTION

Newton's Corpuscular Model

It took a long time for the Western world to evolve a working model for the behaviour of light. To the ancient Greeks light was not thought of as an entity: to Aristotle, the foremost authority, one saw objects by virtue of some sort of emanation from the eye that reached out to the scene and brought back the information. The first recorded model that could be used to predict the behaviour of light was due to Sir Isaac Newton, who postulated that light consisted of rapidly travelling particles or 'corpuscles'. This accounted for the principle of linear propagation, and is still the basis (in the form of ray tracing) of most lens designs. However, the model was unable to account satisfactorily for refraction and diffraction phenomena.

DOI: 10.4018/978-1-4666-4932-3.ch002

Huygens's Wave Model

The corpuscular principle was supplanted by a longitudinal wave model originally suggested by Christiaan Huygens; this model accounted for both refraction and diffraction, though not polarisation. A modification of the model from longitudinal to transverse wave motion rectified this flaw, and this model is now used in basic studies of diffraction and interference. It continued to be employed up to fairly recent times, when diffraction phenomena began to be treated in terms of Fourier theory; this insight resulted in a revolution in lens design.

Maxwell's Electromagnetic Model

The discovery of the connection between electricity and magnetism by Michael Faraday led James Clerk Maxwell to develop the mathematical theory of electromagnetic fields, and put the wave model on a sound basis, as well as showing that light could be considered as a form of electromagnetic energy of the same nature as radio waves and X-rays. It is the basis of what we now call physical optics. However, when used to predict the total energy of a broad spectrum of radiation it gave seriously flawed results.

Einstein's Photon Model

This problem was solved by Albert Einstein, who applied Max Planck's quantum model to light and suggested that it could be considered as consisting of discrete pulses of wave energy, or photons, a photon having an energy related to its wavelength (or, more accurately, its frequency). This accounted

for the photoelectric effect, and, as part of Planck's model, for many atomic phenomena previously difficult to account for. It played a vital part in the discovery of stimulated emission and the subsequent invention of the laser, and became important in the development of the technology we now know as 'photonics'.

LIGHT SOURCES

For many centuries the only source of artificial light was the burning of combustible substances, and right up to the 1930s the international standard for the energy emitted by a light source was based on the 'Standard Candle', to be produced from a specified material and burnt under certain rigorous conditions. Indeed, one still sometimes hears the term 'candlepower' used of powerful light sources.

The spectrum associated with a hot source is linked to its temperature, the emission being higher in the longer wavelengths (infrared) at lower temperatures, moving towards blue as the temperature is increased according to an equation originally derived theoretically by Planck. This group of light sources is called 'incandescent', and their spectral emission is described by a figure known as colour temperature, which is the temperature (in kelvins) to which a so-called black body (i.e. a perfect radiator) would have to be raised to in order to provide the same emission spectrum. Thus, a halogen filament lamp has a colour temperature of 3400 K and midday sunlight at the equinoxes around 5500 K. This definition can also be extended to 'cold' sources such as fluorescent tubes, flashtubes and

LED sources, provided their spectral energy distribution approximates to the Planckian equation.

THE UNITS OF LIGHT MEASUREMENT

Until SI units were universally adopted by the scientific community, units for the measurement of light were, to say the least, haphazard. Even now one sometimes comes across textbooks and even research papers that talk of foot-lamberts, nits and other obsolete units. The SI unit of luminous power is the candela, and as the perception of light is a subjective quantity and energy varies with wavelength, it can only be related to the watt (the SI unit of mechanical and electrical energy) for single wavelengths. Thus whereas the output of a laser can be measured in watts, that of a white-light source has to be measured in candelas. The two quantities are linked at 555 nanometres (nm), the wavelength corresponding to the maximum sensitivity of the human eye.

Power (Intensity)

The candela output of a light source can only be quantified in connection with a plot of its angular distribution, so a more convenient unit is needed, specifying the luminous power output averaged over a specific solid angle. The SI unit is the lumen, which is the luminous power emitted by a completely uniform source of 1 cd over a solid angle of 1 steradian (sr). As there are 4π steradians in a sphere, $1 cd = 4\pi$ lm, provided the source is uniform. As this proviso is seldom true, the lumen is used in preference to the candela as a measurement of luminous power (or 'flux', to use the old terminology), especially for such directed sources as car headlamps. At 555 nm (corresponding to the maximum sensitivity of the human eye), $1 W = 683$ lm. This is a maximum value: at other wavelengths the luminous efficiency is less, becoming zero at the limits of perception. Now whereas the efficiency of a monochromatic source such as a laser can be given as a simple percentage (watts out $\div$ watts in, $\times$ 100), owing to the variation of photon energy with wavelength this conversion cannot be applied directly to a white-light source, so has to be specified as a luminous efficacy, in lumens per watt (lm W^{-1}). The overall value of a broad-spectrum source cannot approach the figure of 683 lm W^{-1}, though efficient sources such as LEDs may have efficacies well over 100 lm W^{-1}, which amounts to something like 40% efficiency over the whole visible spectrum.

Illuminance and Luminance

A uniform source of 1 candela emits 1 lumen per square metre (lm m^{-2}). When this light falls perpendicularly on a surface the illuminance is 1 lux (lx). If the light is oblique at an angle of incidence θ, the illuminance is multiplied by $\cos \theta$ (Lambert's law). The illuminance on a surface from a small source is proportional to the intensity of the source and inversely proportional to the square of its distance from the source (the inverse square law). The luminance of a source is its measured brightness per unit area of emission, assuming this is uniform over the emitting surface. It is measured in candelas per square metre (cd m^{-2}); there is no named unit. The reflectance of a surface has the same units, but measured in a specific direction. The extremes of quality of reflection are *specular* (like a mirror) and *Lambertian* (uniformly diffuse). In between is *partially diffuse*. The best mirrors have a specular reflectance of around 92%, but interference mirrors can be constructed to provide very nearly 100%. The best Lambertian reflector is a block of magnesium carbonate, with a diffuse reflectance of about 90%. This material is used to calibrate diffuse reflecting surfaces.

Luminous Energy

This is the total energy dissipated by a flash source, and is measured in lumen seconds (lm s). If a graphical record is made of instantaneous luminance versus time, the total luminous energy is usually measured between the rising and falling 10% points on the curve.

Spectral Power Distribution

If a source cannot be specified by a colour temperature it is necessary to show power versus wavelength graphically (including IR and UV emission if appropriate).

TYPES OF LIGHT USED IN TV AND CINEMA

Daylight

Natural looking but very variable. The best lighting balance for photography is with bright sun at about 45° with white clouds to fill in the shadows. In the absence of clouds the shadows will be very blue, as blue sky

has a colour temperature of around 16000 K. Full noon sunlight has a colour temperature around 5500 K, which reduces to around 3000 K towards sunset. A fully overcast sky is about 8000 K.

Filament Lamps

When intended for photography, these have a colour temperature of around 3400 K. The most common type used for photographic lighting is the halogen lamp, which uses a compact filament in an atmosphere of bromine or iodine, which prolongs the life of the source by restricting evaporation of the filament.

Fluorescent Lamps

These are compact folded tubes similar to household 'energy saving' lamps. The activating source is a mercury vapour arc emitting UV radiation which stimulates layers of phosphor material coated on the inside of the tube. The phosphor composition is adjusted to give a spectrum corresponding to a particular colour temperature, usually 3400 or 5500 K. However, colours may not always be rendered correctly. The control circuitry in the base consists of an electronic ballast that operates the lamp at a high frequency, thus avoiding flicker when used for high-speed or slow motion filming.

Discharge Lamps

These are small-source arc lamps producing their light from a krypton gas plasma. They give a light close to daylight in quality and colour temperature, and are frequently used in automobile headlamps.

Flash Sources

Xenon gas in a small tube at low pressure is ionised and made conductive by a trigger pulse of high voltage, permitting a bank of capacitors to discharge through the tube and producing a bright flash with a duration of the order of a millisecond. (This can be reduced to microseconds by a quenching circuit.) Both xenon and krypton lamps can be made to produce pulses in rapid succession, and these can be synchronised to cine camera exposures. Flashbulbs, in which a bundle of zirconium ribbon in an oxygen atmosphere is electrically ignited, were once the main source of light for press photographers, but have now virtually disappeared.

Light Emitting Diodes (LEDs)

These are becoming increasingly common as their spectral quality improves and mass manufacture brings the cost down. They can be made to emit light that is a close match to daylight; they run cool and are efficient and economical. They may well become the chief source of illumination in photographic and TV studios, and are already standard for illuminating computer and TV screens, where electron beam tube (CRT) technology is fast disappearing.

Lasers

Diode lasers are a special type of LED in which the two faces form a Fabry-Pérot etalon (a type of ultranarrow-band interference filter) so that they emit a quasi-monochromatic beam that can be collimated by a suitable lens system and used for displays and holography. They are available in output powers of from 0.5 mW to 1 W or more. For higher powers used in large displays gas or solid-state lasers are generally employed. Lasers are not as a rule suitable for general photography as their beam when expanded and reflected from a surface shows a marked speckle pattern.

TYPES OF LUMINAIRE

'Luminaire' is the term used to describe a complete lighting unit for photography. It consists typically of the light source itself, a backing reflector, and some kind of optical device to modify the nature of the beam. Luminaires may use a source that is concentrated, as in spotlights and 'kickers', or extended, such sources being referred to by photographers as a 'flood', 'chip pan' or 'swimming pool', depending on the source size. There are several types of diffuser, from simple ground glass or cloth to a source facing backwards into an extendable reflector or 'brolly', which may be white or aluminised. Spotlights are focused via a Fresnel lens (a large convex lens cut back in steps), and the beam can be restricted by a 'snoot' (a narrow tubular mask) or 'barn doors' (hinged flaps). A useful accessory is a 'gobo', which is basically a small panel on a stand used near a spotlight to give local shading. The name is also often used to describe a light filter attached to the front of the spotlight to serve a similar purpose, or to provide patterned illumination.

LENSES AND IMAGING

The Abbe Sine Condition

Consider a plane wavefront entering a convex lens and emerging as a converging wavefront. Ideally this wavefront should be perfectly spherical, so that it converges to a point; this point defines the image plane (Figure 1a). In practice, if the lens surfaces are accurate to within about one-half of a wavelength, the image will be as sharp as it possibly can be. It is said to be diffraction limited. Ernst Abbe, a specialist in microscopy, found that in order for any optical system to form a good image, it must satisfy the sine condition. This states that an optical system will form an accurate image of an infinitely distant object only if, for each ray in the parallel beam emanating from the source $h/\sin \theta = f$ where h is the radius of curvature of the lens surface, θ is the half-angle subtended by the principal surface at its centre of curvature and f is its radius (Figure 1b).

The sine condition assumes that the air/glass interface is spherical, and thus does not take account of lens aberrations, which are small in very small lens systems such as microscope objectives.

Abbe Theory of Image Formation

In the design of microscope objectives in the early 1870's, Abbe faced a problem that compelled him to abandon the ray model. He had found that with a lens designed using ray tracing methods there was still a minimum resolvable object size, and that this resolvable size was as small as possible when the angle subtended by the lens diameter at the object was as large as possible and the refractive index of the medium in which the system was immersed was as high as possible. The resolution was also greater the shorter the wavelength of the light. These conclusions led (in microscopy) to the use of cedar wood oil to fill the space between the microscope slide and the objective and increase its refractive index. Oil immersion is now standard practice for high power microscope objectives.

WAVE THEORIES OF LIGHT

Christiaan Huygens (1629–95) was one of the first physicists to use mathematics. His wave model for the behaviour of light formed an alternative to Isaac Newton's corpuscular model, in which refraction needed to be explained by a speeding up of the corpuscles on entering a denser medium. Huygens, on the other hand, explained refraction by a slowing of light on entering the medium. Later experiments aimed at finding the speed of light in different media confirmed this view. Augustin-Jean Fresnel (1788–1827) showed that Huygens's principle, together with his own principle of interference, could explain both the rectilinear propagation of light and the visible diffraction effects in the near-field. To do this mathematically he had to make a few assumptions, and at the time there was some doubt about these in the Académie de Sciences. Simeon Denis Poisson (1781–1840) used Fresnel's theory to predict that there would be a bright spot in the centre of the shadow of a small disc and assumed that the theory must thus be incorrect. However, Francois Arago (1786–1853) designed

Figure 1. (a) Focusing of wavefronts; (b) the sine condition

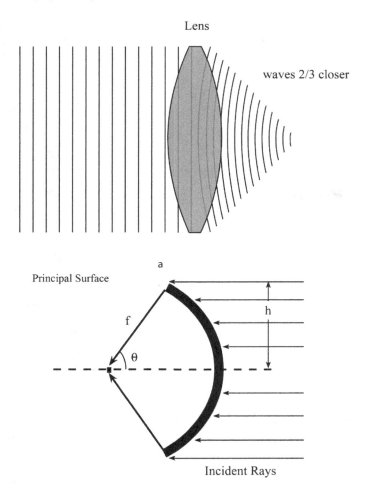

Lens

waves 2/3 closer

a

Principal Surface

f

θ

h

Incident Rays

b

an experiment that showed that there was indeed a bright central spot and vindicated Fresnel's prediction. A century later, James Clerk Maxwell (1831–79) combined Michael Faraday's discoveries in electromagnetism with the newly-minted vector calculus to derive the four splendid equations that clothed the earlier wave theory with the garment of mathematical respectability and showed that light was simply part of a much larger spectrum of electromagnetic waves (see Figure 2).

The mathematician Joseph Fourier (1768–1830) had little to do with optics: he developed his mathematical concepts in connection with his researches on thermodynamics. Fourier series, however exactly describes the diffraction effects observed by Fraunhofer. The so-called Fraunhofer/Fourier laws were prob-

Figure 2. The wavelet principle of diffraction

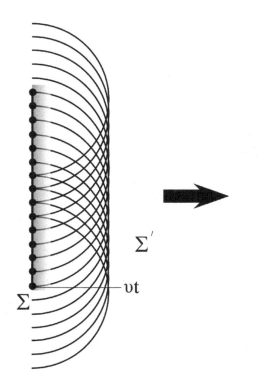

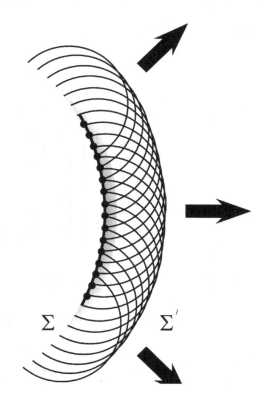

ably formulated by Gustav Kirchhoff (1824–87), who produced a general formula for diffraction.

The Plane Wave

This can be thought of as generating spherical wavelets originating from every point in the gap. The resultant transmitted wave can be calculated as the envelope (Figure 2) of these. The governing rule as to the application of diffraction theory is the Fresnel number F, a dimensionless parameter given by the relationship $F = a^2/\lambda R$, where a is the radius of the aperture, λ is the wavelength and R is the distance of the wavefront from the aperture. If F < 1 we have Fraunhofer diffraction, whereas if F > 1we have Fresnel diffraction.

These are often called respectively far-field and near-field diffraction. Joseph von Fraunhofer (1787–1826) was a professional glass worker. He invented the spectroscope, and was also responsible for the development of ruled diffraction gratings. His name has been given to the absorption lines present in the Sun's spectrum.

LENSES AND DIFFRACTION

The theory of image formation suggests that diffraction is an intermediate step of the image forming process. If an object is situated close to the front focal plane of a lens, every point on the object will emit a diffracted beam, and the sum of these will form a unique diffraction pattern close to the back focal plane (sometimes called the *Fourier plane*) of the lens. From here the light continues on, to form an inverted image of the object (Figure 3). As the finest detail is diffracted by the greatest amount, unless the lens is very large this detail will miss the lens altogether, and this limits the finest detail that can be present in any image. Abbe had thus shown that for a 'perfect' lens its resolution is limited by its diameter.

In the more precise language of the Fourier model for image formation, if the object is in the front focal plane, the lens will form an optical *Fourier Transform* (*FT*) of the object field at its back focal plane. If a second lens is then positioned so that its front focal plane coincides with the back focal plane of the first lens, it will form an FT of this FT at its own back focal plane. Now, FTs possess the mathematical property that if you form an FT of an FT you get back to the original

function (give or take the odd constant), so the system makes mathematical sense. (The math even accounts for the image's being inverted!).

The formula for the Fourier transform $f(x)$ of a function $F(u)$ is

$$f(x) = \int_{-\infty}^{\infty} F(u)e^{2\pi iux} du$$

and its inverse (i.e. the FT of the FT) is

$$F(u) = \int_{-\infty}^{\infty} f(x)e^{-2\pi iux} dx$$

It is unimportant which equation has the minus sign. Note that if the function is symmetrical the exponential terms simplify to $\pm \cos(2\pi ux)$.

If the second lens is omitted the image will be formed at infinity (if this seems anomalous,

Figure 3. Image formation in diffraction terms

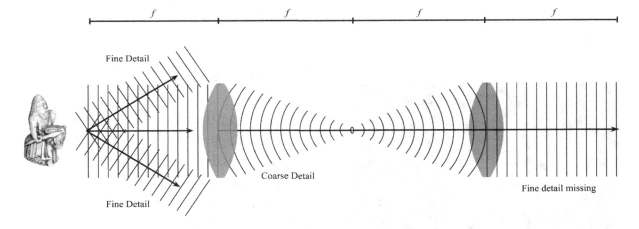

it isn't. Just think of this setup as equivalent to the insertion of a lens of infinite focal length). Incidentally, if the object is farther from the lens than the front focal plane, a transform will still form, though it will not be a true FT, and the inverted image will be formed nearer than infinity.

The transform pattern at the back focal plane is not normally seen, as white light is incoherent (i.e. the wavelengths and phases are all mixed up); but if we use the well-disciplined monochromatic light of a laser to illuminate the object we can actually see the diffraction pattern that indicates the presence of the optical transform. If the object is a simple one, such as a 2-dimensional grating, the pattern will be fairly simple too (Figure 4).

The pattern is perpendicular to the grating slit orientation. The central maximum is called the *zero order*, and represents the light that passed straight through the grating. The nearest side spots are *first order*, and so on. The farther out the spots are, the more the light has been diffracted. High spatial frequencies

correspond to high diffraction orders. For all types of object the frequencies located at the edges of the intensity pattern in the Fourier plane correspond to higher spatial frequencies. The lower spatial frequencies (for which the light was deviated less) are located in the central part of the Fourier plane. If the lens has only a small diameter the higher order diffracted beams will miss it altogether, and the number of outer spots will be reduced.

The reason a whole row of spots appears in the diffraction pattern, rather than a single one each side, is that the transmission function of a simple ruled grating in spatial terms is a square wave function, so that in addition to the fundamental spatial frequency it contains spatial frequencies three times, five times, etc. higher, and these are diffracted proportionally farther out from the centre. If, on the other hand, we transmit our monochromatic beam through a cosine grating (i.e. a grating in which the transmittance/distance relationship is (co)sinusoidal, there will only be one spot each side of the centre spot, because the FT of a cosine function is unity.

Figure 4. Diffraction pattern formed by a grating

IMAGING WITHOUT A LENS: THE ZONE PLATE

Now consider the image formed using a pinhole. Although easy to describe in terms of ray tracing analysis, it cannot in fact produce a sharp image owing to diffraction. (This applies to lenses too, as you may have realised from the analysis above.) What it does produce is the diffraction pattern of a top hat function. As this function is two-dimensional it is more difficult to analyse in Fourier terms than a one-dimensional function, but the associated diffraction pattern turns out to consist of a series of rings, called the *Airy pattern*, after Sir George Airy (1801–92) (see Figure 5).

From the above you might assume that illuminating the pinhole transform pattern with a lens and monochromatic light would get you back to a pinhole. In fact we can do better than that. By blocking off alternate rings we can make a transparency that will focus a beam of light like a lens. It is called a *zone plate*. As its focal length varies with wavelength zone plates are of little use for white light, but where lenses are ineffective,

as in X-ray imaging, they can be very useful (see Figure 6).

To construct a zone plate you simply draw opaque concentric circular rings (zones) with radii proportional the square roots of 1, 2, 3, etc., i.e. $r^2 = nf\lambda$, where r is the radius, λ is the wavelength, f the desired focal length and n = 1, 2, 3 etc.

A more efficient zone plate has a sinusoidal transmittance pattern rather than simple opaque rings (Figure 7). Sinusoidal zone plates are often used in imaging test charts.

With increasing spatial frequency there is a fall-off in contrast. Another important artefact that tends to appear in digital images

Figure 5. The Airy pattern, showing the effective diameter of the central Airy disc

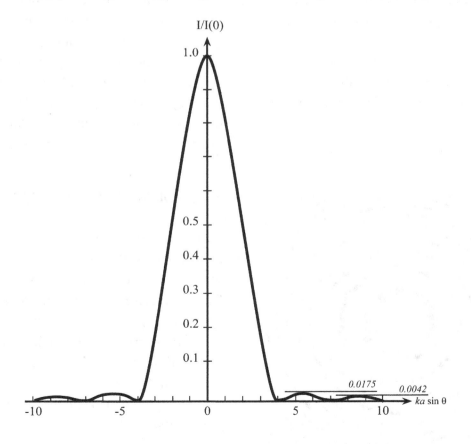

Figure 6. A zone plate (in practice it would be less than 1 mm in diameter)

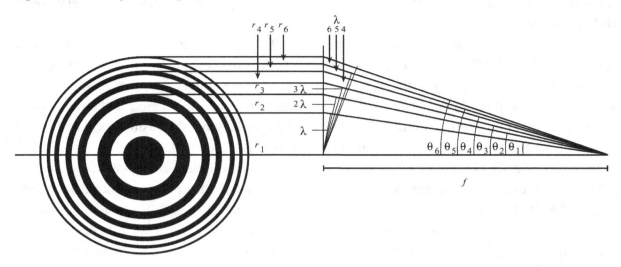

is known as aliasing, a term borrowed from electronics. In optics it means that the pattern of the test object 'beats' with the pattern formed by the pixels of the camera format, forming a moiré pattern. In digital camera images the moiré pattern is often coloured.

Figure 7. A sinusoidal zone plate used as a test chart

A zone plate can be thought of as a trivial hologram, as it can be considered the holographic record of a point object. In fact, one model for a hologram assumes that it consists of the sum of the zone plates corresponding to every point on the object. This principle underlies some of the programs for constructing holograms by computer generation.

THE POLARISATION OF LIGHT

So far we have considered the behaviour of light in terms of rays and of waves. To deal with polarisation we need to treat light as a form of electromagnetic energy that travels by wave propagation. The waves oscillate in a direction perpendicular to their direction of travel, and the plane of the waves in most cases is random. However, it can be constrained to become ordered. This phenomenon is called *polarisation*.

Types of Polarisation

There are several types of polarisation, and these are described by specifying the orientation of the wave's electric field vector at a point in space during the period of a single oscillation. If the vector representing the electric field is orientated in a fixed direction, then as the wave progresses this vector traces a single plane in space. This is termed *linear* or *plane polarisation*. Alternatively, the vector may rotate, performing a complete rotation for every cycle as the wave propagates. This is referred to as *circular polarisation*. The rotation may be either clockwise or anticlockwise (conventionally when viewed looking in the direction of travel), and is described as the *chirality* or *handedness* (as in left- or right-handed) of the wave. Both linearly and circularly polarised light are used widely as a means of viewing 3-D images, especially when projected on a screen (see Figure 8).

When a wave has elements of both linear and circular polarisation it is said to be *elliptically polarised*. This type of polarisation is used for some FM radio broadcasts, as it improves signal reception by receiving antennas that are not always correctly orientated, as in automobiles.

The Generation of Polarised Light

Linearly polarised light can occur as a natural phenomenon. Many crystalline substances (even ice!) are what is termed *optically active*. The effect has been known for many years. In 1669 Bartholinus, a Danish scientist, noted that when a crystal of Iceland spar (calcite) was placed over some words on a sheet of paper two images were seen (Swindell, 1975). He termed the phenomenon double refraction. Such materials are now described as *birefringent*. Over a century later experiments verified that when ordinary light was incident upon calcite and a number of other crystal types, the two beams produced were linearly

Figure 8. Wave oscillations in (a) unpolarised, (b) linearly, and (c) circularly polarised light

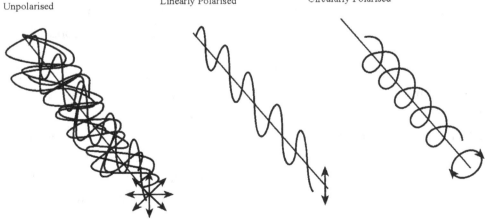

Unpolarised   Linearly Polarised   Circularly Polarised

polarised at right angles to one another. Light reflected from a non-metallic surface is also partially polarised. In 1815 Sir David Brewster was able to show that when the reflected and refracted light rays are at 90° to one another the reflected light is totally polarised in the plane that is orthogonal to the plane containing the incident and reflected rays (this is called *s–polarisation*), and the transmitted light is partially polarised in the plane of the incident and reflected rays (called *p–polarisation*). Total polarisation of the reflected beam occurs when the tangent of the angle of incidence is equal to the refractive index of the reflective material (about 56° for optical glass). This angle is known as the *Brewster angle* after its discoverer. Circular polarisation as a natural phenomenon is much less common, but it can be exhibited by thin sheets of some optically active materials such as mica.

The first application of polarised light was by the geologist William Nicol in 1828 for his researches into the structure of minerals, using a calcite crystal specially cut to eliminate one of the two rays. But the world had to wait until the twentieth century for the use of polarised light for stereoscopy. In 1929 the American Edwin H Land (1909 – 1991) invented the material which he named Polaroid, forming the Polaroid Corporation to manufacture it. Although Land initially studied at Harvard he left early, and instead worked long hours on his own, developing this new material. He did eventually return to Harvard but never graduated, though he later became the recipient of numerous academic distinctions. His work on polarising material began by continuing the 19th century work of William Herapath, who had discovered the optical activity of crystalline quinine iodosulphate (Herapathite), (Herapath, 1852). Land conceived the idea of aligning large numbers of tiny needle-shaped crystals embedded in a transparent polymer film. Modern methods for alignment of the crystals involve a mechanical stretching technique. The final sheet of material transmits only light that is linearly polarised orthogonally to the alignment direction of these crystals.

The Polaroid material finally developed is a material known as K-sheet, which is a polyvinyl alcohol sheet impregnated with iodine. The polymer chains are stretched to produce a structure of parallel linear molecules made electrically conducting by the presence of the iodine, and resistant to heat and humidity. Polaroid material is obtainable in large sheets, and is very effective for producing linear polarisation.

If the polarisation axes of two such filters are crossed at right angles with respect to one another, light passed by the first filter (the polariser) will not pass through the second (the analyser). However, light emerging from the first filter and incident on a second filter with the same alignment will pass through uninterrupted. If the second filter is rotated about the optic axis, the amount of light transmitted will decrease, reaching zero when the two filter axes are orthogonal (Figure 9). In practice polarising filters are not perfect, and a small amount of light is transmitted even in the fully crossed position.

The principle of crossed polarisers forms the basis of viewing stereoscopic image pairs (left and right) projected onto a screen or a

Figure 9. Effect of crossed polarisers

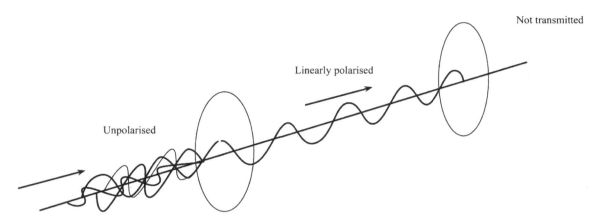

3-D TV display. Such images are too large to view stereoscopically side by side, so the left and right transparencies (or movie films) are projected superposed by the projector optics. Polarising filters are fitted to the projector lenses, their axes orientated at right angles to one another at ± 45°. The screen must be metallised so as not to depolarise the reflected light. The viewer wears similarly orientated polarising spectacles, so the left image on the screen will be seen only by the left eye; the right image, being cross-polarised with respect to the left spectacle filter, will be blocked out; similarly, the right eye will see only the right image. In practice, extinction may not be total, particularly if the viewer's head is tilted slightly so that each eye perceives a ghost of the other eye image.

Filters for Circular Polarisation

Circular polarisation is rather more difficult to achieve than linear polarisation, but it has advantages when used for viewing 3-D im-

ages on a cinema screen or TV. This type of polarisation is achieved from unpolarised light by a two-layer filter. The first layer is a linear polariser, and the second converts linear into circular polarisation. This latter filter consists of a thin layer of a birefringent material such as mica or a cellulose derivative, of such a thickness as to delay one of the beams by exactly one-quarter of a wavelength. The sum of their wavefront equations can be shown to be a rotating vector, which indicates circular polarisation. Such a filter is called a *quarter-wave plate*, or (more pedantically) a *phase retarder plate*. In practice a simple phase retarder plate produces a quarter-wavelength delay for only one specific wavelength, but achromatic retarders have been developed that operate over a range of wavelengths (She et al, 2005). To obtain the effect the incident beam needs its polarisation axis to be at precisely 45° to the axis of one of the birefringent beams. An angle of + 45° produces rotation in one direction, and an angle of – 45° produces rotation in the other. Figure 10 demonstrates this.

Figure 10. Clockwise and anticlockwise helices

Left Handed   Right Handed

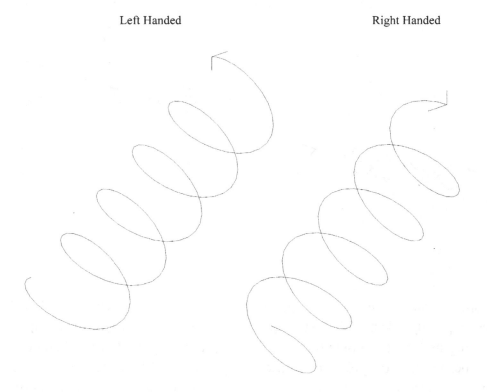

A right-handed (clockwise) polarised beam will not pass though a left-handed (anticlockwise) polarising filter and vice versa, so if the stereo pairs are projected through circular polarising filters of opposite chirality and viewed through spectacles containing the appropriate quarter-wave filters, each eye will see only its appropriate image. This type of stereo projection is free from crosstalk even when the viewer's head is tilted, and is fast becoming the norm for stereo projection systems.

Other Applications of Polarisation

The use of polarisation has already been mentioned in connection with the examination of geological specimens. In addition to the quarter-wave plate there is also a *half-wave plate*, which, as its name suggests, retards one of the beams by one half-wavelength. This has the effect of rotating the plane of a linearly polarised beam. Rotating the half-wave retarder in the beam by a given angle rotates the beam's plane of polarisation through twice this angle. This device is useful when two beams that are linearly polarised at different angles are required to interfere, as in holography, and for adjusting the planes of polarisation in the type of stereo projection that uses linearly polarised beams.

In general photography, polarising filters are used regularly for eliminating unwanted reflections from glass and water, and for adjusting the intensity of recorded skylight (skylight at right angles to the sun is strongly polarised). Polarisation has many scientific applications (some substances rotate the plane of polarisation, and this is a valuable analytical tool), and in engineering it is employed in stress analysis, using acrylic models which become optically active when stressed. Edwin Land's original purpose in developing Polaroid material was to eliminate glare from automobile headlamps, but his scheme was never adopted.

VISUAL PERCEPTION

Evolution of the Eye

The eyes, and their associated neural pathways in the brain, represent such a complicated system that creationists often point to it as evidence against the principle of evolution, saying 'What use would an only half-developed eye be?' In fact, such rudimentary eyes do exist: every stage in the evolution of the eye is present even today, from the single light-sensitive skin patch of an earthworm to the crude eye-pits of the limpet, and on to the sophisticated telescope that is an eagle's eye. Each organ has developed just as far as is necessary to do its job. In the course of evolution several different types of eye have evolved, the mammalian eye being only one of these. Many sea animals have evolved excellent vision using quite different optical principles, and almost all insects have a compound eye consisting of many tiny lenses each focusing a small area of the visual field. Neither is the animal kingdom confined to two eyes: many arachnids have up to eight, and some shellfish possess even more.

But in the mammalian (and avian) eye the positioning of the two eyes is crucial. Depending on whether you are likely to belong to the hunted (pigeons, antelopes) or the hunters (falcons, leopards) your eyes may be set at the sides of your head, for all-round vision, or in front, for concentrating on prey. Humans, having been at some time both hunters and hunted, have a compromise: with our eyes set forward but fairly far apart, we have excellent forward vision, but also a wide angle of view (around 210°).

Structure of the Eye

The structure of the human eye is shown, somewhat simplified, in Figure 11. Optically it resembles a camera. The *cornea* forms the first surface of a convex lens which does most of the focusing of the image onto the *retina*, which consists of light-sensitive nervous tissue that has the job of relaying the image to the parts of the brain that do the interpretation. Fine focusing is achieved by the *lens*, which can change its curvature to adjust the focus under the tension of the *ciliary muscles*. The eye is filled with fluid of refractive index 1.34, the portion in front of the lens being watery (*aqueous*) and in the rear chamber jellylike (*vitreous*). The amount of light entering the eye is controlled by an opaque muscular ring (the *iris*); its aperture forms the *pupil*. The refractive index of the lens varies from 1.38 at its rim to 1.41 at its centre; this variation helps to mitigate the aberrations present in a simple lens.

Figure 11. Schematic representation of the eye

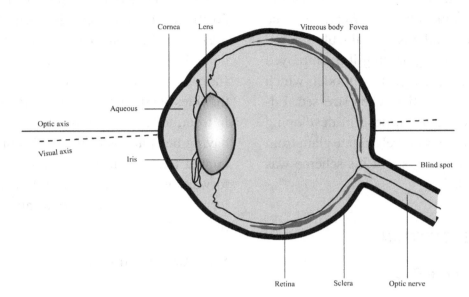

The Retina

This is a complicated structure, as it has to deal with far more information than can pass down the optic nerve. The light-sensitive cells are of two types: *rods*, which function only in dim light, and *cones*, which respond only to bright light. Cones operate efficiently at light levels around 10 cd m^{-2}, and cease functioning altogether at a level of about 0.1 cd m^{-2}, at which point the rods take over. They contain a dye called *rhodopsin*, which is bleached by light and provides the operating mechanism. Rods are connected together in parallel, sometimes in quite large numbers, and this increases low-light sensitivity almost to the point of detecting single photons, at the cost of low resolution (try reading a newspaper by moonlight!), and they are not colour sensitive. Cones, on the other hand, are activated by a different dye, *iodopsin*, which comes in three

forms, ρ, γ and β (rho, gamma and beta), sensitive respectively to red-orange, yellow-green and blue-violet. So the cone system can detect colour, operating in an additive manner much the same as a colour TV. In the retina, on the optic axis of the eye there is a large concentration of cones, and rods are absent: this is the centre of vision and of maximum resolution, the *macula lutea*, within which the centre, the *fovea*, is the most sensitive. It is only a tiny area covering about 2 square degrees, but it seems larger, as in normal vision the eyes are in constant movement. Farther out from the centre there are fewer cones, and none at all towards the periphery, where the rods are very large and can only detect movement. The neural impulses from the rods and cones do not feed directly into the optic nerve, but pass through other types of cell where the first stage of processing takes place.

Ocular Acuity

This is usually assessed by looking at a white board on which are printed lines of letters of regularly diminishing size from the top line downwards. At one time these letters were in block-sérif form, called 'Snellen letters' after their original designer, but this face was inherently difficult to read, and a simpler sans-sérif face is now employed, though the old name seems to have stuck. The limit of resolution is assessed from the size of lettering such that the letters can only just be identified. In the UK, at a standard distance of 6 metres, the ability to identify 6 mm detail (6/6) is rated as normal vision (20/20 in the US). This is a statistical norm, and if you have 6/6 vision, you should remember that half the population can see better than this! An angular resolution of 6 mm at 6 m represents 1 milliradian or about 0.5°, and while this figure is perhaps a little pessimistic it is convenient for calculating optimum screen distances etc. – but when constructing displays it is as well to remember the 50% who can see better than this, and to go for a higher standard.

Visual Pathways

Most of the functions of the right side of the body are controlled by the left side of the brain, and vice versa. In most mammals this is also true of the eyes, but in anthropoid apes, including humans, it is just the right half of the visual field that is processed by the left half of the brain, and vice versa. So half of both fields goes to each lobe of the brain, crossing over at the *optic chiasma* (Figure 12). These then pass into the *lateral geniculate nuclei* (one each side) where the signals undergo further analytical processing, and finally to the hindermost part of the brain, the *occipital cerebral cortex*. It is here that the composite image is finally analysed and the matter in front of the eyes recognised. Some receptors are programmed to match signals from the left eye with those from the right eye, and it is these that are responsible for stereoscopic perception (*stereopsis*).

How Our Perception is Fooled

All of us have at some time been taken in by a visual illusion. Some of these illusions are very powerful. They can often reveal much about our visual processes. Some of them plainly originate in the retina, and are the effects of afterimages (where the opsin components have become exhausted a negative image remains and interferes with subsequent vision); some originate in the lateral geniculate nuclei (figures that appear distorted when seen against a confusing background) and some originate in the cortex (self-inverting cubes and staircases, as exemplified by some of M C Escher's vertiginous woodcuts). Perhaps the most aesthetically compelling illusion, though, is the impression of depth conveyed by viewing a stereoscopic pair of images. This is undoubtedly a cortical construction, and even now is not completely understood. It does not even develop until a child is several years old, and is poorly developed in as many as one person in five.

The recognition of a perceived scene is a complicated matter, and there are conflicting theories. The work of Hubel and Wiesel

Figure 12. Schematic of optical pathways

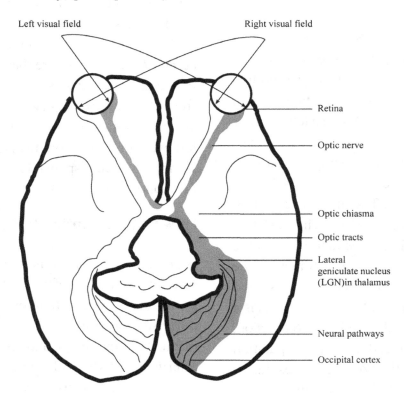

(1979) showed that specific areas of the cerebral cortex are programmed to recognise image elements such as sloping lines, curves, colour, movement in specific directions, etc. This supports the recognition by components theory developed by Biederman (1987), which suggests that the image of the input is segmented into an arrangement of simple geometric components such as cubes, cones and cylinders that can be assembled into almost any shape. This theory does not account for the recognition of the more complex images such as faces, clearly an acquired ability; this may be the job of small discrete groups of cells (ironically known as the 'grandmother cell' theory, which suggests that a single group of cells becomes specialised in recognising the face of one's grandmother). A complementary theory, developed by Treisman and Gelade (1980) suggests that there are two stages of attention: in the first, primary visual stimuli are processed and represented by feature maps that become integrated into saliency maps that direct attention to the most conspicuous areas; the second stage involves the focusing of attention on the information presented in the image and comparing the whole with prior knowledge of comparable images. The preliminary feature search identifies such primitive features as colour and shape, and the conjunction search phase, which is slower and conscious, integrates these features and matches them to stored images.

THE INTRODUCTION OF STEREOSCOPIC IMAGING IN FILM AND TELEVISION

The idea of stereoscopic moving pictures followed rapidly on the invention of the zoetrope in the 1850s. There were many attempts to modify this device to show stereo pairs. When cine arrived around 1900 it was not long before there were attempts to make stereoscopic film sequences, some more successful than others. The early history of stereoscopic cinema is told in detail by Zone (2007). The first really successful public showing of stereoscopic films was in 1951, when a dedicated cine theatre (the Telekinema)was erected on the South Bank of London for the Festival of Britain. In the main exhibition, at the optical technology display site the television company Pye had rigged up a 3-D television display using two monitors set at right angles, using a 45° mirror on the Pepper's Ghost principle, showing the L/R images superimposed, and viewed with the same polarising eyewear used for the Telekinema displays. There was no suggestion of such an unwieldy device's having any commercial prospects, and the idea languished. During the 1990s there were some experimental colour stereo TV broadcasts using the anaglyph principle, with the red image for the left eye and the blue and green images combined for the right eye, but public interest soon flagged and the transmissions were quietly dropped.

The story would be taken up again when the battle for supremacy between cinema and TV produced high-definition film, and stereo images of high quality became possible in both media. By the 1980s information technology (IT) had reached the kind of performance level that would permit the display and manipulation of 3-D images in near real time. At that time the models for display were based on wire mesh type virtual frameworks, viewed (in 2-D) by rotation and translation of the viewpoint. Ray-tracing calculations were then performed and a 2-D pixellated image generated. Today we use wire models less, as fractal-generated images permit the production of more realistic scenes at lower bitrates. This development was quickly taken up by the movie industry, using computer-generated imagery (CGI), in which high profile animation and special effects (SFX) film makers employed both wire frame and fractal methods in stereo image generation. Other industries such as medical imaging, chemical molecular modelling, architecture and engineering eagerly adopted 3-D imaging, resulting in rapid development of the technology. (One of the more esoteric examples of its application in the engineering world was the design of compact engine compartments in automobiles; these new configurations were introduced during the 1990s and owed much to associated advances in technical drawing.) Given a view from a single viewpoint, IT can now generate a two-eye pair of viewpoints, viewable as a stereoscopic image. In 1997 the twelve-minute film 'Marvin the Martian' was produced in this manner, although it was 2009 before advances in the quality of computer-generated imagery (CGI) and the added stability of digital projection resulted in the public's awarding 'Avatar' the hypothetical laurel for 3-D features. The creation of 3-D objects for film and computer driven games has become a huge commercial success.

Production in the 2-D world today involves a great deal of electronic manipulation, the solution of which in turn has made it possible to cut the initial costs of location production. 2-D production methods, however, still treat recorded material in a manner analogous to that of an audio recording, the post-production stage becoming effectively a mastering process. Unfortunately, it cannot always be assumed that a less-than-perfect sequence can be economically rescued at this stage. Further progress is still following a path beaten by the audio recording business, but the increasing processing capabilities of generic computers now enables much of this work to be carried out in real time. In audio recording, file formats were created to allow for the need for low bitrate proxy files in the editing process. Once finished the original high quality audio data could be processed overnight if necessary. But soon the need for proxies disappeared, IT systems now being capable of handling the high-quality data at the full rate. This advance was repeated with stereoscopic data, from online editing and formatting, via colour grading and title design, all the way to mastering and creation of duplicate copies. In a contemporary digital imaging suite there are four main operations that can be performed interactively, namely the correction of stereoscopic, colour and convergence errors, and the introduction of appropriate optical effects. The evolution of tools capable of this kind of stereoscopic film finishing has created commercial opportunities to perform new services as part of the 3-D post-production process.

STEREO VIEWING CONDITIONS

Public Viewing

In the cinema (except for outdoor venues), viewing and listening conditions generally follow the guidelines given in SMPTE standard EG-18-1994, which recommends a minimum viewing angle of 30° for movie theatres, amongst other requirements. THX also publishes standards requiring that the back row of seats in a theatre should have at least a 26° viewing angle, and recommending a maximum angle of 36°. An interesting observation is that it was the introduction of sound in the late 1920s that initially established quality requirements, and this remains today where commercial cinema sound systems own quality interests. The original spaces where silent films had been shown suffered from such poor acoustics that the reproduced dialogue could be almost unintelligible. The only real solution to this problem was to produce new spaces specifically built for both good picture display conditions and good sound reproduction. A 3-D film display brings additional problems, with lower image luminance, and possible crosstalk between L/R images with some types of display system (in particular linear polarisation when viewed off-axis). Consequently, for 3-D screenings some cinemas keep the side front seats vacant.

Home Viewing

Television home viewing distances vary, but there are recommendations, usually based on the ITU-R 'Methodology for the subjective assessment of the quality of television pictures'. The figure of three times the picture height (3H) is assumed here, although the EBU recommendations contained in Tech 3276 (supplement), which, like cinema recommendations, were primarily intended to define reference listening conditions, also mention 4H and 6H, which are probably more representative of most homes. Certainly 3H viewing enables the optimum viewing of 1080 line pictures, but this is very close to the screen, especially for a wide-screen feature. Also, any transmission and display impairments will be less visible at greater screen distances, and social arrangements may be easier to meet. Table 1, an extract from EBU Tech 3276 (supplement), shows the relationship between some viewing and listening arrangement parameters (see Figure 13 and Table 2).

Consumers may seldom consider the impact that the size of the viewing screen might have on any 3-D content, and the director will

Table 1. Viewing and listening parameters

Object Distance (Metres)	Parallax (Minutes of Arc)	Number of Depth Steps	Size of Depth (Metres)
0.3	742	1036	0.0002 (i.e. 0.2mm)
1	224	224	0.002 (i.e. 2mm)
2	112	112	0.009 (i.e.9mm)
4	56	56	0.036 (i.e. 36mm)
8	28	28	0.145 (i.e. 145mm)
16	14	19	0.592 (i.e. 592mm)
50	4.5	4.6	6.227
100	2.2	3.28	28.71
400	0.56		3314

Figure 13. Viewing angles for different television systems

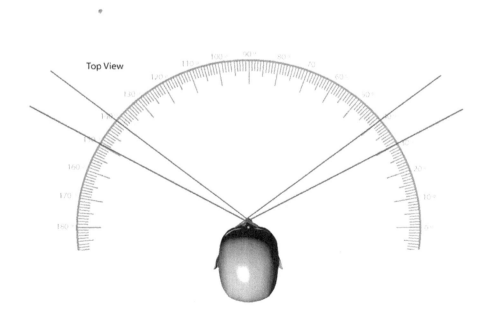

Table 2. 3-D implications of television screen size

3D TV Size (Diagonal – inches)	Optimal Viewing Distance 720p (inches) – 2.3x	Optimal Viewing Distance 1080p (inches) 1.56x	Closest possible Viewing Distance 70 Degree FOV (inches)
42	97	66	26
50	115	78	32
55	127	86	35
60	138	94	38
65	150	101	41

of course have assessed the programme content in the studio control gallery on a 50-inch monitor screen viewed at the recommended 3H distance. Although the current range of 3-D-enabled television sets follows a similar limited range of sizes, larger screen sizes are likely to become available, and this will increase the potential viewing range. When a stereoscopic 3-D image is viewed, the apparent position of objects in the 3-D space is controlled by the distance of the viewer from the screen and the L/R separation of the respective images. Changing the size of the screen affects both these factors. The effect is comparable to varying the spacing of your eyes with the screen size, and unless the parameters are carefully chosen the result can be disturbing.

REFERENCES

Biederman, I. (1987). Recognition-by-components: A theory of human image understanding. *Psychological Review*, *94*(2), 115–147. doi:10.1037/0033-295X.94.2.115 PMID:3575582

Elheny, V. K. (n.d.). *The life and work of Edwin Land*. Retrieved 11 November 2011 from www.nap.edu/html/biomems/eland.html

Herapath, W. B. (1852). Quoted. *Philosophical Magazine*, *3*, 161.

Hubel, D. H., & Wiesel, T. N. (1979). *The brain: A scientific American book*. San Francisco: Freeman.

Swindell, W. (1975). *Polarized light – Benchmark papers in optics/1*. Stroudsburg, PA: Dowden, Hutchinson & Ross.

Treisman, A. M., & Gelade, G. (1980). A feature-integration theory of attention. *Cognitive Psychology*, *12*(1), 97–136. doi:10.1016/0010-0285(80)90005-5 PMID:7351125

un She, S. S. & Wang, Q. (2005). Optimal design of a chromatic quarter-wave plate using twisted nematic crystal cells. *Optical and Quantum Electronics, 37*(7), 625-34.

Zone, R. (2007). *Stereoscopic cinema: The origins of 3-D films, 1838–1952*. Lexington, KY: Univ. of Kentucky. doi:10.5810/kentucky/9780813124612.003.0010

ADDITIONAL READING

Gregory, R. L. (1998). *Eye and brain* (5th ed.). Oxford, UK: Oxford University Press.

Hecht, E. (2002). *Optics*. San Francisco, CA: Addison Wesley.

Kingslake, R. (1989). *A history of the photographic lens*. San Diego, CA: Academic Press.

Chapter 3
Principles of Binocular Stereoscopic Imaging

Geoff Ogram
Stereoscopic Society, UK

ABSTRACT

The relation between visual perception and the recorded image is discussed in this chapter, emphasising the historical growth of the understanding of depth perception and its visual cues. The stereoscopic principle is explained in detail, and figures are given for comfortable viewing of stereoscopic images.

SPACE: MAPPING OUR SURROUNDINGS

Our world is three-dimensional. We were born into this three-dimensional space, we grew up in it and we handle it as part of our daily lives without conscious effort. We rarely contemplate its characteristics in detail or philosophise on its significance; we take it for granted. Nevertheless, if we wish to un-derstand the principles underlying 3-D vision (stereopsis) or 3-D imaging (stereoscopy), we need to examine three-dimensional space in detail.

The word 'space' is not easy to define, although we may use it freely in common parlance. A typical dictionary definition is 'Space, noun: that in which bodies have extension; a portion of extension; room; intervening distance; an interval; an open or empty place.' A more helpful interpretation of 'space' can

DOI: 10.4018/978-1-4666-4932-3.ch003

be achieved by mathematics. One of the many functions of mathematics is to describe the physical world in a succinct and accurate way. The dictionary definition covers both 2-D and 3-D space, but this chapter concentrates only on the latter. Using mathematical techniques we can map out our space to locate key points and show the relationships between various features within that space, such as the distance between two points or the angle between two lines. All such mapping enables us to understand the structure of that space in terms of the objects within it, their sizes, orientations and locations. These details can be established accurately with reference to some fixed point or scale. This principle will be familiar to anyone conversant with map-reading. Britain, for example, can be mapped onto flat sheets

of paper despite the curvature of the Earth and the topography of the landscape. The space is two-dimensional and the reference lines are the N-S and E-W axes on which the Ordnance Survey references are based. Thus the map reader can locate places and estimate distances from the scale of the map. Such maps are based upon the principle of a Cartesian graph plotted on a flat sheet with orthogonal x- and y-axes. Flat objects can be mapped in terms of width and depth (x- and y- coordinates). In a three-dimensional world we can apply the same principles, but we need an extra axis, the z-axis, perpendicular to both the x- and y- axes (Figure 1).

Whereas in 2-D representations the y-axis represents depth and the x-axis width, in 3-D space height is now represented by the z-

Figure 1. (a) Three axes (x, y & z) mutually at 90°, representing width, depth, and height; (b) one of the conventional labelling methods for axes

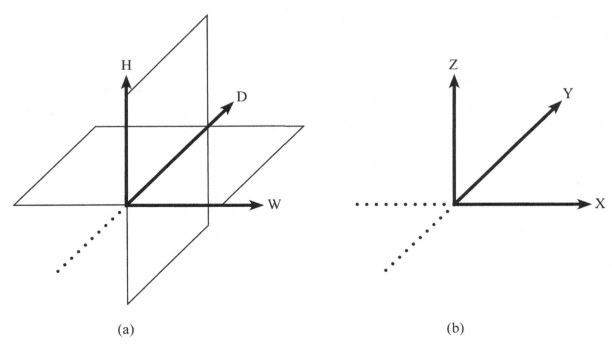

(a) (b)

axis. In this convention, distances in the *y*-direction are measured away from the observer. (Note, however, that in many mathematics textbooks the convention is to have the *y*-axis pointing out of the page.)

VISION: SEEING OUR SURROUNDINGS

Sight is arguably the most valuable of the senses, because it enables us to explore and understand our three-dimensional world to a greater extent than do hearing, taste, smell and touch. From the effects of visible light entering the eyes we are able to interpret information about our surroundings. The result of looking at our environment is usually described as 'creating an image' of that environment. The term 'image' is also applied to photographs and to what we see on a cinema or TV screen. An image matches what we see; it is an echo of reality. But the image we 'see' is not like that projected onto a screen or printed on paper. Light passes through the cornea and lens of the eye, then falls on the retina, and light-sensitive molecules within the rods and cones of the retina absorb the energy, resulting in a change in electrical potential. The detected signal is sent to the brain via the numerous nerve fibres. These signals are processed by the brain to produce what we call the 'image'. Like all real images produced by a convex lens, the retinal image is inverted, though the brain interprets it as being erect. The physiology of vision was discussed in Chapter 2.

RECORDING IMAGES

The images produced on the retina and processed by the brain are transitory, changing continuously as we look around and move about. Humankind seems to have longed to produce permanent images for thousands of years, from cave paintings to present-day sophisticated art forms. For centuries before this the principles of light propagation and image formation had been known, even in ancient Greece: as early as the fourth century BC Aristotle noted that sunlight passing through small apertures produced discs of light on the ground. This phenomenon was probably first described by the Chinese philosopher Mo-Ti in the fifth century BC, according to the sinologist Joseph Needham (1986).

The earliest demonstration of the projection of an entire outdoor image onto a screen indoors by a pinhole is attributed to Abu Ali Al-Hasan Ibn al-Haytham, known in the West as Alhazen, who described his work in his Book of Optics in 1021 (Crombie, 1990). He is credited as being the first person to appreciate that light passing through a small aperture produces an image of whatever is in front of it. A box with a pinhole aperture at one end and a translucent screen at the other, would project an inverted image of the view on the screen . This discovery marked an important contribution to the understanding of the way images formed in the eye, and demonstrated that light travelled in straight lines.

The image quality in a pinhole camera depends upon the size of the aperture: as its diameter is reduced the image becomes sharper (and less bright). But at a certain point the image begins to become unsharp again, this time owing to diffraction, a spreading of the light rays due to the wave nature of light. The principle of the pinhole camera was eventually scaled up using a darkened room with a larger aperture and a focusing lens. (Later workers included a mirror to render the image erect.) This construction became known as a *camera obscura*, which means literally 'dark chamber'. It has been shown that that several artists, notably some of the Dutch Masters, including Vermeer, must have used portable versions of the camera obscura to produce images that could be traced in order to achieve accurate detail and convincing perspective, the rules of which were only imperfectly understood at the time. However, evidence of extensive use of the device is not conclusive. No doubt the artists who made use of the device wished that the beautiful images it produced could be made permanent; and there were many attempts to achieve this using materials known to be affected by light. But it was only in the early nineteenth century that scientific (and amateur) research at last led to the invention of photography.

THE PERCEPTION OF DEPTH

The disadvantage of these images was that while they represented the three-dimensional world they were themselves only two-dimensional. They were the equivalent of a one-eyed view, whereas most of us see the world through two eyes. This binocular vision gives us an enhanced ability to perceive the depths of objects and their spatial relationships in our immediate environment. The scientific name for this phenomenon is stereoscopy, ('seeing solid' from the Greek), and it signifies the ability to perceive depth as well as height and width. A more formal definition, by Chibisov (Valyus, 1966) was 'the science of the visual perception of the three-dimensional space surrounding us'. Valyus (1966) also states that the word 'is usually used to convey appreciation of relief, plasticity, roundness or spatial qualities of the visual images of the objects viewed.' Pinker (1997)discusses the interpretation of two-dimensional images in his book How The Mind Works. He points out initially that there is an element of illusion about various kinds of image and cites an example of the Victorian novelty in which a peephole in a door reveals a well-furnished room, but when the door is opened the room is seen to be empty. The furnished room is a doll's house nailed over the peephole. Pinker also describes the work of Adelbert Ames Jr., a painter who turned to psychology. He hung rods and slabs from the ceiling of a room apparently at random. However, when seen through a peephole in the door there appeared to be a kitchen chair present. In his discussion of two-dimensional images, Pinker states:

Now, a picture is nothing but a more convenient way of arranging matter so that it projects a pattern identical to real objects. The mimicking matter sits on a flat surface, rather than in a dollhouse (sic) or suspended by wires, and it is formed by smearing pigments rather than cutting shapes out of wood. The shapes of the smears can be determined without the twisted ingenuity of an Ames.

The trick, Pinker said, had been stated succinctly by Leonardo da Vinci, who asserted that the perception of perspective was the same as would be obtained by a flat drawing seen behind a glass pane. It is our experience of the real world that enables us to view a flat (2-D) picture and from it gain a sense of the missing third dimension. In most photographs there will be visual clues that aid our understanding of the relative locations in space of the various objects within the picture. Some of these are to be seen in Figure 2.

We subconsciously compare this view with images of real landscapes stored in our memory, and conclude that the scene stretches into the distance from our vantage point. We deduce this from the following observations:

1. Various objects are partly obscured by others (one mountain by another, or part of the wall by a standing figure), so in each case the obscured object must lie farther away than the one that obscures it. A distant object cannot mask a closer one.

2. The pathway appears to get narrower as it reaches the tower near the centre of the picture. We know that parallel lines appear to converge as they recede so we assume that the path leads away from us and that the tower is some way distant.

3. The size of the people in the photograph varies greatly. We know most adults are 1.6–2 m tall, so we deduce that the 'larger' man on the right is nearer to us than the 'smaller' man in shorts on the left. The same reasoning applies to other figures.

Figure 2. Illustrating visual clues to gauge depth (Great Wall of China)

4. The mountains have a gradation of colours, some being very pale. We see this effect in life as a result of haze; the phenomenon is referred to as 'aerial perspective'. The mountains that are paler (and bluer) in colour are more distant, and the fact that in this scene the darker ones partly obscure them confirms this.

5. The patterns of light and shade on objects reinforce the impression of solidity.

6. The closest objects are usually located near the bottom edge of the picture and the most distant ones nearer to the top (though there may be exceptions).

From the above observations and others we can compile a list of visual clues that help us to interpret the spatial aspects of the 2-D images. Not all such clues are necessarily present in every image, but they certainly exist in various combinations. In summary they are:

- **Masking (or occlusion):** Nearer objects partially or wholly mask more distant ones, but not vice versa.
- **Perspective:** Parallel lines appear to converge in a direction away from the observer towards a point on the horizon (the vanishing point).
- **Relative size:** If two objects of the same size are at different distances from the observer the farther one appears smaller as the distance increases.
- **Aerial perspective:** The presence of haze in a landscape lowers the contrast of distant objects less distinct and produces a bluish hue.
- **Lighting and shadows:** Side lighting produces shadows on objects and suggests their shape and solidity. Thus we can distinguish between a flat disc and a sphere of similar diameter and colour. If we view a photograph taken with the sunlight to one side of the camera, we may see an object such as a pole or tree casting a shadow along the ground and onto the front wall of a building.
- **Colour:** So-called aerial perspective is partly associated with colour, as mentioned above.
- **Relative movement (dynamic parallax):** A classic example can be seen in the cinema when a scene is filmed (in 2-D) through the window of a moving train with the camera aimed at right angles to the direction of travel. Nearby objects such as trees appear to pass by rapidly; those farther away pass more slowly, while objects on the horizon appear to be almost stationary. Dynamic parallax is a key factor in depth perception. We rarely look at any real scene without some small involuntary head movements. Such movements change the relative positions of objects in the image we see, giving an immediate impression of depth.

There is also a psychological phenomenon associated with colour that may have some influence on depth perception of a 2-D image. It is that hues towards the red end of the spectrum, especially if saturated, can seem to stand out, whereas those towards the blue end, and desaturated colours, seem to recede.

This effect can best be observed in projected images: bold lettering in a title, especially when seen against a plain blue or dark grey background, can often appear to float in front of the screen, while the background recedes into it. The effect resembles that of looking into an empty box with words painted on a window in front. It may enhance the appearance of depth in a 2-D picture if the bright colours are on the nearer objects, but it can sometimes be disturbing if distant objects are in bolder colours than nearer ones.

All the above visual clues are important in judging spatial distribution of the various components in a flat image such as a photograph, painting or the image on a TV or cinema screen. They are the only factors that contribute to our relating of the perceived image to the real spatial world in which we live, and this ability develops in early childhood. The visual image alone is insufficient; it has to be assessed in conjunction with other stimuli. The classic scenario of the infant reaching out to grasp the moon is a valid one. Over the first few years of life, the child, by touching and feeling objects and by moving around the real world, will gain the experience required to relate the visual data to real space. A little later it learns to interpret the images representing that space more fully. In addition the child is all the time learning these spatial skills with two eyes, each with a slightly differing viewpoint .Now, the visual depth cues discussed above are relevant to our perception of three-dimensional images as to two-dimensional ones. But with binocular (two-eyed) vision we have the ability to gain additional information that would be denied to us if we were restricted to a view through one eye only. Stereoscopy enables us to recreate a visual experience of the real world with uncanny accuracy, provided we follow certain rules. Stepping too far outside these rules may introduce distortions to mar the illusion, as we shall see later. Fortunately, the eye–brain combination is fairly tolerant, and may ignore these distortions.

BINOCULAR VISION AND THE PERCEPTION OF DEPTH

If you move your head slightly when viewing a scene, the apparent movement of objects relative to one another gives the essential clue to true stereoscopy, the ability to perceive the depth of a scene by being aware of the third dimension, the y-axis of Figure 1. Moving the head laterally from one position to another causes the view of the scene to be changed because our viewpoint has shifted. When you view a scene with, say, your right eye closed, your left eye simply sees a flat image. If you now move your head about 65 mm to the right, so that your left eye now occupies the point previously occupied by your right eye, your eye will see a slightly different flat image, the 'parallax' described earlier. In this case you see the two images one after the other; but if you now view the same scene with your head still and both of your eyes open, you see the two flat images simultaneously, one with your left eye and the other with your right eye. The receptors in your cerebral cortex combine these images to provide you with a sense of depth that is lacking in a monocular image. The distance between the two viewpoints, i.e.

your eye separation, is the stereo base. It differs from one person to another. but of course remains constant for any individual. In making stereoscopic images by photographic or other means the stereo base can sometimes be greater or less than this 'interocular distance', as long as this is appropriate for the conditions. In normal vision we are not usually aware of two separate images. Instead we seem to be looking out as if from a single eye positioned at the bridge of the nose. This is the so-called cyclopean image (Julesz, 1960), named after a race of giants in Greek mythology, who possessed only a single eye in the centre of their foreheads.

A simple experiment to illustrate parallax is as follows. Close your right eye and hold a pencil or similar object upright at arm's length and, viewing with your left eye alone, align the pencil with a vertical line such as the edge of a window frame, about 3 m or more away. Without moving the pencil or your head, close your left eye and open your right eye. The pencil will appear to have jumped to the left relative to the window frame. If you repeat the experiment with the pencil closer to you, the apparent leftward movement of the pencil will be greater. Your left and right eye images make up what is termed a stereoscopic pair, and the pair of drawings in Figure 3 is a representation of what you have seen. It shows the difference between your left and right eye images. (Note that in this diagram the eyes are assumed to be focused on the pencil.)

In this experiment, the images are formed on the retinas of your eyes and interpreted by your brain; but we can create similar images as drawings or as photographs made from the two viewpoints occupied by the eyes. If these drawings or photographs are viewed in such

Figure 3. A pencil-and-window experiment

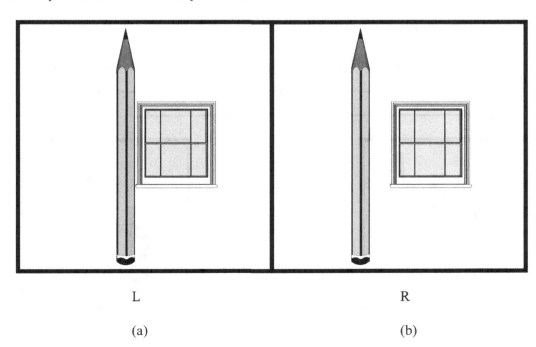

L   R

(a)   (b)

a way that the left image is seen only by the left eye and the right image by the right eye, your brain combines them to reproduce the scene in depth as a three-dimensional sketch or, with colour photographs, a full-colour three-dimensional replica of the original scene. This is the basis of stereoscopy, and there are many techniques for accomplishing it. It is a simple concept, but to achieve the best results it requires a deep understanding of the subject.

LATERAL DISPLACEMENTS IN STEREO PAIRS

The pencil experiment described above employs a simple arrangement of two objects in which the nearer one (the pencil) appears to be laterally displaced leftwards relative to the background. With the left and right views placed side by side as in Figure 3, the window

frame images are seen to have moved relative to the pencil images. As both of your eyes are focused on the pencil, it will not itself appear to have moved. When your eyes are focused on a near object any far object will appear to have a greater separation between the two images. These separations between corresponding points in the two images of any one object are known as homologous points or homologues.

Figure 4 shows three post-like objects at different distances, the one on the left being closest and the one on the top right farthest away. In this diagram $s_f > s_m > s_n$. This illustration forms a true stereoscopic pair and can be viewed as such, verifying the relative locations of the objects as described. These separations are not true parameters because their absolute values depend upon how far apart the two images have been placed in whatever format is to be used for viewing them

Figure 4. Image separations for three objects at different distances

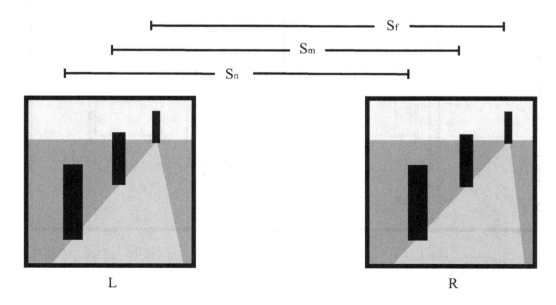

L   R

as a stereoscopic pair. The measurements are the differences $(s_f - s_m)$, $(s_f - s_n)$ and the like; these are independent of the left and right image spacing. Even so, the left and right images of a stereoscopic pair have to be a distance apart that suits whatever device is to be used for viewing, if the result is to be satisfactory.

If these differences in separation are to be of any practical use, they need to be compared against an appropriate reference value. The way in which your eyes function in normal vision provides a logical one. If you view any real scene, your eyes adjust in two ways: they re-focus (accommodation) and they rotate towards each other (convergence) as they focus on closer objects. In normal perceptual situations accommodation and convergence work together. However, the physiological mechanisms of accommodation and convergence are not fully linked, and can function independently. In viewing stereograms in a typical viewer they do so. Your eyes have to be focused on the surface of the photograph for every part of the image, but they converge as normal for objects at different distances, just as they would in a real scene. Thus, your eyes converge while maintaining a constant focus on the plane of the photograph; this does not occur in normal vision. (The eyes can even be made to diverge, but this is not a natural action and may result in eyestrain.)

The sense of depth resulting from binocular vision is not an absolute one. It enables us to understand the relative positions of objects but it does not tell us the precise distances; our eyes are continually adjusting their convergence by various amounts as we survey a photographed scene; but since we are not consciously aware of such movements or their magnitudes we do not relate these variations in convergence to actual distances. Nevertheless, with practice one can develop a certain degree of skill in estimating distances, using convergence as a subconscious clue.

In theory, the axes of the eyes would be parallel when viewing an object at infinity, and when you view distant objects your sight-lines are so nearly parallel that for practical purposes they can be regarded as parallel. It therefore seems logical to take the separation of infinity homologues as a convenient standard, as it is essentially a natural limit. So, in defining the differences in separations of homologues, we can use the terms $(s_\infty - s_m)$, $(s_\infty - s_n)$, and so on, where s_∞ is the separation of infinity homologues. The differences are now referred to as parallax deviations or simply deviations.

PERSPECTIVE DIFFERENCES IN STEREO PAIRS

So far this chapter has treated objects as consisting simply of a number of points (or flat cardboard cut-outs) situated at various distances from the observer. In reality, most objects are themselves three-dimensional and this will give rise to other differences in a stereo pair. Whereas in a flat object, photographed head on, the left and right images in a stereo pair are identical, for an object with depth the left eye sees more of its left side and the right eye sees more of its right side. Consequently each eye sees part of the object that is not visible to the other eye. Figure 5, representing a plan view of a cylinder, illustrates this point. L and R represent the eyes. In Figure 5 (a)

Figure 5. Perspective differences in stereo pairs: (a) near view, (b) far view

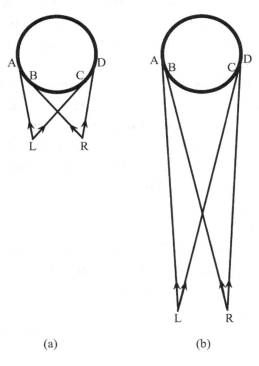

(a) (b)

both eyes can see the portion BC, but AC is visible only to the left eye and CD only to the right. For the same object seen at a greater distance as in Figure 5 (b), the portions AB and CD are both smaller, but BC and AD are both larger. While more of the object is seen with both eyes together the object does, of course, appear smaller because it is farther away. A further difference, which appears when you view an object such as a cube is in the shape that your two eyes perceive. Figure 6, another stereo pair, shows left and right images of a cube, each drawn in correct perspective. Consider the left face marked as HKLM in each image. The edge HK is farther away than the edge ML, and from our discussions of parallax deviations it follows

that s_r is greater than s_n. So the shape of the face HKLM is different in the two images: in the right image, it is slightly narrower, so the lines HM and KL are slightly steeper. When viewed stereoscopically this left face of the cube appears as a receding plane, giving visual depth to the image of the cube.

ACCEPTABLE RANGE OF DEPTH

The pencil and window experiment described above provides another important aspect of depth perception. Try setting up the experiment as before, aligning the pencil with the edge of the window frame, but now view the result with both eyes. First focus on the window frame, and then refocus on the pencil. In the first instance you can become aware of two images of the pencil (somewhat out of focus), while in the second the pencil will be in focus but you will see two out-of-focus window frames in the background. This effect may be noticeable once you have become aware of it, although it should not normally be disturbing.

Now try a variation on the experiment, which may be enlightening. First, still holding the pencil at arm's length and viewing with both eyes, gradually move closer to the window frame. Then fix the pencil in some kind of holder and place it successively in several positions closer to the window frame, each time viewing with both eyes from your original vantage point. In each case, as the distance between the pencil and window frame decreases, the lateral distance between the relevant double images decreases, and eventually the double vision effect seems

Figure 6. Left and right images of a cube

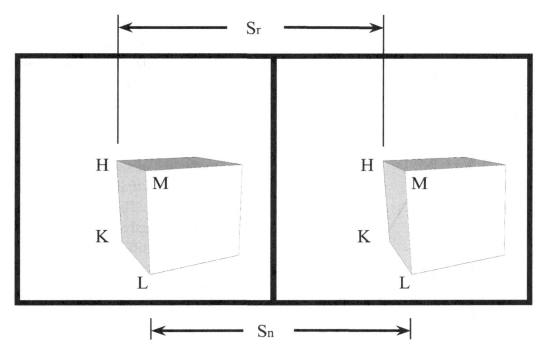

to disappear. In fact it is still present, but at some point your eyes change focus repeatedly between the pencil to window frame and back without your being aware of it. This suggests that if the distance between the nearest and farthest objects in a scene is excessive, then the double imaging effect becomes too prominent for comfortable vision, and the image is described as having too much depth. There is thus a limited range of depth in a 3-D image that allows it to be viewed comfortably.

There is another difference between viewing reality and a viewing a 3-D photographic image. In a real scene, when an object is so close that it leads to double imaging, it is usually a simple matter to move away to a more favourable position. But when a still 3-D image includes too much depth there is no way of avoiding the discomfort. In the cinema or on 3-D TV the effect may be tolerable as long as it is short-lived, as when objects move swiftly across the screen and out of sight, or if the camera pans around a particular location.

It is generally agreed that the eye–brain combination will accept the view of an outdoor scene containing items ranging from the far distance to about two metres with no obvious double imaging. as long as the nearest object is no closer than this. This is not a precise physical law, but is a fair description of the acceptable depth range for comfortable viewing. By the same reasoning it follows that subjects for stereoscopic imagery should not exceed this range. Consider the geometry of Figure 7, which represents an overhead view of the eyes (marked L and R) looking at a scene that stretches to infinity, with a near object

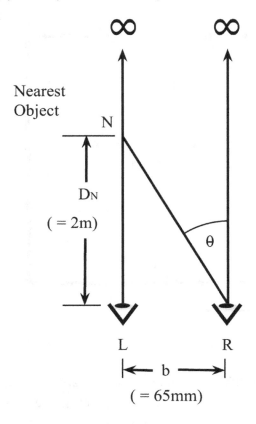

Figure 7. Maximum depth range (2m to infinity) for comfortable viewing

situated at N. For simplicity N is located on the left eye's direct line of sight.

The optic axes of the eyes are effectively parallel when they are focused on infinity; but when the focus is changed to the nearest object N at a distance D_N from the observer, the right eye swings through an angle θ. In adult humans the interocular distance (eye spacing) varies between some 55–75 mm. Taking a mean value of 65 mm (the dimension b in the diagram) we can calculate the value of θ. Plainly, angle LNR is equal to θ, so tan $\theta = b/D_N$.

Since $D_N = 2000$ mm and $b = 65$ mm, tan $\theta = 65/2000 = 0.03$, giving $\theta = 1.87°$.

For convenience, we can round this value up to 2°. This angle represents approximately the maximum eye convergence that avoids obtrusive double imaging. As long as the nearest object is at 2 m distance it can be anywhere in the scene, and need not be confined to the line of sight of either eye. If it is located more centrally, say, then both eyes will turn inwards. The total eye convergence will not have exactly the same value. but for our purposes the difference will be negligible.

This 2 m-to-infinity restriction does not mean that we cannot view objects that are closer than 2 m; that would be absurd. We can view closer objects perfectly well, but in so doing we have to be aware of having too much depth for comfort. This suggests that some other restriction has to be imposed to prevent this occurring. The 2° eye convergence limit provides the answer. To avoid too much depth we must ensure that the permissible far point must be only so far from the nearest point that the 2° eye convergence difference is not exceeded. Thus if the nearest object is closer than 2 m then clearly we may not include in our view any objects at infinity. Instead, for every near point closer than 2 m we can calculate a corresponding permissible far-point distance based on this angle. Table 1 gives some values. Notice that there is a reduction in the acceptable depth range as the near-point distance becomes less. We are already familiar with the 2 m–infinity depth range, but for a near point just 1 m closer, i.e. at 1 m distance, the permissible far-point distance is reduced to as little as 2 m.

Table 1. Permissible far point distance D_F for a given near point D_N

Near distance D_N	200 mm	400 mm	600 mm	800 mm	1m	1.2m	1.4m	1.6m	1.8.m	2m
Far distance D_F	222 mm	500 mm	857 mm	1333 mm	2m	3m	4.67m	8m	18m	inf

Another way to represent this information is to calculate the total permissible depth range ΔD that extends beyond the given near point distance D_N, i.e. $\Delta D = D_F - D_N$. Using the distances given in Table 1 we obtain Table 2.

This information is useful when recording images by photography as camera viewpoints can be selected on the basis of the near point/depth range combinations so that the final stereoscopic image can be viewed comfortably. In a scene too much depth can lead to eyestrain. These acceptable depth ranges can be expressed as deviation values in images. Consider the two objects M and N in Figure 8, which lie on the axis of the left lens of a stereoscopic camera. From any object, light rays that pass through the optical centre of a lens are undeviated, enabling us to define in the simplest way where their images are formed. Thus, the points m, n and i on the film are the images of M, N and an object at infinity respectively. The on-film deviations are shown as d_m for object M and d_n for N, measured as the separation differences between pairs of homologues (mm – ii) and (nn

– ii). These are identical to the lengths im $(=d_m)$ and in $(=d_n)$ both on the right image.

Note: In the diagram the separations mm and nn are greater than the infinity separation ii, which is at odds with what is stated above, namely that separations of near points are smaller than infinity point separations. However, because the two optical images are inverted, when the uncut processed film is viewed from the non-emulsion side with the images the right way up, the left image will be on the right of the right image, so they have to be cut from the film, separated and transposed for correct viewing as a stereo pair. Thus the true values of the separations will be restored.

If point N represents the nearest permissible object (at 2 m in this example, which includes infinity) then d_n becomes the largest permissible on-film deviation. The same maximum deviation value will apply to any acceptable depth range, such as the examples given in Tables 1 and 2. Nevertheless, the value of maximum deviation as defined here is not fixed. If the camera lenses in Figure 8 were of longer focal length, the lens–film

Table 2. Permissible depth range ΔD for a given near point distance D_N

D_N	200 mm	400 mm	600 mm	800 mm	1m	1.2m	1.4m	1.6m	1.8m	2m
ΔD	22 mm	100 mm	257 mm	533 mm	1m	1.8m	3.27m	6.4m	16.2m	inf

Figure 8. Parallax deviations d_m and d_n produced on film by point objects M and N, respectively

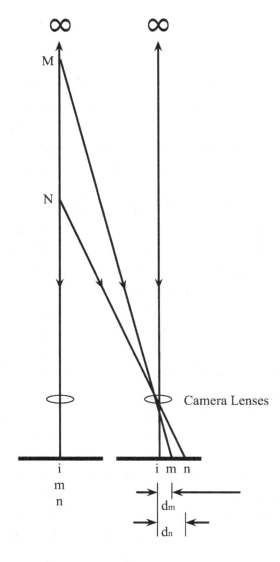

midway between the two parallel sight lines, there would be a deviation of $d_n/2$ on each of the left and right film images, so the value of the total deviation would be unchanged. The same value of total deviation will apply to any other object lying on a line through N parallel to the film plane, but it will be divided unequally between the two film frames. For objects left of centre, the greater portion of the deviation will appear on the right film frame, and vice versa. The same applies to the deviations of all points lying on a line through M parallel to the film plane, in this instance with a smaller total value of d_m rather than d_n. And the same principle applies to all points along any other similar line parallel to those through N and M; the more distant the line, the smaller will be the value of the total deviation.

EXTENT OF BINOCULAR VISION AND DEPTH PERCEPTION

The clue to stereoscopic vision lies in the way in which nerve fibres connect the retinal rod and cone cells to cells within the cerebral cortex of the brain. In a simplified model, we can think of each brain cell being connected to two visual nerve fibres, the impulses in which have come from two corresponding eye areas, one in each retina. These are referred to as corresponding points. When the gaze is fixed on a stationary object, the impression of a single object is generated. Strictly, this applies only to an individual point on the object; the whole object is perceived by a replication of this effect as the eye scans the scene. However, returning to the fixed-gaze situation, the part

distance would be larger in order to achieve correct focus, and the deviation values would consequently be greater in magnitude.

For objects that do not lie on one of the lens axes the deviation will be split between the two film images. For example, if N were to be to the right of its position in Figure 8, exactly

of the object on which the eyes are focused forms images on corresponding points of the two retinas. Other objects either in front of or behind the primary object will form retinal images in different locations in the two eyes (Ackermann, 1992). Certain nerve cells in the brain, known as binocular disparity neurons, are activated only when this image difference occurs. Such interpretative skill is for a static scene gained through experience. The brain estimates distance using two main clues, namely the differences in retinal images and the size of the image, and if either the content of the scene or the viewer's head moves, the dynamic parallax becomes important.

Two parameters are relevant in defining the ability of the individual to distinguish depth clues and establish object locations in space. The first of these is known as stereo acuity, and represents the smallest depth difference that can be resolved by the eyes of the observer. It is also referred to as the stereopsis threshold or stereo definition. The second factor is the maximum distance at which any depth differences can be identified, what might be regarded as a limit to stereoscopic vision, or stereo infinity. The two are linked.

In Figure 9 the two eyes are shown fixated on a point A situated at a distance D. If A is off centre, the lengths LA and RA will differ, but provided they are much larger than the eye spacing b they can be taken as equal. The circle passing through L, R, and A represents the section of the horopter in the plane of the diagram. This horopter is the locus of points in space that form images on corresponding points in the retinas, and is spherical. The geometric properties of the circle are such that the angle subtended by a chord (LR.) at the perimeter, angle θ, is constant for all points on the perimeter. If the eyes are focused on a more distant point B, then there will be a similar horopter of greater diameter, representing other object points that produce images on corresponding retinal points, and a smaller angle θ_1. In other words, each horopter relates to a particular eye convergence.

Referring to Figure 9, which shows the eyes focused at A, consider point B, lying just outside the horopter. The retinal images formed by B will not lie on corresponding points in the eye, and the image disparity will trigger the brain into registering the image of B as farther away than point A, provided ΔD is large enough. However, if B is too close to A, the two points will be seen as one. For the observer to be able to distinguish the depth difference between A and B, and to see two

Figure 9. Horopter: Point A, on which the eyes are focused, gives retinal images on corresponding points in the two eyes

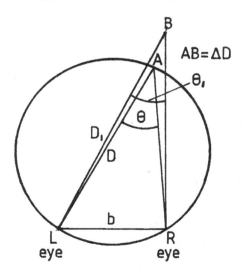

separate points, the difference between the two angles LBR and LAR has to be sufficiently large. For angles LAR and LBR (θ and θ_1 respectively), the parallax angle is given by $\Delta\theta = \theta - \theta_1$.

Stereoscopic acuity is defined by the minimum discernible parallax angle $\Delta\theta_0$. Assigning a value to this is not straightforward, as it varies between individuals, and depends on external factors such as subject brightness and duration of observation. A value as low as 2 arcseconds has been reported (Klooswijk, 1978); but Valyus (1966) earlier suggested that 30 arcseconds was a representative value for use in calculations. Assuming that $D > b$ (Figure 9), then all parallax angles can be written in the form $\theta = b/D$ in radian measure. Accordingly, the expression can be used to determine the limit of stereoscopic vision. Rearranging and taking $b = 65$ mm and $\Delta\theta_0$ $=30$ arcseconds or 0.000145 radians (i.e. 30 arcseconds or 1/120 degree) then

$D = b/\Delta\theta_0 = 65/0.000145 = 448256$ mm or just over 448 m.

The value of 30 arc seconds for stereo acuity takes into account the various factors that may diminish an individual's ability to detect small depth increments. The stereo acuity of most people is in practice better than this, in the region of 5–10 arc seconds, which suggests that the limit of stereoscopic vision is likely to be greater in magnitude. If the value 0.000048 radians is used in the calculation, we have $D = 65/0.000048$ mm $= 1354$ m or just over 0.8 miles.

It should now be apparent that one cannot be dogmatic about the extent of binocular vision as it depends upon so many factors. 1.35 km is probably as good a value as any other quoted distance, though in many circumstances this figure is perhaps optimistic. On the other hand, figures as low as 200 m, which are sometimes quoted, seem pessimistic. Stereo acuity is undoubtedly likely to be better in an actual location (in a good light) than when viewing a stereoscopic picture of the same scene through a viewer. Under projection conditions stereo acuity will be probably be even lower.

To recapitulate, stereo acuity is defined as the parallax angle that just enables an observer to see that two objects lie at different distances. In Figure 10, A and B are two such points as determined by the stereo acuity angle $\Delta\theta_0$ as shown. The length AB is the detectable depth step, labelled as ΔD_1. When the eyes focus onto a more distant point A1, the nearest point that can be detected as farther away is now B1. Because of the change of viewing angle the depth step is greater, with a value ΔD_2. It should be clear that the smaller the stereo acuity value, the smaller is the depth step that can be detected for a given set of circumstances.

The conditions that affect the value of stereo acuity referred to above when outdoors include such factors as the quality of lighting (sunny or cloudy) and the presence of haze or rain; when viewing stereograms indoors variables such as the brightness and lens quality of viewers or projectors, and the quality of the photographic image itself, will be important. But, for a particular observer in specific conditions the stereo acuity will be fixed for the whole scene being viewed, and we can analyse the variation of depth step

Figure 10. Variation of depth step ΔD with distance

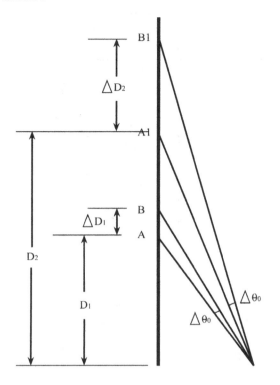

with distance more quantitatively. The exact relationship between object distance and the size of the minimum depth step can be determined from Figure 9. The distance AB normal to the horopter is given by

$$\Delta D = D_1 - D$$

Since in practice D and D_1 are very much greater than b, the angles are very small. For the same reason we can assume that the sides of the triangles LA and RA are equal, as are LB and RB equal to D and D_1 respectively, so

$$D_1 = b/\theta_1 \text{ and } D = b/\theta.$$

hence

$$\Delta D = b(1/\theta_1 - 1/\theta) = b(\theta - \theta_1)/\theta_1\theta$$

i.e. $\Delta D = D^2\Delta\theta/(b - D\Delta\theta$

The minimum depth step ΔD_0 is the value of ΔD when $\Delta\theta$ is equal to the stereo acuity $\Delta\theta_0$. Thus

$$\Delta D_0 = D^2\Delta\theta_0 /(b - D\Delta\theta_0)$$

This expression, which follows the analysis by Valyus (1966), shows that the minimum discernible depth step is roughly proportional to the square of the object distance, measured from the observer. The significance is that our perception of depth falls off rapidly for distant objects and that the stereoscopic effect is more pronounced at relatively close range. As a general guide, the photographer should ensure that the important elements of a stereoscopic image lie within the 2–200 m range, though more distant objects need not necessarily be excluded. It is worthwhile to calculate the number of discernible depth steps that occur with various distance ranges. For the purposes of this exercise the stereo acuity is taken as 30 arcseconds (0.000145 radians). An observer will thus just be able to discern two objects as lying at different depths if the parallax of one is 30 arcseconds greater than that of the other.

This can be seen from the following, where $b = 65$ mm and $\Delta\theta_0 = 0.000145$ radians.

From the above equation the minimum discernible depth step at $D = 10$ m will be given by

Table 3. Depth step values

Object Distance (metres)	Parallax (minutes of arc)	Number of Depth Steps	Size of depth step (metres)
0.3	742		0.0002 (0.2mm)
		1036	
1	224		0.002 (2mm)
		224	
2	112		0.009 (9mm)
		112	
4	56		0.036 (36mm)
		56	
8	28		0.145 (145mm)
		28	
16	14		0.592 (592mm)
		19	
50	4.5		6.227
		4.6	
100	2.2		28.71
		3.28	
400[#]	0.56		3314

 # With a stereo acuity of 30", the limit of stereo perception is calculated to be 448m

$\Delta D_0 = (10^2 \times 0.000145)/[0.065 - (10 \times 0.000145)]$ m $= 0.228$ m or 228 mm

So, if an object is located at a distance of 10 m, a second object will just be distinguished as lying farther away than the first if it lies at 10.228 m, but not if it lies at, say, 10.2 m. The 228 mm depth step value is shown in Table 3 together with those calculated for object distances of 2 m and 200 m.

We can now calculate the number of depth steps that lie between two points at different distances from the observer. To do so, we need to consider the parallax angles as illustrated in Figure 11. The calculation simply involves determining the parallax angles θ_1 and θ_2 for the relevant distances D_1 and D_2. The angular difference $(\theta_1 - \theta_2)$ divided by the stereo acuity gives the number of depth steps. Of course, within a given distance range the step size is not constant but increases with distance, as shown in Table 3. For example, the parallax angle for 1 m distance, as seen by the eyes, is 224 arcminutes, and that for 2 m is 112 arcminutes. Since the stereo acuity of 30 arcseconds can be written as 0.5 arcminutes,

Figure 11. Variation of parallax angle with distance

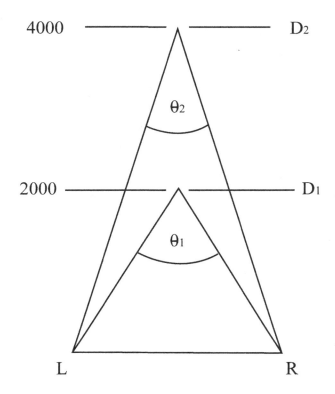

the number of depth steps between 1 m and 2 m will be (224–112)/0.5 = 224 steps.

Table 4 gives the results of similar calculations for object distances from 0.3 m (representing approximately the nearest distance of distinct vision) to 400 m, just short of the stereo infinity value of 448 m for the chosen stereo acuity value of 30 arcseconds. The table shows the number of depth steps in each distance range, and the size of the step at each key distance.

The significance of this brief analysis is that the human eye can discern finer depth

Table 4. Variation of depth steps with distance

D	ΔD_0 for $\Delta \theta_0 = 30''$	ΔD_0 for $\Delta \theta_0 = 10''$
2 m	8.65 mm	2.96 mm
10 m	228 mm	74.4 mm
200 m	160 m	34.7 m

detail at close range, and that successful stereoscopic image making is likely to gain the maximum effect with the inclusion of objects in the foreground.

REFERENCES

Ackermann, U. (1992). *Essentials of human physiology*. St Louis, MO: Mosby Yearbook.

Crombie, A. C. (1990). *Science, optics, and music in medieval and early modern thought*. New York: Continuum International Publishing Group.

Julesz, B. (1960). Binocular depth perception of computer-generated images. *The Bell System Technical Journal, 39*(5), 1125–1163. doi:10.1002/j.1538-7305.1960.tb03954.x

Klooswijk, A. I. J. (1978). Natural and photographic stereo acuity. *Stereoscopy, 5*.

Needham, J. (1986). *Science and civilization in China: Physics and physical technology*. Taipei, Taiwan: Caves Books Ltd.

Pinker, S. (1997). *How the mind works*. London: Penguin Books.

Valyus, N. A. (1966). *Stereoscopy*. London: Focal Press.

Chapter 4
Analogue and Digital Stereo–Photography

Geoff Ogram
Stereoscopic Society, UK

ABSTRACT

The origins of stereoscopic imagery are discussed briefly, and a practical method of producing stereoscopic pairs of images with a single-lens camera is explained in this chapter. The history of stereoscopic cameras is summarised, and the models and formats listed. The various formats for 35 mm formats are discussed and the fundamental geometry reviewed. Instructions for the correct mounting of stereo pairs of images are given. Equipment for digital stereoscopic photography is discussed, with descriptions of available models and their control devices and associated software listed for each model. Specifically designed programs for stereo image processing are available.

BACKGROUND TO STEREO-PHOTOGRAPHY

Stereo-photography is almost as old as photography itself. In 1826, the French inventor Nicéphore Niépce produced the first image on a pewter plate coated with a layer of bitumen of Judea, a white substance that hardened on exposure to light. He exposed it for a whole day, then washed away the unaffected bitumen with oil of lavender to leave a positive image. This could be coated with ink and pressed onto paper to give a permanent image that he called a 'heliograph'. He discovered that silver compounds were more amenable, and threw his lot in with Louis Daguerre, whose images, based on silver as the recording

DOI: 10.4018/978-1-4666-4932-3.ch004

medium, became known as *daguerreotypes*; these were also positive images. Around 1839 Henry Talbot in England produced negative images on silver chloride impregnated paper, from which any number of prints could be made; he called these images *calotypes*. The standard size was 85 x 170 mm. Contributions by John Herschel led to the chemical techniques still used today for (analogue) photography. Later developments included the wet collodion process, in which glass plates coated with a solution of nitrocellulose in ether were sensitised and exposed wet, being prepared and processed on location by the photographer. George Eastman in the US had not yet introduced gelatin-silver halide coated cellulose nitrate film and small cheap cameras, so large cameras constructed of wood with brass fittings producing images on glass plates were the rule, and exposure duration was regulated by removing and refitting a cap over the lens.

It was not long before stereoscopic cameras, with two lenses side by side to record simultaneous left- and right-eye images on wide plates, were introduced. Single lens cameras were also used to make stereoscopic pairs of negatives by exposing two plates, moving the camera a few inches to one side between shots. By the 1850s stereophotographic image pairs were widely available to the public. These were viewed using an optical device invented by Charles Wheatstone called a stereoscope. Stereoscopic photographs became very popular in Victorian homes, as Roger Taylor explained in Chapter 1. There is a fascinating modern study by Brian May and Elena Vidal of a set of stereo cards produced in the 1850s in a book entitled A Village Lost and Found (May & Vidal, 2010).This book includes reproductions of the complete original set of stereo card images and a specially designed viewer.

Successful stereophotography relies to a large extent on the sharpness of the images. Anything in a scene that moves during an exposure and produces a blurred image will look odd when viewed through a stereo viewing device, and any object that changes position in between two sequential shots, or is missing from one of them, will give rise to strange visual effects. To obtain overall sharpness the camera should normally be focused on a distance that will give maximum depth of field.

Stereophotography was the natural successor to the 3-D drawings of geometric shapes produced in 1838 (as side by side left and right views) by Wheatstone in a scientific paper presented to the Royal Society (Wheatstone, 1838). In his paper, Wheatstone also described stereopsis and gave details of his invention, a mirror device for viewing such drawings as a three-dimensional image. It was he who coined the term *stereoscope* from the Greek 'seeing depth'. The name is now used to describe any device used for the same purpose. Production of stereograms (also known as stereographs) was then almost entirely confined to the professional photographers of the day.

David Brewster was the first to design a stereoscopic camera; he also invented an improved design of stereoscope, still in use today (Brewster, 1856). Towards the end of the 19th century smaller cameras began to be marketed, an example being the French Verascope, designed and patented in 1891 by Jules Richard (Perrin, 1997). The first model appeared in 1893, and many improved models were made and sold over the next fifty years or so. The availability of such equipment might have been expected to result in a flurry of amateur activity in stereophotography, but interest in the subject tended to fluctuate, as it does to this day. After 1900 interest in stereophotography waned, although it assumed importance as a tool in the hands of the RAF photo-interpreters who examined aerial reconnaissance photographs taken over enemy territory using simple stereoscopes. Interest reawakened in 1947 when the American Stereo Realist camera, using 35 mm colour film, was launched. The 35 mm camera soon became the standard for amateur stereophotography, but in general the technique remained a niche interest, strictly for enthusiasts. Then early in this century, public interest in stereoscopy began to increase as a result of the production of 3-D movies and the promotion of 3-D television and computer games. Present-day stereo cameras use two lenses separated by around 65–75 mm, the average eye spacing in human adults. Almost all the new models are digital, though film is still popular.

Stereophotography with a Single Lens Camera

This method is restricted to static subjects because it relies on taking sequential exposures, moving the camera sideways keeping the optic axes parallel (i.e. not toed in), with a horizontal spacing of 65mm between the two exposures. It is important to keep the camera level (Figure 1). The camera may be tilted downwards when taking a picture looking down from a high vantage point, provided the lens axes remain parallel.

For standard stereoscopy, the stereo base of 65–75 mm will produce satisfactory stereo images as long as the nearest object is no closer than about 2 m. The most distant object can, of course, be nearer than 'infinity'. With close subjects (nearer than 2 m) the stereo base may need to be reduced (hypostereoscopy), whereas for very distant subjects it could be increased to advantage (hyperstereoscopy). This is discussed later. For standard 35mm film cameras the individual image size is 36×24 mm (so-called full-frame), 50% wider than the images produced in the so-called 5P format (5 perforations per frame); this does provide a pleasing rectangular format, almost panoramic by comparison. There is also the option of mounting the images in 5P or 7P format mounts if the subject is better portrayed in a narrower frame, so the full frame images give more flexibility in this respect.

One technique in hand-held shots is to stand perfectly still and make the two exposures, the

Figure 1. Making sequential exposures with a single lens camera

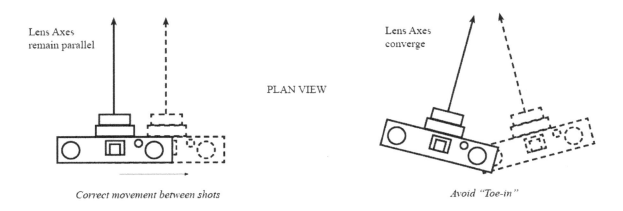

Lens Axes remain parallel

Lens Axes converge

PLAN VIEW

Correct movement between shots

Avoid "Toe-in"

first using your left eye in the viewfinder and the second using your right eye. Alternatively, and perhaps more reliably, take the first shot with your weight on your left foot and the second with your weight on your right. Working with a tripod, you can fix some sort of slide bar that allows the camera to be shifted sideways by the required distance, and this is the most reliable method. Some simple designs are shown in Figure 2.

Twin Rigs with Two Single Lens Cameras

When there is movement in a scene to be photographed the two exposures must be made simultaneously. It is fairly easy to make a rig with two identical single lens cameras mounted side by side on a bar. Although the shutter buttons can be fired together using both hands, it is difficult to synchronise them precisely. It is preferable to use either a double cable release or to have the cameras connected together electrically so that operating the exposing button on one camera activates the shutters on both. Most single lens reflex (SLR) 35 mm film cameras mounted side by side give a stereo base well in excess of 65 mm because of their physical size. This increased base enhances the depth of the image, but this may not be desirable. One possible solution is to mount the cameras vertically, base to base. The images will be in portrait format and may have to be trimmed to fit conventional stereoscopic viewers.

Although compact cameras are much smaller than SLRs it may still not be easy to reduce the inter-lens spacing sufficiently. However, in many compact cameras the lens is not central, but is closer to one side. If this is so, a convenient solution is to mount the two cameras on a Z bar. This is a bar bent twice at right angles to form a central pillar adjoining two horizontal arms at different heights, one to the left and one to the right. One camera is inverted and fixed to the upper arm, while the other is fixed right way up to the lower

Figure 2. Simple slide bars and pantograph: (a) recessed tray, (b) tray with spacer block, (c) simple slide bar with scale, (d) pantograph

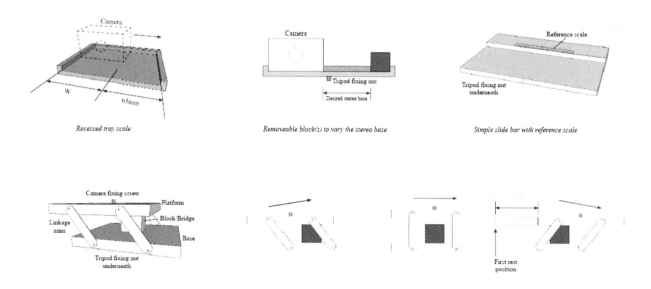

one. The bar should be designed so that the two lenses are at the same level. The stereo base can be changed as desired by varying the camera separation.

At one time most professional photographers worked with medium format cameras using 120 film to produce images of 60× 45 mm, 60× 60 mm, or 60×70 mm, depending upon the camera design. Since the images were larger than those on 35mm film, the image quality was usually superior. Some stereophotographers made twin rigs of linked medium format cameras, but expensive film stock – and fewer images per film – made this format less popular.

Analogue Stereoscopic Cameras

The most convenient method of taking stereo-photographs is to use a purpose-built twin-lens stereo camera with a single linked shutter mechanism giving perfect synchronisation. Probably the best-known amongst many that use 35mm film is the Stereo Realist (Figure 3), made by the David White Company in Milwaukee. This is regarded as a classic, and is still widely used; it may be considered typical of its era. From 1947 to 1972 many thousands were made, with some variations

Figure 3. Stereo realist 35mm film camera

such as faster (f/2.8) lenses, double exposure prevention and an improved film advance mechanism.

Its appearance in 1947 prompted other manufacturers (mainly in the USA and Germany) to produce a large range of 35mm stereo cameras. Interest in stereophotography blossomed over the next decade – until (as usual) the novelty wore off and only die-hard enthusiasts kept up the momentum. At first sight the Realist appears to have three lenses. Actually, the central one is the front lens of the viewfinder system. This is linked via internal mirrors to the viewfinder eyepiece at the bottom of the rear face. This central location of the viewfinder window ensures accurate framing with no vertical parallax errors. The following is a brief specification:

- **Lenses:** Two matched lenses, f/3.5 3-element (Model ST- 41) or f/2.8 4-element (ST- 42) focal length 35mm:
- **Lens separation:** 70 mm
- **Frame separation:** 71.25 mm
- **Shutter speeds:** ST- 41: 1– 1/150 s + T & B; ST- 42: 1–1/200 s + T & B.
- **Apertures:** f/2.8 (or f/3.5) to f/22

The camera range focuses down to 760 mm, and has double exposure prevention (can be over-ridden), auto-exposure counter and cable release socket. With manual exposure and focusing, it is a versatile camera. Unusually, the lenses do not rotate when focusing. Instead, the focusing knob controls the position of the film plane. The built-in rangefinder is coupled to the camera focusing mechanism; with a base of 120 mm it is accurate and fairly sensitive. There is a depth of field scale. The flash shoe is an obsolete design and requires modification if it is to accept modern flashguns. Spares for this camera may be difficult to find, but the camera itself still appears regularly in the second-hand market.

The camera has been served well by the publication of two books. The Stereo Realist Manual (Morgan & Lester, 1954) is a comprehensive guide to the camera and associated equipment (viewers, projectors). It is a useful handbook for any stereo photographer. How to use and maintain your Stereo Realist (Themelis, 1999) gives much new information about the camera and its production history; it lists the numerous accessories available, and is a valuable guide to its use. The sections on maintenance and modifications are particularly helpful.

Cameras made by other companies at this time were of much the same size, appearance and specifications as the Realist, though without the central viewfinder lens. They were designed to be operated manually; cameras with auto-exposure and autofocus appeared much later. Examples of well-known makes include Kodak, Wray, Iloca, Edixa, Revere, Wollensak, and Third Dimension Company (TDC). The individual images were 23 × 24 mm, corresponding to five perforations. This image size is usually referred to as the Realist (or 5P) format, and produced 27 image pairs in a standard 35 mm 36-exposure film.

A few companies made cameras that used a slightly larger image size of 30×24 mm based upon a seven-perforation length of film; this became known as the European or 7P format. Examples are the German-made Belplasca and the French Verascope F40, both of which gained good reputations. These cameras were designed with a stereo base of 63.4 mm. Much later, in the 1990s, Russia marketed the FED, which is optically excellent, but its exposure system in auto mode is a curious design, the

shutter doubling as the aperture diaphragm. It is not the most reliable of cameras, but when it works well it is capable of achieving good results. Certain more specialist cameras were made, such as the Macro Stereo Realist, which had a smaller lens separation and was designed for close-up work. These cameras had only limited appeal.

Some manufacturers, and mechanically-minded individuals, have constructed stereo cameras by combining two mono cameras, usually SLRs. This involves intricate work cutting the cameras, joining them, redesigning the film advance system, linking the aperture settings and the focusing mechanisms, and making additional parts or modifying existing ones to create a 35 mm stereo camera with two full-frame images (8P format). Such cameras are often referred to as 'siamesed' for obvious reasons. A few companies marketed such cameras but only in small numbers. The best-known is the German company Raumbildtechnik (RBT), who until the end of 2010 made a range of quality cameras of this type which were eagerly sought after. One such camera built from two Ricoh SLRs is shown in Figure 4.

Figure 4. RBT 'siamesed' camera, model X3, built from two Ricoh XR-X 3PF mono cameras

RBT also made stereo projectors and many accessories. The company has now ceased manufacturing but remains active to provide repair and servicing facilities. The rapid advance of digital imaging has caused film-based photography to dwindle. One of the advantages of 'siamesed' cameras over those of the Realist era is that they possess much more sophisticated technology, such as autofocus, auto-exposure and different programme modes. Medium format cameras (e.g. the Russian-made Sputnik) also appeared from time to time, but the greater cost per image made them less popular than those using 35 mm film.

In 1938, The US Company Sawyer's Inc. brought out the familiar Viewmaster stereo viewer taking circular cardboard mounts each containing seven 11 x 12mm stereo pairs. These covered a wide range of subjects, entertaining and educational. In the early 1950s the Stereocraft Engineering Company began to manufacture the Viewmaster Personal Stereo Camera for Sawyer's; this development allowed amateurs to produce their own Viewmaster format stereograms. The camera used 35 mm film which was transported twice through the camera. During the first transport the film advanced by 8P each time to give 36 pairs of images on the lower half of the film. Operation of a lever shifted the lenses and diaphragms upwards, and a further 36 exposures could be stored on the upper half of the film, wound back in the opposite direction. A later German-made camera, the Viewmaster Color Camera, had a different film advance reduced to 3.5 perforations, with the film running diagonally to give a unique arrangement of interlinked pairs. The stereo bases in both of these cameras were 61.5 mm (personal model) and 65 mm (color model).

Stereo Attachments

Another method of taking stereophotographs with a single lens camera is to fit an attachment, commonly called (incorrectly) a beamsplitter, which produces left and right images simultaneously side by side in a single frame. Fig 4.1.5 shows a typical design made by Pentax, which uses mirrors, for use with colour reversal (slide) film. It forms part of a kit with its companion piece, a specially designed viewer which takes 50× 50mm slide mounts each containing an image pair.

The mirrors are front-surface silvered. The two images (each approximately 18×24 mm) appear side by side within the standard 36×24 mm frame, so the stereo image is seen in portrait format, which tends to limit the usefulness of such attachments. They are designed to work best at a certain aperture (f/5.6), which again restricts their usefulness. For portraits and medium close-up work they can be effective, but they are not really suited to landscape photography. The optical design is not perfect and leads to some convergence distortion, though this is not obtrusive.

Figure 5 indicates the way the optical paths from the left and right views cross over so the images appear to be transposed. However, because the image in the camera is inverted, the two images are in the correct configuration for viewing. The dedicated viewer is similar in appearance to the attachment, but has the optics reversed: the front apertures become eyepieces, and are fitted with lenses, and the mounted slide is inserted in a slot at the back.

An unusual version of the 'beamsplitter' design is the American Tri-Delta attachment, which contains a prism and two mirrors that rotate the two beams separately: 90° clockwise for the left half of the full frame, and anti-clockwise for its right half. Thus the two

Figure 5. Pentax stereo adaptor

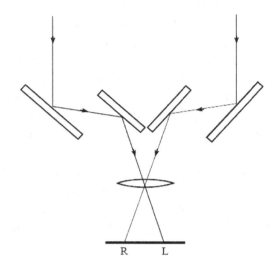

images abut, forming a vertical centre line in the frame. One advantage of this attachment is that the stereo image is in landscape rather than portrait format, and avoids the convergence of the Pentax type beamsplitter.

A more recent attachment of similar design made by the Hong Kong company Loreo Asia Ltd is advertised as a '3-D Lens in a Cap'. It is intended to be used for creating prints for viewing via a dedicated Loreo viewer taking 150 ×100 mm prints. As with the Pentax adaptor, the left and right half frame images lie side by side in a single full frame. It is a simple system aimed at the popular market rather than the serious stereophotographer. Again, the portrait format limits its usefulness; also, the half-frame width of the individual images means that the horizontal angle of view is restricted to half that of the camera lens alone. It can be used to view image pairs taken with the Pentax adaptor.

Stereoscopic Image Formats

Medium format stereo pairs produced on 120/620 film stock in a stereo camera such as the Chinese 3-D World TL-120-1 camera no longer represent a popular format. The pairs of images, each 58×56 mm, are on adjacent frames. The lens separation is 63.5 mm, and the winding mechanism advances the film by two frames for each exposure. As images of each pair are inverted in the camera, when the film is viewed the right way up the left image is to the right of the right image, so they need to be transposed for correct viewing. This necessity for transposition applies to most medium formats.

As discussed above, the standard formats for 35 mm film cameras are 5P and 7P, with the occasional full frame 8P for 'siamesed' SLR cameras. These designations (5P etc) not only indicate the individual image size, but also determine the arrangement of the L/R frames on the film, and dictate the width of the stereo base. The perforations on 35 mm film are spaced 4.75 mm apart, so 5P gives 23.75 mm, which a single image of 23 mm width will just fit with a small margin. This value represents the width of the rectangular film aperture. Three frame widths, totalling 15 perforations (15P) measures 71.25 mm, a convenient size for the stereo base. Consequently the left and right images need to be three frames apart. This arrangement can be seen in Figure 6a, as viewed from the back of the camera. The first exposure records images R1 and L1, inverted. Thus when the film is advanced by ten perforations (10P), the second exposure produces images at R2 and L2. Likewise, the third exposure, after a further film advance of 10P, gives R3 and L3; and so on. The left and right images are interwoven with a blank frame between R1 and R2 and another at the other end of the film. This 10P film advance is known as the Colardeau progression, after Colardeau (1923),

who devised it and applied the principle to an early Jules Richard camera, the Homeos.

If the film advance were to be only 5P, then the second exposure would place its images in the locations labelled 'blank frame' and 'R3' in Figure 6a without any overlap, and successive 5P film advances would cause a succession of double exposures.

Because of the nature of the interlacing of the images in this format, wherever the film is cut between any two adjacent frames the left and right images of one stereo pair will have been physically separated. With the European

Figure 6. (a) 5P, (b) 7P formats

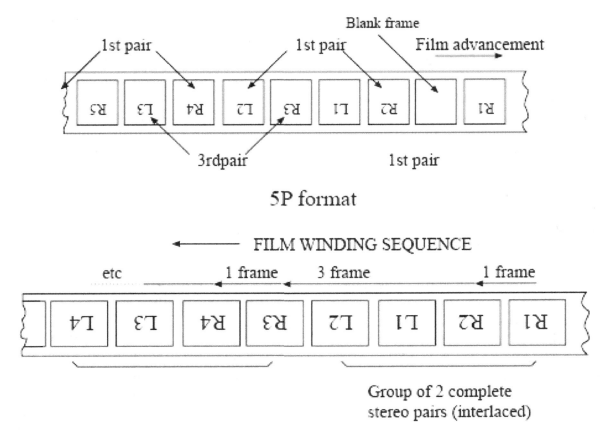

5P format

7P format

format (7P), the left and right images of a pair are separated by only one frame rather than two as with the 5P format. The reason for this is based upon geometrical factors. The extra width of 7P images compared with 5P (30 mm rather than 23 mm) will not allow a two-frame separation because the stereo base (at three frame widths) would then be 96 mm, i.e. too large. Two frame widths (64 mm) is a more satisfactory layout, but requires a different film advance sequence, as shown in Figure 6b. If after the first exposure the film is wound on by one frame (32 mm) the new images L2 and R2 will be neatly interlaced with the first pair L1 and L2. The next advance requires a three-frame shift (96 mm) to give L3 and R3 in order to avoid a double exposure. The configuration is now similar to that at the beginning, so a one-frame advance is now all that is required to interlace the third and fourth pairs. Hence the film transport mechanism in 7P format cameras is designed to produce a one- and three-frame advance alternately. There is a slight advantage in this layout because the film can be cut into lengths of four images each containing two (interlaced) stereo pairs; this makes it easier to keep track of the separate images when mounting them as stereo pairs for viewing. This 1-3-1… frame advance is also used in full-frame 'siamesed' cameras.

Although referred to as the 7P format, the true description of the frame width should be 6.75P. A seven-perforation film length measures 33.25 mm, whereas the frame separation in 7P cameras is smaller, (e.g. only 32 mm in the Russian FED). There are slight variations in these dimensions in different 7P cameras. The frame width is, of course, the basic unit on which the film advance mechanism is designed. The decision by manufacturers to use a 32 mm unit for film advance fits an extra pair of images into the film.

MOUNTING FILM (ANALOGUE) TRANSPARENCIES AND PRINTS

Transparencies or prints have to be combined to form a viewable stereogram. Probably the most common layout is with the images mounted side by side in a format to suit the viewing method. Amateur stereophotographers have largely favoured transparency film for their images, though the principles of mounting apply equally to prints. Card or plastic mounts are available for 5P, 7P or 8P formats. The ideal mounting system:

1. Enables comfortable viewing, displays a 3-D image that is an optimum match for the original subject matter. The viewer (with or without an optical aid) can fuse the two images without eyestrain.
2. Locates the image appropriately with respect to the frame. This represents a hypothetical ideal situation in which the original scene and its recreated 3-D image are spatially identical. It is both acceptable to refer to mounting as being "correct" when the stereo image matches certain criteria and appears to reproduce the original subject faithfully.
3. Embodies the concept of a stereo window (explained below), and can form part of a convenient storage system.

The Stereo Window: A Key Factor in Mounting

Each of your eyes covers a horizontal field of view of about 100°; with both eyes it is somewhat more than 180°. However, you are not as a rule aware of these boundaries. As you are continually moving and changing your viewpoint, space appears continuous and three-dimensional. On the other hand, a photograph or slide has distinct borders, so when viewing a stereogram there is a sudden transition at the edges from the 3-D image to the 2-D of the surrounding frame. These borders form a stereo window in space. When viewing a stereogram, preferably with the area surrounding the scene neutral and unobtrusive, the 3-D image is as though you were looking at the scene through a window. The rectangular apertures in the mount are responsible for creating the stereo window, and these are deliberately made smaller than the frame size to allow for alignment and adjustment of the film images. The mounts themselves are usually rectangular, made to fold down over the transparencies (see Figure 7). When these have been mounted in the lower half of the mount, the top half is folded down with the apertures aligned, and the two layers are sealed.

The separation of the apertures (62.2 mm in Figure 7) is designed to create a stereo window at about 2 m distance in space. As in Figure 7(b) the stereo image is typically aligned so that it lies behind the window, in the maximum case extending from 2 m to infinity. This conforms to the ability of the eye to view this depth range without discom-fort. The two shaded regions in Figure 7(b) are monocular areas; the one on the left (A–A´) is seen only by your right eye and the one on the right (B–B´) only by your left eye. It is the central region (B–A´) that exhibits the stereoscopic effect. This is normal: exactly the same impression will be gained by looking through a real window. Your left eye sees part of the view to the right of the window that is hidden from the right eye, and vice versa.

Now imagine that the stereogram in Figure 7(b) is removed and the stereo window replaced by a real window, so that we are looking at the original subject instead of its stereo image. If your eyes are in the same positions all the sight lines will be unchanged. Each eye still sees its respective monocular region, and the central region appears in three dimensions as. If the eyes were to be replaced by the lenses of a stereo camera it would record exactly the same left and right views, and each image would include the relevant monocular area. The presence of these monocular regions in the images is perfectly correct, and they should not be masked out.

Figure 7. (a) Layout of a typical 5P stereoscopic mount, (b) The field of view of a mounted stereogram (for simplicity the focusing lenses are omitted)

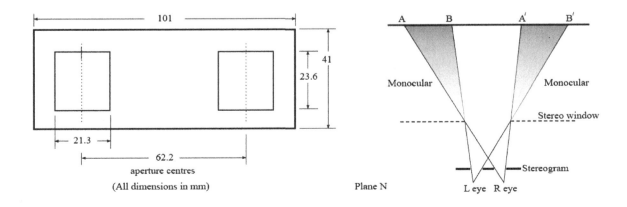

Some stereo cameras have what is popularly called 'a built-in stereo window'. Strictly, this is a misnomer. The term really means that the camera is designed to create a stereo window at about 2 m distance. To understand the significance of this it may be helpful to study the formation of the images in a stereo camera with no 'built-in stereo window'. This is shown in Figure 8a, which also represents the images formed by a pair of single lens cameras used as explained earlier. Taking the line ABXCD as representing an arbitrary reference plane, it is clear that the region ZBC and beyond is covered by both lenses and this will be seen stereoscopically. The shaded area extending to AB and beyond is imaged only by the left lens and that on the right extending to CD and beyond only by the right lens. These are recorded as monocular areas on the films at a–b and c–d respectively. (These distances are exaggerated in Figure 8a for clarity.)

Now when the film offcuts are inverted to erect the images, the region a–b will lie on the left side of the left image and the region c–d will lie on the right of the right image. Both monocular areas are therefore on the wrong sides (compare this with Figure 7). This means that these monocular areas have to be masked out when mounting, or there will be floating edges (see Figure 8; the term is explained fully later). Part of the original frame width is thus redundant. These incorrect positions of the monocular areas arise because the optic axes of the lenses are aligned with the centres of the image frame. In this example the stereo window formed from the frame border is at infinity, since the sight lines from corresponding edges in the two images are parallel.

If the camera lenses are placed slightly closer together than the frame separation, as shown in Figure 8, the problem is solved. This design feature leads to the built-in ste-

Figure 8. (a) Images produced by two single lens cameras or a stereo camera without a built-in stereo window, (b) images formed in a stereo camera with a built-in stereo window

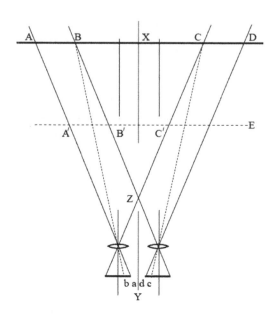
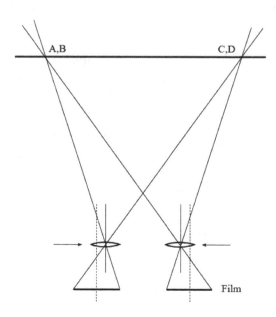

reo window mentioned above. The reduced inter-lens distance means that the sight lines from the corresponding frame edges of the two images now converge, and what were two separate points A and B in Figure 8a will now coincide at a specific distance from the lenses, as will the points C and D. The line A–B to C–D forms a stereo window just as in Figure 7. Its actual position in space, i.e. its distance from the lenses, depends upon how much closer the inter-lens spacing is when compared with the frame separation. It is generally set to be located at around 2 m to match that produced by the aperture separation in the slide mounts.

It is perhaps appropriate at this point to introduce a fundamental equation derived from the image geometry in the camera as depicted in Figure 9. An object lying centrally at distance D produces images at P and S on the relevant film frames. From an object at infinity the light rays are parallel and give images at Q and R. The total deviation (PQ + RS = d) is split equally between the two images.

Assuming that the lens to film plane distance is equal to the focal length of the lens we have:

b = lens spacing (stereo base)
f = focal length of the lens
D = object distance
d = deviation

From similar triangles OLX and LPQ:

PQ/QL = LX/XO

so

PQ = (QL× LX)/XO = f $(b/2)/D$

i.e.

Figure 9. Determination of deviation d in camera images for an object at distance D, with a lens of focal length f

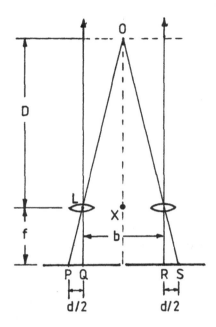

$d/2 = f \times b/2D$

so

$d = f \times b/D$

This tells us that the deviation is directly proportional to focal length and stereo base and is inversely proportional to object distance. Now the lens–film distance is equal to the focal length f only when the lens is focused on infinity; for closer objects it will be greater than f, but the differences are small and errors in using f in the equation can be ignored.

From this formula the maximum deviation, which occurs for a nearest object at 2m, can be calculated for different stereo formats to give the following results:

- **5P format:** Focal length typically 35 mm, stereo base 70 mm: $d = (35\times70)/2000 = 1.225$ mm. Similar calculations give:
- **7P format:** Focal length typically 35 mm, stereo base 63.4 mm: $d = 1.11$ mm.
- **8P format:** Focal length typically 50 mm, stereo base 70 mm: $d = 1.75$ mm and for the 5P and 7P formats, the maximum deviation is standardised at 1.2 mm. The original standards were based upon the old Imperial measure of 7 feet, which is 2.134 m rather than 2 m.

If you refer back to the specifications of the 5P Stereo Realist camera you will see that the frame separation is 71.25 mm (equal to the 15P frame advance) whereas the lens spacing is 70 mm. It is no coincidence that the 1.25 mm difference matches the maximum deviation value. Infinity homologues in the two images will be separated by 70 mm, the lens spacing. However, the more widely spaced frame edges will produce a stereo window at 2 m, because the maximum deviation of 1.25 mm represents the shift from infinity to 2 m.

Mounting Slides

To mount a stereo pair of transparencies, lay them over the apertures in the mount, align them carefully, and fix them in position. Take care over the alignment, as this is where the 3-D image is set in space, and you need to produce a realistic result. Although here we are concentrating on slide mounting, this section is equally relevant to the mounting of any image pairs on film, or as silver or digital prints. It is preferable to work from the rear of the film images, i.e. emulsion side up as in Figure 10a, to protect the image from scratches as you move the transparencies around. The image should be illuminated by using a mounting jig based upon a slide-viewing light box, or a light panel, with additional features as shown in Figure 11. You can then view the transparencies in 3-D while they are being adjusted; this makes accurate adjustment much easier.

Mounting a Stereo Slide Pair

Assuming you are mounting two film images in the type of mount shown in Figure 7, there are some rules you need to follow if the result is to be successful:

1. Homologues must be at the same height. Any given pair should lie on a horizontal line parallel to the top and bottom edges of the mount. The numerous parallel lines printed on a typical mounting gauge make this easy to achieve.

2. Infinity homologues should be set at a specified distance apart, usually around 65 mm, the exact distance depending on the mounting technique. This separation is basically close to the average eye spacing so that when the image of a distant object is viewed the optic axes of the eyes are parallel. For the 5P Realist format this separation will be 62.4 mm.

3. Near point homologues should not be set closer than a specified distance, which will also depend upon the particular mounting format. This separation will be smaller than that for infinity homologues by an amount equal to the required maximum deviation for the image concerned. So it will also be equal to the aperture separation of the mount. For the 5P Realist format this will be $62.4 - 1.2 = 61.2$ mm. This ensures that the closest object in the image does not appear in front of the stereo window created by the mount apertures.

Figure 10. (a) Mounting from the back of the mount, (b) mounting gauge

(a)

(b)

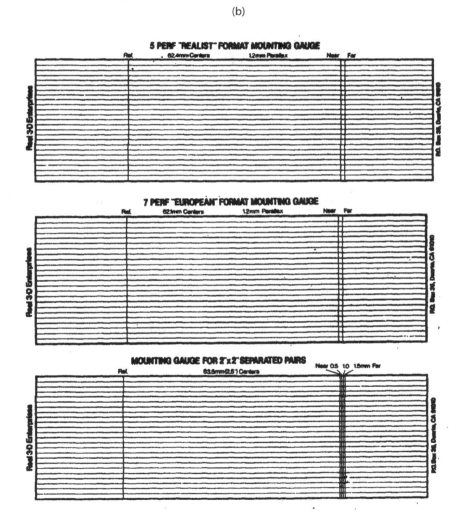

Figure 11. Mounting jig based upon a slide-viewing light box

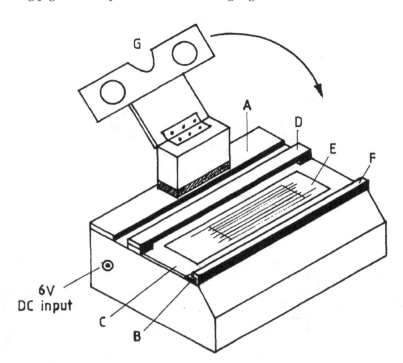

4. There must be no rotational differences between the two images with reference to an axis perpendicular to the image plane. (This relates directly to Figure 7.)

The jig is designed for use mainly with four-aperture double card mounts that fold in half to enclose the film offcuts. The additions to the light box are as follows:

1. A thin steel strip (A), about 20 mm wide, running across the back edge from left to right. This acts as a track for the lens mount, which incorporates a magnet in its base.

2. A thin steel strip (B), about 5 mm wide, running adjacent to the front edge. The sloping front of the light box projects upwards past the top surface to form a natural stop. The metal strip allows the use of magnets to hold the films temporarily in position over the mount apertures.

3. Clear acetate sheet (C), to cover the rest of the top surface. It abuts the metal strip B and should be of the same thickness to avoid a step where they meet.

4. A rigid metal or plastic strip (D) fixed at either end and raised slightly to leave a gap underneath. One long edge of the opened mount can be wedged under this strip so that it is held in position.

5. The mounting guide (E), attached to the clear acetate sheet. Its exact position must be determined, to ensure correct alignment.

6. A narrow plastic strip (F), glued to the raised ridge to form a flange to retain the front long edge of the mount when it is in place.

7. The lens assembly (G), consisting of a block with a strong rectangular magnet as its base. The lenses (focal length about 40 mm) are fixed to a plate hinged to the top back edge of the block, to allow it to be tilted back out of the way. The lens assembly as a whole can be positioned anywhere along the strip A.

Mounting of images is easier if they have been made using a camera with a built-in stereo window. As the frame edges for a stereo window appear at about 2 m distance, placing the films centrally over the mount apertures should place the image in the correct position in space.

Once the mount is in place on the mounting jig, it is probably simplest to place the right image, emulsion side up, over the left mount aperture and fix it with a strip of adhesive tape along the top edge. Then, while viewing the images through the lenses, manoeuvre the left film offcut into the desired position over the right aperture using tweezers. Do this with reference to the gridlines on the mounting guide, which you can see through the film images. In addition to the horizontal lines used for height adjustment there are three vertical lines on the left, marked 'Ref', and two to the right. The first of these is 62.2 mm to the right and marked 'Near'. The second is 1.2 mm (the maximum allowable deviation) further to the right and marked 'Far'. It thus lies 62.4 mm from the reference line. The reference and near lines used as a pair give the correct separation for homologues of an object at 2 m; the reference and far lines do likewise for infinity homologues.

The mounting procedure adopted also depends partly on the details within the images. Assuming again that the transparencies are being mounted from the back as in Figure 4, there are three possible routes to the making of a correctly mounted stereogram. They are:

Infinity Referencing

If the images include a mountain peak or the edge of a cloud, or any object sufficiently distant to be taken as at infinity, the mount should be moved laterally on the mounting jig until this feature, in the right image (already fixed in position) is aligned with the Ref line on the mounting guide. Next, the left image should be located so that the same feature is aligned with the Far reference line. You will need to tease the image into place; if you apply a little pressure to the film with a finger you can move it accurately by a fraction of a millimetre either horizontally or vertically until the stereoscopic image snaps into place when you examine it through the lenses of the mounting jig. All other objects will then fall into their correct locations in space.

Near Point Referencing

If the near object is at 2 m, the mount should be moved laterally and the left image adjusted to align homologues on this object with the Ref and Near vertical lines on the mounting guide. This will place the object exactly at the stereo window. In practice it is best to set these homologues a fraction farther apart than these two reference lines to ensure that the image of this particular object lies slightly behind the stereo window.

Intermediate Point Referencing

We might call this 'intelligent guesswork'. It applies to those cases where there are no objects as close as 2 m, and none so far away that they can be regarded as at infinity. When one homologue on the nearest object is set on the Ref line, the other should lie between the Near and Far lines. Homologues from the most distant object will act similarly, but will be farther apart. Your own recollection of the location of key objects in the original scene will help in the adjustment of the images.

The lateral separation of the films has a marked effect on the appearance of the 3-D image, as will be clear from Figure 12. In both diagrams the film offcuts are denoted by F. Infinity homologues are marked as 'i' and near point homologues as 'n'. In Figure 12a the transparencies are mounted with the infinity homologues separated by a distance equal to the eye spacing. Thus the relevant sight-lines are parallel and the corresponding image correctly located at infinity.

At the same time the near point homologues, being closer, produce an image at N (assume this is at 2 m), coinciding with the stereo window position. The stereo image clearly extends from 2 m to infinity and duplicates the original position. The stereo image also extends from 2 m to infinity, and duplicates the original scene in size and scale.

In Figure 12b, the images have been moved closer together so that the infinity homologues occupy the positions originally occupied by the near point homologues n in Figure 12a. This now moves the 'infinity' image to point M (identical with point N in Figure 12a) so that it coincides with the stereo window position. The near point has now become closer and lies at N´, close to the window. The stereo

Figure 12. Effect on scale and position of the stereo image of changing the separation of the 2-D images

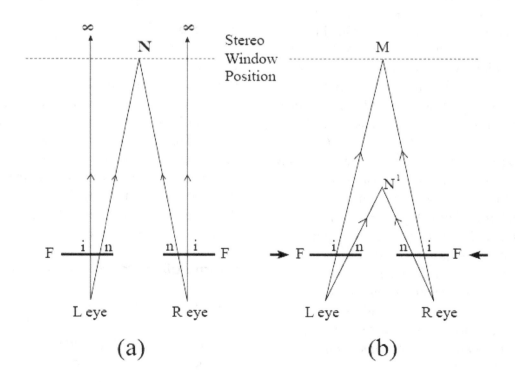

image has now been severely compressed, and extends only from N´ to M. As a result, it looks much smaller and lacks the visual impact of a correctly mounted image. This is an extreme example, which is further exaggerated if the images are moved still closer. With the whole of the stereo image close to the window the configuration is sometimes called the puppet-theatre effect.

To summarise, moving the film offcuts towards each other causes compression and pulls the image forward; moving them apart causes stretching and the image retreats away from the observer. Separating them by too much will lead to the infinity homologue sight-lines diverging, and the resulting eyestrain makes viewing uncomfortable or even impossible. Being aware of compression and stretching will enable you to adjust the position of the stereo image to match the original subject matter as perfectly as is possible. Under the right viewing conditions the stereo image can approach parity with the real scene, both in apparent size and in the locations of the individual objects within the frame.

Too close a spacing of the film offcuts can not only cause a near object to appear to protrude right through the window; it can also result in the monoscopic area being, for example, larger on the left side of the left offcut than on the left side of the right offcut, and correspondingly for the right sides of the images. This can make it difficult for the eyes to fuse the frame edges; this effect is referred to as *floating edges*. A height difference between the two mounted images can cause a similar problem with the top and bottom frame edges.

Problems can also occur if the pictures are taken without reference to the acceptable depth ranges . When mounting by infinity referencing the nearest object will protrude through the stereo window. If the images are separated to

correct this the infinity homologues will move too far apart. But even if the total depth range is satisfactory, ill-chosen film separation can still cause part of the image to jut through the window. With the appropriate subject matter this can be quite effective and dramatic: indeed, there are some stereophotographers who delight in this kind of display and deliberately arrange their images to exhibit the effect.

However, although something like a sword or stick protruding through the window might be acceptable, anything that is partly obscured by the image frame borders will look odd if it projects in the same way. In a head and shoulders portrait, for example, if the upper torso is floating in front of the window it can appear as if the lower part of the body is cut off, or even masked by something actually behind it. This effect is known as *window violation* and is unacceptable. There are some situations where a small amount of window violation may be unavoidable, but there is no excuse for blatant disregard of the principle. Any object that protrudes without intersecting the frame edges in any way can of course be allowed to do so for effect; but if overdone this becomes a tiresome gimmick.

There is an alternative method of mounting full frame images produced either as sequential pairs from one camera, or from two cameras linked in a twin rig, or from a siamesed camera. Instead of fixing the two images into a Realist type 5P card or plastic mount, they can be retained as two separate standard 50×50 mm mounts. Since it is unlikely that any of the camera set-ups will have a built-in stereo window, the left and right images will each display monocular areas on the wrong side of the frame, as shown in Figure 8. The slides will require a fair amount of masking to avoid including these unwanted monoscopic areas. Fortunately, a whole range of masks can be inserted into the mount to crop the image as required; these are available in various shapes and sizes.

Mounting Prints

As a rule, prints are mounted side by side. There is some freedom, since there are different stereoscopic viewing aids that enable the prints to be more widely separated (and therefore larger) than the 65 mm norm. The principles for mounting slides apply equally to prints. Card or plastic mounts are rarely used. Prints will usually have been made from monochrome or colour negatives taken on single lens cameras in twin rigs or 'siamesed' cameras and, like the full frame slides discussed in the last section, may suffer from wrongly located monocular areas. The prints will require a fair amount of trimming in order to create a 3-D image with a correctly located window. The procedure is outlined below with reference to Figure 13.

1. Position the prints temporarily as shown to a predetermined s_i value (say 65 mm).
2. Locate the nearest object (i.e. the gate in Figure 13) and measure its distance from the right hand side (in this case) in each print.
3. Trim the right hand edge of either the left or right print so that the distance a_L is greater than a_R. Usually it will be the right hand print that is trimmed, but this depends on the techniques that were used in taking the stereo pair. The difference between the two measurements should be no more than a few millimetres, in accordance with the monoscopic areas within the image.

4. With the prints separated as shown ($s_i = $ 65 mm) measure s_n, the separation between near point homologues. Trim the left hand side of each print to an equal width s_w, where s_w is slightly less than s_n.
5. Fix the pictures permanently in place according to the s_i value selected (see Figure 13).

Assuming that the nearest object has been correctly selected as a reference, making s_w smaller than s_n will result in the image of the subject lying entirely behind the window. On viewing the image, other objects present may appear nearer than that originally used and they may even protrude through the window. As long as there is no window violation, nothing more need be done. If the images of some objects intersect the frame edges, then more must be trimmed from each print from opposite sides (i.e. left side of left print and right side of right print). This will bring the window forward in relation to the subject.

Another method of mounting involves two 150×100 mm prints aligned laterally but mounted (touching) one above the other on the page with the right image above the left image. The vertical separation is 100 mm, and the format is usually described as 'up and over'. A special viewer marketed under the name ViewMagic is required.

DIGITAL STEREOPHOTOGRAPHY

Although this technology is so young, it is already the front runner in this field. There are still (at least in the amateur world) many stereophotographers who claim to be too old to learn new tricks and are likely to keep to

Figure 13. Mounting of prints to produce a correct stereo window s_i value

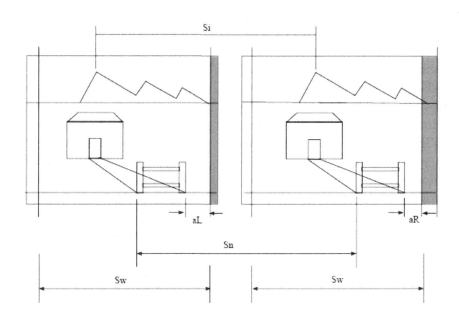

film until it disappears – if ever. Digital photography began to become popular with both amateur and professionals since about the year 2000, and has now completely taken over in the amateur field. The quality of digital imaging now matches the performance of film in regard to image quality. The main advantages of digital imaging over film imaging are: are:

1. Images can be viewed immediately after being recorded, and deleted if unsatisfactory.
2. Processing is minimal and cost effective: prints can be made using home computers and printers.
3. Digital memory devices can store large numbers of images.
4. Successful images can be produced in conditions that would otherwise require flash or complex lighting.

5. Images can be modified or creatively using specialised software programs.

All of this looks favourable for stereoscopic applications, but development in this area has been much slower, because until late in 2009 digital stereo cameras did not exist, and twin rigs were not widely used because of the problem of synchronisation. Most digital cameras suffer from shutter lag, a delay between triggering the shutter and recording the image. Many photographers have taken shots only to find that the subject has moved from its original position. Shutter lag is usually present in film cameras too, but to a much lesser extent. Shutter lag is due to the time required for the camera to set the exposure and focus. Advances in camera design have reduced this delay, but it still has to be overcome for two-camera stereoscopic photography, as no two cameras can be guaranteed to fire at precisely the same time without the aid of external electronic circuitry. This problem does not arise in cameras designed or adapted for use with a 'beamsplitting' device to take two images side by side. One such device was the 3-D Advantage V1200 model devised by Larry Heyda (Website, undated) in the USA, based upon the Tri-Delta beamsplitter principle. In this system a digital single-lens reflex (DSLR) camera is mounted on the beamsplitter horizontally, with the lens pointing downward onto an arrangement of mirrors that produce the two images side by side. This device allows various lenses to be used, and the stereo base can be varied from about 63–100mm. It does, however, need to be customed to fit specific cameras, and is a costly solution to the problem.

Digital Twin Camera Rigs: Achieving Synchronisation

A simple design by Aldous (2004) for synchronising two digital cameras has, however, achieved some success. A beam located above the camera operates the two shutter buttons simultaneously when pressed and achieves synchronicity to within 0.01 s, which is acceptable for most general photography, though perhaps not for fast-moving subjects. In the Netherlands, van Ekeren (2007), a specialist in stereo equipment, makes and sells a large number of digital camera twin rigs, the on–off shutter (half-press and release) and zoom functions of the lenses permanently connected to improve the synchronisation, though this method needs cameras to conform to particular dimensions.

A slightly different approach has been devised by Rob Crockett (2008) in the USA. He has designed electronic circuitry to monitor the timing signals of the two cameras in a rig. This is built into a separate control device which he names the LANC Shepherd. It has two cables, one for each camera; these plug in to the ACC ports. Originally the device was made to work with various Sony cameras, also with twin Sony and Canon video cameras. To quote from his website: 'It coordinates synchronized camera power-up and power-down, focus lock, shutter, and zoom functions, and displays the synchronization of the cameras' internal electronics. It also has a 10 second self-timer, allowing the photographer to get into the picture, and includes a formal interval timer with intervals from 2 seconds to 24 hours, for time lapse photography in stereo.' Crockett has also developed models of the controller with additional features, for example to control an external flash. The LANC Shepherd controller powers up the two cameras simultaneously, so the cameras are in close synchronisation. The lag is shown on the controller's LCD display. A similar device marketed by Pokescope (2006), also designed mainly for certain Sony models, claims that the cameras are synchronised to within 0.2 ms, which is adequate for a wide range of action photography.

A breakthrough has occurred with the appearance on the Internet of a software program entitled StereoData Maker (SDM) developed by Suto and Sykes (2007), a program that can be freely downloaded for amateur use. This was a development of the Canon-Hack Development Kit (CHDK) that was published in 2006 on a Russian website by a programmer 'VitalyB' who had studied the firmware for a Canon Ixus camera in order to understand the procedure for updating, and as a result was able to write his own program that installed itself via the standard 'Firmware Update' menu option. This new program made extra features such as uncompressed RAW images as well as JPG versions available to the user. Further developments by others added even more features, of which SDM was one. The programmers have devised software that diverts the firmware upgrade or 'autoboot' function in a range of Canon compact cameras to their own programs downloaded onto the memory cards. The cameras do not need any modification. A published list of the Canon cameras needs to be consulted as there are versions of the SDM software to suit each model. Even two examples of the same camera may need different versions, because the manufacturer sometimes makes subtle changes to the internal software at a late stage in the production.

Having identified the correct version, the appropriate software is downloaded onto two memory cards, one for the left camera and the other for the right (the two versions differ), and the cameras themselves have to be linked with further equipment. A small pack has to be made, containing batteries (4.5V to-5V total) connected to a press switch and two leads connected to the mini-USB sockets on the cameras. Figure 14 shows four modern digital stereo cameras. (a) and (d) are dedicated stereo cameras; (b) is a standard camera fitted with a stereoscopic lens system.

Figure 14c is a twin-camera rig using two Canon Ixus cameras with the right hand camera inverted to reduce the inter-lens spacing, and linked by the SDM software described above. When the cameras are first switched on they remain independent and will operate as normal with their own controls. But if the battery pack switch is pressed and held down, each camera receives the external voltage and this activates the software on the memory cards which take over from the cameras' firmware. The blue LEDs on the cameras light up and the LCD screens blanks out. The exposure can be made at any time during the next ten seconds, by releasing the switch to give a highly-synchronised pair of images. If the shutter is not released within this period the photograph will be taken automatically

after ten seconds. As an example of the effectiveness of the synchronisation, the measured desynchronisation for a Canon A460 compact camera is between 1/5000 and 1/20000 s (Aldridge and Sykes, 2008).This tiny discrepancy is more than adequate for the use of flash. The system is not 100% reliable, and on occasion the synchronisation fails and only one camera takes the shot, but most of the time everything works correctly.

The SDM program software avoids the nuisance of having a pair of cameras firing two full flashes simultaneously. It sets the cameras so that in one of them the flash is fired at a much reduced power to enable the camera to set the white balance. Then, a fraction of a second later, while the first camera shutter is still open, the flash on the second camera is fired at full power to illuminate the subject. The program also adds a number of menu options to give added on-screen information and useful aids such as alignment grids to indicate if a subject has an excessive depth range.

Not long after the success of the SDM solution for digital synchronisation, which had been taken up by stereophotographers worldwide, there came the welcome news that the Fujifilm Company of Japan was to market a true digital stereo camera. It appeared in the latter part of 2009 as the Finepix Real 3D W1 (Figure 14a), renamed the Fuji W1 for convenience. It proved a popular choice amongst 3-D aficionados. Styled in black, and measuring approximately 124(W) x 68(H) x 26(D) mm, it is fitted with two Fujinon 3× optical zoom lenses and produces two images with up to about 10 megapixels each, depending on the choice of format and image quality. The novel LCD screen enables the image to be viewed in 3-D without the need for any optical aid. It is not lenticular, but operates on the principle of two light beams, one for the left

Figure 14. Digital stereo cameras: (a) Fujifilm W1, (b) Panasonic Lumix with stereo lens, (c) twin-camera rig (Canon Ixus), (d) Fujifilm W3

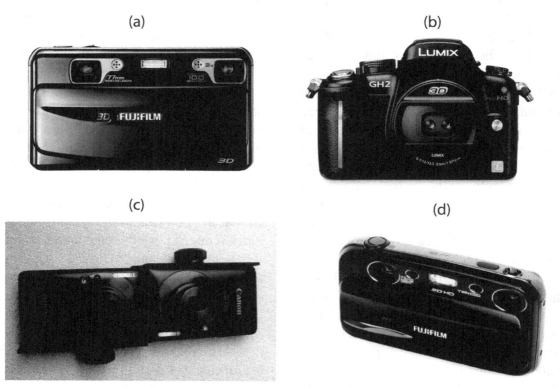

image and the other for the right, directed to the corresponding eyes when viewed centrally.

The camera has a number of operational modes including auto, manual, aperture priority, and Scene Position (SP), which gives the user a choice of automatic settings suitable for portrait, sport, landscape, night shots, snow, sunset, etc. It can also record video images in 2-D or 3-D. There are also advanced 3-D modes that allow a pair of sequential exposures to be taken, both using the one lens but with the images stored as a normal image pair. This facility allows macro shots with a reduced stereo base, or distant landscapes with an increased base. The first shot remains visible on the screen as a ghost image to assist the alignment of the second shot. The camera

can be switched to 2-D mode, and there is an advanced 2-D setting under which two shots can be taken simultaneously with different settings, for example 'Tele/Wide simultaneous' or 'Two-Color simultaneous' shooting. The latter can display differing colour tonalities, or one image in colour and the other in black and white.

A very useful feature for stereo work is a parallax control button. When operated it allows the two images to be moved laterally relative to one another (akin to altering the lateral spacing of film offcuts in a mount) to move the image forward or back relative to the stereo window. This can be performed when viewing the image after recording, but it can also be set as desired before taking the shot.

Also if a recorded image is readjusted in this way, the new version can be copied into the memory card as an extra image, the original image being retained unaltered.

The built-in flash is typical of those in most compact cameras, not over-powerful but adequate. In any case, a digital camera can be used without flash in many situations where film cameras would require it. The only problem with the Fuji W1 flash is that it is located centrally between the lenses and on the same level. This can lead to a double-shadow effect for relatively near objects close to the background: the left lens records a shadow to the left of the object and the right lens one to the right of it. Viewed in 3-D, the effect gives the appearance of a floating halo, which looks odd. It is less noticeable or absent when the background is either farther away or complex so no clear shadows are visible. There is no socket for an external flashgun to be connected.

The only way to overcome this omission is first to set the flash to 'forced flash' mode (without any pre-flash for red-eye reduction) and mask the camera flash window with a layer of exposed film to act as a filter, then mount the camera on a flash bar with an external flashgun triggered by a small flash slave unit. You will need to make a few trials to ensure the external flashgun is successfully triggered by the slave unit. Without this ancillary equipment you will simply have to resort to Adobe Photoshop (or similar) to remove the undesirable shadow and save an otherwise good picture.

The images are stored in MPO (multi-picture object) format files as side-by-side pair images together with, by default, a JPEG image in 2-D, though the latter can be excluded when saving images. In the MPO format the stereo pair is treated as a single unit, though there is a software program to separate the two images if required.

A year after the Fuji W1 was launched its successor, the Fuji W3, appeared (Figure 14 d). Although slightly smaller than its predecessor, it has virtually the same specifications; but has a large lenticular LCD screen with a very fine pitch. This gives a high quality image that is brighter than that in the W1. The control buttons and dials are redesigned and rearranged to be more compact. Another improvement is the inclusion of HD video (1280 × 720 pixels at 24 pps) as well as two lower definition modes. The flash is still central, so the double shadow effect is still present; and the battery capacity is lower than that of the W1, but a problem of haze that appeared under some lighting conditions in the W1 seems to have been eliminated.

Panasonic is also involved in 3-D imaging, mainly in TV and video, but their Lumix range contains some hybrid cameras that have interchangeable lenses, but are not DSLRs. The GF2, shown in Figure 14b, can be fitted with a special 3-D lens consisting of a normal size lens ring containing two small lenses side-by-side, about 12 mm apart. Thus two images are produced as a stereo pair. The close separation of the lenses restricts stereo usage to close-up subjects (around 60 mm distance). Normal landscapes do show some very restricted depth.

A number of Sony Cyber-shot cameras can be used in Sweep Panorama Mode. This produces a number of sequential pairs of images in a novel way. In the appropriate operating mode the shutter is released and the camera is panned from left to right. As the camera is swept sideways up to one hundred images are taken from the different angles. The camera's processor then analyses and combines these images to produce a left and right panoramic pair, giving a full panoramic stereoscopic picture when viewed on a 3-D TV with glasses.

Stereo Synthesis

Some models of Cyber-shot Sony cameras claim to produce stereo still pictures by exposing two images rapidly in succession at different focus settings. The differences between these images are processed by the camera's firmware to produce the 3-D version. It is not clear exactly how this is achieved from the two images. There are computer software programs that can be used to convert a single 2-D picture to a 3-D; these involve identifying where the various parts of the scene should be located in space in the z-axis direction. By working in layers, parts of the image are shifted laterally by amounts that relate to their intended position in the z-direction and filling in the gaps. These programs require only a single 2-D image as the starting point so presumably the Sony camera software must work in a slightly different way.

The Panasonic Company markets a ZS10 compact camera featuring a 3-D shooting mode which also operates by panning the camera in order to obtain multiple shots. In this case, however, only two of the images are selected automatically by the camera to create an MPO format image. This is a kind of truncated version of the Sony 3-D sweep panorama mode.

**PROCESSING OF DIGITAL
IMAGES FOR VIEWING**

`Processing of digital images is the equivalent of the mounting of film-based images, whether slides or prints, but with important differences. With slides the images are fixed, and all that is required is to locate them into a mount for the best effect. Print images obtained from processing laboratories are treated in essentially the same way. But prints can now be produced at home by virtue of the personal computer and associated printers and scanners. This means that the negatives can be scanned and modified in many ways with the aid of sophisticated software programs,: brightness and contrast, enhancement or reduction of colours, distortions to convert images to abstract forms among others, before producing a positive or maybe even a negative print. Here, though, we are concerned only with the basic processing of digital images necessary to produce a satisfactory stereogram for viewing as a digital 3-D image. Programs like Adobe Photoshop can be used successfully for this purpose, but most other programs do not cater for stereo images. Correctly separating the two images and aligning them accurately are key factors in producing correct stereoscopic images, as was explained for film-based images; and these are equally important for digital images. In photo processing programs, cropping the images as you would do for prints is not as simple as it might seem if you want precision, and correcting a twist is a trial and error operation.

StereoPhoto Maker

Fortunately there is a program designed specifically for stereo image processing and, like StereoData Maker, it is available on the Internet and can be freely downloaded for personal use. It is called StereoPhoto Maker (SPM). This software program was again created by Suto and Sykes (2007), prior to their production of StereoData Maker, discussed earlier. From the initial published version SPM has been developed and improved to cope with technological changes, such as the handling of MPO files generated in the Fujifilm W1

Table 1. StereoPhotoMaker features. Items in capitals are likely to be the most commonly used.

1. OPEN STEREO IMAGE	15. Real size or zoom (various settings)
2. OPEN LEFT/RIGHT IMAGES	16. INTERLACED
3. Open single image	17. GRAY ANAGLYPH
4. Open image file list	18. COLOR ANAGLYPH
5. SAVE STEREO IMAGE	19. SIDE-BY-SIDE
6. SAVE LEFT/RIGHT IMAGES	20. OVER/UNDER
7. FREE CROPPING	21. Sharp 3DLCD
8. RESIZE	22. Page flip (for shutter glasses)
9. SWAP LEFT/RIGHT	23. EASY ADJUSTMENT
10. Show/hide navigator	24. AUTO ALIGNMENT
11. Show/hide border	25. ALIGNMENT MODE
12. Full screen	26. AUTO COLOR ADJUSTMENT
13. Resample on/off	27. ALIGNMENT RESET
14. Fit to window size	

and W3 cameras. The program's strength lies in its ability to align images accurately; but it also includes many useful features for enhancing them and preparing them for various methods of display.

To give some idea of the versatility of SPM, Table 1 is a list of the features available on the toolbar of one of the latest versions when the program is running. The displayed icons, reading from left to right, have the following functions:

All of the above functions and additional features are also available via the menu bar on the computer screen when the program is in use.

When working with SPM, the separate left and right images from sequential exposures or from digital twin rigs are first downloaded to the computer into separate files. SPM has a facility for adding 'L' and 'R' tags to each image. The left and right images can then be assigned to a single folder (named, for example, Source Images) when they will pair up. Clicking the Open left/right images opens a box on screen which allows you to locate the appropriate folder. After selecting the left image a Confirmation box appears on screen; clicking 'Yes' will place the image automatically with its right partner into the SPM program as a side-by-side pair on screen, ready for processing.

MPO format images from Fuji W1 and W3 can be downloaded into a single folder (marked, for example, MPO images) when they will appear by default as a 2-D JPEG image next to an icon with the same title, with the suffix MPO. The camera can be set to download only the MPO file without the JPEG image if desired. In the SPM mode you simply select Open Stereo image, and by clicking on the MPO marked icon (not the JPEG image) the stereo pair will appear side by side and ready for processing.

REFERENCES

Aldous. (2004). Retrieved Nov 2012 from www.aldous.net/photo/project07

Aldridge, R., & Sykes, D. (2008). StereoData maker. *Journal of 3D Imaging. Stereoscopic Society, 180,* 25.

Brewster, D. (1856). *The stereoscope: Its history, theory and construction.* Paris: Triage Général de Stéréoscopie.

Crockett, R. (2008). *Ledametrix.* Retrieved Nov 2012 from www.ledametrix.com/lanc-shepherd/index.html

Heyda, L. (n.d.). *Freewebs.* Retrieved Nov 2012 from www.freewebs.com/larryeda/

May, B., & Vidal, E. (2009). *A village lost and found.* Bristol, UK: Canopus Publishing.

Morgan, W. D., & Lester, H. M. (1954). *Stereo realist manual.* New York: Morgan and Lester.

Perrin, J. (Ed.). (1997). *Jules Richard et la magie du relief.* Paris, France: Prodieux.

Pokescope. (2005). Retrieved Nov 2012 from www.pokescope.com

Suto, M., & Sykes, D. (2006). *Stereo.* Retrieved Nov 2012 from www.stereo.jpn.org

Suto, M., & Sykes, D. (2007). *Stereo.* Retrieved Nov 2012 from www.stereo.jpn.org

Themelis, G. A. (1999). *How to use and maintain your stereo realist.* Cleveland, OH: Author's Publication.

Van Ekeren, J. (2011). *Ekeren3d.* Retrieved Nov 2012 from www.ekeren3d.com

Wheatstone, C. (1838). Contributions to the physiology of vision: Part 1: On some remarkable, and hitherto unobserved, phenomena of binocular vision. *Phil. Trans. Royal Society, 1,* 371–394. doi:10.1098/rstl.1838.0019

Chapter 5
Head Tracked Auto-Stereoscopic Displays

Philip Surman
De Montfort University, UK

ABSTRACT

This chapter covers the work carried out on head tracked 3-D displays in the past ten years that has been funded by the European Union. These displays are glasses-free (auto-stereoscopic) and serve several viewers who are able to move freely over a large viewing region. The amount of information that is displayed is kept to a minimum with the use of head position tracking, which allows images to be placed in the viewing field only where the viewers are situated so that redundant information is not directed to unused viewing regions. In order to put the work into perspective, a historical background and a brief description of other display types are given first.

HISTORICAL BACKGROUND

Historically, 3D display systems were often considered an obscure research interest whose fruits were of dubious practical or commercial value. Recent technological advances, coupled with the demands of sophisticated methods of interaction (for example virtual reality) have reawakened interest in the techniques and applications of 3D imagery. So we have to consider the terminology, as in most discussions of 3D displays. Historically, authors have often used many definitions and descriptions ambiguously, often serving only to confuse the reader. For the purpose of this discussion a 3D display system is defined as one that provides additional information to the observer as a consequence of the observer's possessing stereoscopic vision. In order for any technological advancement to gain widespread

DOI: 10.4018/978-1-4666-4932-3.ch005

acceptance it should be capable of something new and previously unachievable, or perform its task so much better than the existing alternative that it naturally supersedes it.

Our primary interest in stereoscopy involves its application to eliciting depth and detail in an otherwise flat image. However, there are many other applications, including the detection of counterfeit currency and documents, and the investigation of convergence disorders in ophthalmology, to name two important examples. It seems natural to utilise stereoscopy in the display of electronic imagery. We see the world stereoscopically, so presumably we should endeavour to recreate the visual experiences to which we are accustomed in other images. This argument was no doubt applied when colour was introduced into electronic imagery, although the universal acceptance of colour was achieved with some opposition. Indeed, at its inception colour TV was considered by many as an extravagance, receivers costing as much as a small family car. Nowadays most households own more than one colour TV receiver.

The visual perception of depth depends upon image disparity. Our eyes are horizontally disposed and receive disparate images. A synthetic image must include that disparity. This is generated in computer graphics by geometric transformations. However, when viewing the world we are also receptive to other kinds of disparity. The surface reflectance of some materials can result in the images formed on each of the retinas differing in colour and brightness. Binocular mixture of the hues occurs, and a physiological phenomenon known as retinal rivalry accepts and rejects the binocular mixture in rapid alternation, and as a result imparts a lustre-like characteristic to the perceived image of, for example, metals and crystals. Although it may be useful if a 3D

display system can portray a likeness of such materials more realistically, the effect may appear as a defect. Points of correspondence in computer-generated images may appear to be differently coloured, and this can give rise to a distracting 'twinkling' at these points.

Binocular vision is only one member of a set of properties commonly referred to as 'visual depth cues'. These properties may be psychological or physiological or both (Graham, 1965). These cues include: static and dynamic parallax, linear and aerial perspective, size constancy, accommodation, and, of course, stereopsis. An examination of these terms is worthwhile, particularly for those involved in computer graphics, where the objectives may be clear but expense dictates a compromise. For example, given the choice between stereoscopic wire-frame images and monoscopic shaded images, which should we use? There is no clear answer, but an understanding of the factors involved may be useful.

Imparting the illusion of depth to a two-dimensional image can hardly be claimed to be a new idea. For hundreds of years artists have been familiar with the techniques. However, there are two aids to depth perception that have not been fully exploited, particularly in the field of computer graphics, namely motion effects (dynamic parallax) and stereopsis.

Although not possible in a still picture, dynamic parallax can be readily represented in modem animated art. For example, any depth ambiguities between two objects are resolved if one object moves in front of another and thus obscures the view of the latter. This effect is used in computer-generated images but is usually limited. In a real-life situation the effect is present at all times, from the relative motion of both the viewer and the object viewed. Stereopsis has been employed with varying success in computer graphics, the graphic arts, and in television systems, but while the effect is well understood it can be fairly difficult to implement. The exploitation of binocular and motion depth cues is the goal of many 3D displays, and there have been a number of approaches.

Stereoscopic imagery has been appreciated for many centuries, but it was Charles Wheatstone who invented the first practical display device in 1832 (Figure 1a). A form of Wheatstone's stereoscope is still used in survey photography (photogrammetry). The essence of Wheatstone's system and many subsequent systems is the presentation of left and right views, each to the appropriate eye. Many of the techniques available to us today are derived from earlier photographic systems. Not all are compatible with electronic display devices and some are suited only to specialised applications.

In 1849 David Brewster demonstrated a stereoscope of the design in Figure 1b where the left and right images are viewed through a pair of convex lenses. The lenses enable the viewer's eyes to focus at a distance further than the actual distance of the image pair thus enabling fusion. By the time the stereoscope was introduced, photography was well established with stereoscopic photographs being very popular during the Victorian period.

The use of anaglyph coloured images and spectacles to select images appropriate for each eye was first demonstrated in 1858. This system had the advantage that several viewers could observe the same image. The Lumière brothers produced the first effective stereoscopic cinema films. This approach has since been used many times in the cinema and on TV, but the two differently coloured filters can cause retinal rivalry and discomfort and the colour images are not very convincing.

The polarisation method has been used almost exclusively in recent 3D cinema productions. Linear polarising filters set orthogonally, or, better, circular polarising filters polarised in opposite senses, provide a means

Figure 1. Wheatstone and Brewster Stereoscopes

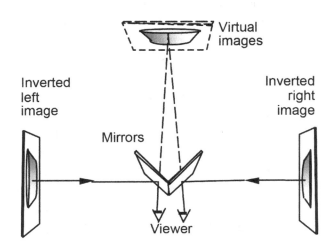

(a) Wheatstone Stereoscope (1833)

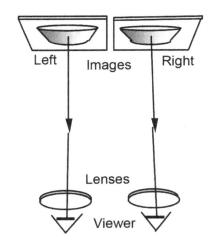

(b) Brewster Stereoscope (1849)

of maintaining separate channels of visual information. The stereo pairs to be viewed are projected through appropriately orientated polarising filters. The screen must be metallised, to avoid destroying the polarisation characteristics of the two beams. The plane polarised display is somewhat compromised by the crosstalk, especially if the viewer's head is tilted (Balasubramonian et al., 1982). It also suffers from low light transmittance of the filters. Circular polarisers avoid the problem of crosstalk but have an even lower transmittance.

The parallax barrier method of 3D display is probably the oldest free-viewing technique; it does not require any kind of optical apparatus for viewing and preserves horizontal (static) parallax. Such displays are termed autostereoscopic. The original proposal by H Ives in 1903 consisted of a fine grid of vertical slits superimposed upon a specially prepared photograph comprising narrow interlaced vertical strips of left and right eye views. The principle of operation is very simple: assuming the geometry is correct' each eye can see only one set of strips and is prevented from seeing the other by the barrier effect of the grid. A major limitation of the Ives parallax stereogram was its restricted viewing position. A later improvement was a reduction in the pitch of the grid, permitting a somewhat wider viewing angle while preserving horizontal parallax. But apart from some experimental work by Ives there was little further development of this approach for many years. Meanwhile, Kanolt in 1918 had overcome the limitation of limited viewing position with his 'parallax panoramagram' (Kanolt, 1918).

In parallel with the development of parallax barrier techniques was that of integral photography. The various systems have much in common. Lippmann (1908) proposed a system for integral photography comprising a 'fly's-eye' lens sheet focused on a photographic plate. A lens and a pinhole have similar optical properties and may be equally applicable in many situations. The first practical system of integral photography used a pinhole sheet, owing to difficulties at that time in producing lens sheets. Integral photography preserves both horizontal and vertical parallax.

It has been argued that as our eyes are horizontally disposed it should be possible to dispense with vertical parallax information, as with the parallax stereogram and panoramagram. This led to the development in the late 1930s of what has come to be known as lenticular sheet photography. A very simple and effective 3D display system can be produced using lenticular screens; many of us will no doubt be familiar with such autostereoscopic images. Any commercial 3D image that does not require special viewing devices is likely to be employing this technique. It is characterised by a lenticular plastic screen on the surface of the image. The lenticules are focused on the surface of the image so that they are retrotrflective; this technique is common in novelty images. The American Nimslo system has attempted to introduce the effect into serious photography by using four lenses rather than just two. This is was a step on the way to a full multi-image stereogram; but the lenses were too close together to give a wide angle of view.

John Logie Baird, the British inventor who pioneered 3D television before the Second World War developed a 3D television in the late 1920's. The apparatus used was an adaption of his mechanically-scanned system. As in his standard 30-line system the image was captured by illuminating it with scanning light. In this case two scans were carried out sequentially, one for the left image and one for the right. This was achieved with a scanning disc having two sets of spiral apertures. Images were reproduced by using a neon tube illumination source as the output of this can be modulated sufficiently rapidly. Viewing was carried out by observing this through another double spiral scanning disc running in synchronism with the capture disc.

Possibly of greater interest is Baird's work on a volumetric display he referred to as a 'Phantoscope'. Image capture involved the use of the inverse square law to determine the range of points on the scene surface and reproduction was achieved by projecting an image on to a surface that moved at right angles to its plane (Brown, 2009; Baird, 1945).The advent of electronic displays in the form of cathode ray tubes (CRTs) set off a revival of interest in many of the techniques outlined previously. In 1938 Vladimir Zworykin proposed a television system with alternate vertical scan lines keyed respectively to left and right images (Zworykin, 1938). This was merely a repetition of Ives's original parallax stereogram proposal. In 1953 Bradshaw made

a similar proposal (Bradshaw, 1953) which appears to be based upon Kanolt's parallax panoramagram.

Robert Collender published details of a display system in 1967 using a different type of parallax barrier (Figure 2a). He called the system a Stereoptiplexer (Collender, 1986). Instead of a grid he used a single rapidly moving vertical slit (a slot cut in a cylindrical drum). This system could not use a composite image as did the stereogram and panoramagram described earlier, because the scanning slit could not move between the relevant grid positions in discrete steps. Instead, a large number of separate photographs of the subject matter (96 in total) were taken from different points along a horizontal line, and subsequently processed to form a continuous loop of cine film. The film was then projected into the drum. The rapidly changing images and the moving slit then produced a 3D reconstruction which could be seen within the drum. Being an entirely mechanical system the Stereoptiplexer had serious limitations and was best suited to displaying still images; also, the use of film

Figure 2. Stereoptiplexer and Parallactiscope

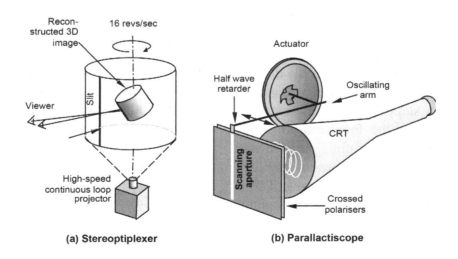

(a) Stereoptiplexer   (b) Parallactiscope

as the imaging medium was not compatible with television or computer graphic applications. Even if it had been practical the use of 96 separate frames per displayed image would certainly be unacceptable in terms of bandwidth.

In 1968 the Parallactiscope (Tilton, 1987) inspired by the work of Collender was invented by Homer Tilton who set about producing a 3D display using electronic rather than photographic images (Figure 2b). His approach was to use a Cathode Ray Oscilloscope (CRO). Although this was still unsuitable for TV applications it was a step in the right direction. The use of a CRO heralded other departures from Collender's work; for example, the Stereoptiplexer allows the viewer 360° horizontal freedom of movement around the display volume (the slotted drum), and without elaborate optics this would not be possible with a CRO. So in Tilton's device a different arrangement is used for the scanning slit; it comprises a vertical strip of half-wave phase-retarding material sandwiched between crossed linear polarisers. The strip is across the open ends of a horizontally arranged 'U' shaped torsion pendulum driven by a loudspeaker.

In some ways the Parallactiscope is the ultimate scanning slit parallax barrier system. It is an analogue instrument in every sense: neither the image nor the scanning slit change discretely, so that within the display envelope the horizontal parallax is continuous and not a stepped approximation. But despite its virtues the Parallactiscope has serious limitations. Its original purpose was to apply 3D imaging techniques to oscillography, i.e. to display 3D waveforms (Tilton calls them 'spaceforms'). This device cannot display images of the type we would expect to see on television or computer displays.

Subsequent research in this area involves the use of advanced ferroelectric liquid crystal devices to implement the scanning slit in the hope of applying this technique to television and computer-generated imagery (Sexton & Crawford, 1989). Another innovation in this area involves an 'inversion' of the parallax barrier arrangement. Carefully structured backlighting of a transmissive liquid crystal display (LCD) panel constrains each eye to see different sections of the display device (Eichenlaub, 1990).

3D DISPLAY TYPES

In order to determine the advantages of the head tracked approach it is useful to investigate other display types and then consider their advantages and disadvantages, in particular with respect to currently available enabling technologies. A classification of the different categories of available display provides a good starting point for this study. Figure 3 shows the principal display methods of stereoscope or Head Mounted Display (HMD), glasses and autostereoscopic. Within the autostereoscopic methods there are the generic types of holographic, volumetric and multiple-image. There are also other 3D display classification schemes, for example that of the 3D@Home Consortium (McCarthy, 2010). Methods requiring stereoscope equipment or head mounted displays are not considered in this chapter and only the most recent developments in holographic video displays are covered.

Glasses

Some glasses displays were described in Section 5.1 but there have been some recent developments that should be mentioned

Figure 3. 3D display types

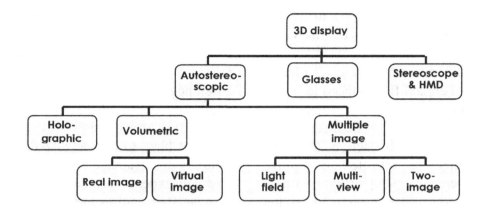

here. Since around 2009 3D television has become widely available, in a large part due to the availability of 120 Hz LCD displays that enable the left and right eye views to be presented sequentially. 3D cinema has been around for a little longer than this and these systems in general use either linearly or circularly polarised glasses.

The first 3D televisions employed active shutter glasses where the filters in the glasses switch in synchronism with the images being presented on the screen (Figure 4). When the display is being addressed, both of the cells in the glasses must be switched off otherwise parts of both images will be seen by one eye thus causing crosstalk. The synchronisation is usually achieved with the use of an infrared link. In order to avoid image flicker images, must be presented at 60 Hz per eye; that is a combined frame rate of 120 Hz. The earliest shutter glasses displays used a CRT to produce the images as these can run at the higher frame rate required.

The simplest method of projecting 3D is with the use of two projectors with orthogonally aligned polarizers in front of each lens. When viewed on a metallized screen through correctly matched linearly polarised glasses

Figure 4. Shutter glasses timing

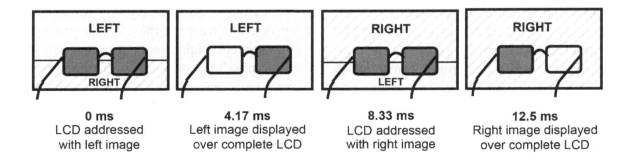

3D is seen. This was first demonstrated by Edwin Land, the inventor of Polaroid film and the Polaroid instant camera, in 1936. Analog IMAX 3D also operates on this principle.

The RealD cinema system uses a circularly polarized liquid crystal filter placed in front of the projector lens. Only one projector is necessary as the left and right images are displayed alternately. The active ZScreen polarisation filter was invented by Lenny Lipton of the Stereographics Corporation and was first patented in 1988 (Lipton, 1988). In the RealD XLS system used by Sony a 4K projector (4096×2160 resolution) displays two 2K (2048×858 resolution) images; this uses only a single projector as the two images are displayed one above the other in the 4K frame. These are combined by a special optical assembly that incorporates separating prisms and the lenses for each channel are located one above the other at the projector output. .

Masterimage produce a mechanical version of the switched filter cell where a filter wheel in the output beam of the projector running at 4320 rpm enables 'triple flash' sequential images to be shown; triple flash is where within a 1/24 second period the left image is displayed three times and the right image displayed three times.

LG produce an LCD display where left and right images are spatially multiplexed with the use of a Film Pattern Retarder (FPR) where alternate rows of pixels have orthogonal polarization. Some observers claim that the perceived vertical resolution is halved but LG suggest that as each eye sees half the resolution, the overall effect is of minimal resolution loss (Soneira, 2012).

An advanced anaglyph display is made by Infitec (Jorke & Fitz, 2012). This is a projection system where narrow band dichroic interference filters are used to separate the channels. The filter for each eye has three narrow pass bands that provide the primary colors for each channel. In order to give separation the wavelengths have to be slightly different for each channel however the effect of the slight differences in the primary colors is relatively small. One advantage of this system is that polarized light is not used so that the screen does not have to be metallized but can be white. A disadvantage is that the glasses are relatively expensive and are therefore not disposable. This requires additional manpower at the theatre as the glasses have to be collected afterwards and they also have to be sterilised before reuse. The Infitec system has been adopted by Dolby where the projector can show either 2D or 3D. When 3D is shown one with two sets of filters replaces the regular colour wheel used, one set for each channel.

Cave Technology

Cave Technology refers to a large theatre sited within a larger room. The walls of the CAVE are fabricated from rear-projection screens, and the floor is made of a down-projection screen. High-resolution projectors display images on the screens via a complex array of mirrors. The user wears polarized glasses inside the CAVE to see 3D graphics. People using the CAVE can see objects apparently floating in the air, and can walk around them, getting a better realistic view of the scene. This is made possible by electromagnetic sensors. The frame of the CAVE is made of non-magnetic stainless steel to interfere as little as possible with the electromagnetic sensors. These sensors track movements and the video adjusts accordingly. Computers control both this aspect of the CAVE and the audio aspect. There are multiple speakers placed at multiple angles in the CAVE, providing 3D sound to complement the 3D video. This form of virtual reality is as close as we can get to the science fiction dream of the Holo-Deck made popular in the science fiction series 'Star-Trek'.

Volumetric

Volumetric displays are a broad and diverse collection of various methods, technologies and ideas. Most of them are interesting only for academic purposes and it appears unlikely that they will ever become large scale or mainstream. Volumetric displays produce the surface of the image within a volume of space. The 3D elements of the surface are referred to as 'voxels', as opposed to 'pixels', on a 2D surface. As volumetric displays create a picture in which each point of light has a real point of origin in a space, the images may be observed from a wide range of viewpoints.

In principle, volumetric displays could provide realistic images that do not exhibit Ac-commodation/Convergence (AC) mismatch where the eyes converge at a different distance to their focus (Lambooij et al, 2009) and they also have the ability to display motion parallax in both the vertical and horizontal direction. They do however generally suffer from image transparency and they cannot readily portray surfaces that have a non-Lambertian distribution. In real-world scenes it is possible for a large proportion of the area of the surfaces in the scene to have non-Lambertian emission, which is generally caused by specular reflection. If a display is incapable of reproducing this, the reproduced image will lack a 'natural' look.

Volumetric displays can be either real image or virtual image (Traub, 1967) and the real image types can use a moving screen (Blundell & Schwartz, 2000) or a series of static screens (Sullivan, 2004). Examples of the different types of display are shown in Figure 5.

When an image is produced by any of the three basic methods described the light distribution from the voxels will be isotropic as

Figure 5. Volumetric display types

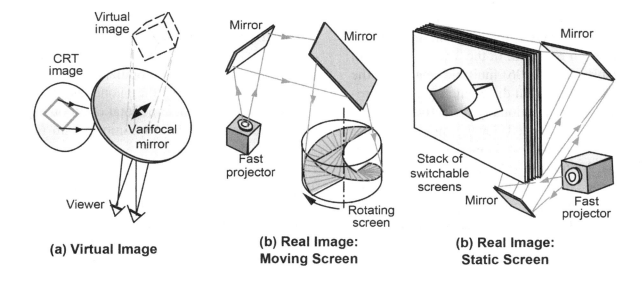

(a) Virtual Image

(b) Real Image: Moving Screen

(b) Real Image: Static Screen

the real surfaces from which the light radiates are diffusing have a Lambertian distribution. Images will therefore lack the realism of real-world scenes where many of the surfaces exhibit specular reflection. This isotropic distribution is also the cause of image surface transparency. Making the display hybrid so that it is effectively a combination of volumetric and other display types can overcome this deficiency (Favalora, 2009).

A conventional volumetric display is referred to as an IEVD (isotropically emissive volumetric display) and its limitations can be partially alleviated by rendering using a technique employing 'lumi-volumes'. A team comprising groups from Swansea University in the UK and Purdue University in the US (Mora et al, 2008) has applied this technique.

Isotropic emission limitations can be overcome with a display developed at Purdue University that is effectively a combination of volumetric and multi-view display(Cossairt et al, 2007). This uses a Perspecta volumetric display that gives a spherical viewing volume 250 millimetres diameter. The principal modification to the display hardware is the replacement of the standard diffusing rotating screen with a mirror and a vertical diffuser that provides a wide vertical viewing zone.

Two displays employing rotating screens have been reported. In the Transpost system (Otsuka et al, 2006) multiple images of the object, taken from different angles, are projected on a directionally reflective spinning screen with a limited viewing angle. In another system called Live Dimension (Tanaka & Aoki, 2006) the display comprises projectors arranged in a circle. The Lambertian surfaces of the screens are covered by vertical louvres that direct the light towards the viewer.

Multi-View

In multi-view displays a series of discrete views is presented across the viewing field. One eye will lie in a region where one perspective is seen, and the other eye in a position where another perspective is seen. Current methods generally use either lenticular screens or parallax barriers to direct images in the appropriate directions.

Lenticular screens with the lenses running vertically can be used to direct the light from columns of pixels on an LCD into viewing zones across the viewing field. The principle of operation is shown in Figure 6(a) The liquid crystal layer lies in the focal plane of the lenses and the lens pitch is slightly less than the horizontal pitch of the pixels in order to give viewing zones at the chosen distance from the screen; this is the distance where uninterrupted 3D is seen across the complete width of a group of viewing zones. In this case three columns of pixels contribute to three viewing zones.

The simplest multi-view displays show two images and provide only limited viewer movement. Sharp introduced the Actius RD3D notebook in 2004 that incorporated a 2D/3D switchable display that used an active parallax barrier to provide an exit pupil pair. They also made a 3D mobile phone that sold for a short time in Japan and was so popular that at one point in the mid-2000's the phone had produced higher sales than the total of all other auto-stereoscopic displays combined up to that time.

A simple multi-view display where the lenses of the lenticular screen are aligned vertically suffers from two quite serious drawbacks. Firstly, the mask between the columns of pixels in the LCD gives rise to the appearance of vertical banding on the image

Figure 6. Multi-view display

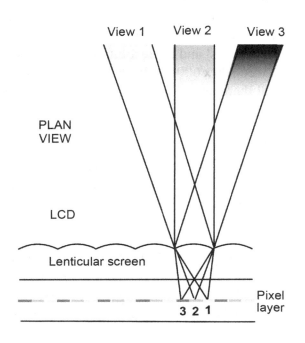

(a) Lenticular Screen

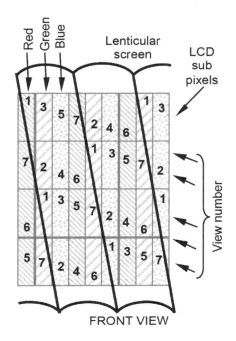

(b) Slanted Lenticular Screen for 7 Views

known as the 'picket fence' effect. Secondly, when a viewer's eye traverses the region between two viewing zones, the image appears to 'flip' between views. Philips Research Laboratories in the UK originally addressed these problems by the simple expedient of slanting the lenticular screen in relation to the LCD. An observer moving sideways in front of the display always sees a constant amount of black mask, therefore rendering it invisible and eliminating the appearance of the 'picket fence' effect which is a moiré-like artefact where the LCD mask is magnified by the lenticular screen.

The slanted screen also enables the transition between adjacent views to be softened so that the appearance to the viewer is closer to the continuous motion parallax of natural images (Figure 6b). Additionally it enables the reduction of perceived resolution against the display native resolution to be spread between the vertical and horizontal directions. For example, in the Philips Wow display the production of nine views reduces the resolution in each direction by a factor of three. The improvements obtained with a slanted lenticular screen also apply to a slanted parallax barrier.

The most common form of multi-view display in current use employs the slanted lenticular view-directing screen. Philips (Ijzerman et al, 2006) was active in this market area for around fifteen years but pulled out in 2009. The work is being continued by a small company called Dimenco (2012) that consists of a few ex-Philips engineers. It appears that Philips has again taken up marketing this display.

A slanted parallax barrier (Boev et al, 2008) can also be used where the result of the slanting has the same effect as for the lenticular screen; that is the elimination of the appearance of visible banding and the reduction in resolution spread over both the horizontal and vertical direction.

Light Field

In light field displays, discrete beams of light that vary with angle radiate from each point on the screen. These can take several forms and integral imaging, optical module and dynamic aperture types are considered here.

In integral imaging, an array of small lenses is used to produce a series of elemental images in their focal plane. If the lens array is a two-dimensional fly's eye lens then motion parallax in both the horizontal and vertical is provided. A fundamental problem with this technique is the production of images with reversed depth known as pseudoscopic images; these are caused by the virtual surface being produced in the same position as the real surface that was captured, however when this reproduced surface is viewed it is seen effectively from the 'inside' and hence reversed. The effect is unnatural and normal orthoscopic images are required; there have been methods devised to overcome this problem (McCormick et al, 1992; Arai et al, 2006) without degradation of the images.

NHK in Japan has carried out research into integral imaging for several years including one approach using projection (Okui et al, 2007). In 2009, they announced the development of an 'Integral 3D TV' achieved by using a 1.34 millimetres pitch lens array that covers an ultra-high definition panel.

Hitachi has demonstrated a 10" 'Full Parallax 3D display'; that has a resolution of 640

x 480. This uses 16 projectors in conjunction with a lens array sheet (Hitachi, 2010) and provides vertical and horizontal parallax. There is a trade-off between the number of 'viewpoints' and the resolution; Hitachi uses sixteen 800 x 600 resolution projectors and in total there are 7.7 million pixels (equivalent to 4000 x 2000 resolution).

The Institute of Creative Technologies (ICT) Graphics Lab at the University of Southern California has developed a system that reflects the images off an anisotropic spinning mirror (Jones et al, 2007). The mirror has an adjacent vertical diffuser in order to enable vertical movement of the viewer. The display is inherently horizontal parallax only but vertical viewer position tracking can be used to render the images in accordance with position thus giving the appearance of vertical parallax.

The Holografika display (Baloch, 2001) uses optical modules to provide multiple beams that converge and intersect in front of the screen to form real image 'voxels', or diverge to produce virtual 'voxels' behind the screen. The screen diffuses the beams in the vertical direction only therefore allowing vertical viewer movement. As the projectors/optical modules are set back from the screen, mirrors are situated either side in order to provide virtual array elements each side of the actual array.

A dynamic aperture type of light field display is one that uses a fast frame rate projector in conjunction with a horizontally scanned aperture. An early version of this was the Tilton display with the mechanically scanned aperture. More recently this principle has been developed further at Cambridge University with the use of a fast digital micro-mirror device (DMD) projector and a ferroelectric shutter (Moller & Travis, 2004).

Holographic

Holographic displays may be some years away but research is being carried out that is leading the way to its inception. The University of Arizona is developing a holographic display that is updatable so that moving images can be shown (Blanche, 2010). In order to be suitable for this application the material displaying the hologram must have a high diffraction efficiency, fast writing time, hours of image persistence, capability for rapid erasure and the potential for large display area; this is a combination of properties that had not been realized before. A material has been developed that is a photorefractive polymer written by a 50 Hz, nanosecond pulsed laser. A proof of concept of demonstrator has been built where the image is refreshed every two seconds.

In the 1990s, MIT carried out work in a horizontal parallax-only holographic display that incorporated Acousto-Optic Modulators (AOMs). It appears that work on this system has been revived by the Object-Based Media Group. In this display a multi-view holographic stereogram is produced in what is termed a Diffraction Specific Coherent (DSC) Panoramagram (Barabas et al, 2011). This provides a sufficiently large number of discrete views to provide the impression of smooth motion parallax. Frame rates of 15 frames per second have been achieved. So far horizontal-only parallax has been demonstrated but the system can be extended to full parallax.

SeeReal is developing a method of reducing the resolution requirement of a holographic display; this combines the advantages of head tracking and holography (Haussler et al, 2008). When a holographic image is produced a large amount of redundant information is required in order to provide images over a large region that is only sparsely populated with viewers' eye pupils. Normally holography requires a screen resolution that is in the order of a wavelength of light and even if vertical parallax is not displayed, the screen resolution is still too high to be handled by current display technology. The solution developed by SeeReal is to produce holograms with a small diffraction angle that serve only a small viewing window and therefore require a resolution that is as low as around ten times that of HDTV. In conventional holography the complete screen area contributes to each image point; the SeeReal display uses small proprietary sub-holograms that reconstruct the point that is seen over the small viewing window region. By reconstructing the points in this fashion, the A/C conflict problem is overcome.

INTRODUCTION TO HEAD TRACKED DISPLAYS

Head tracked displays provide a means of supplying to 3D to several viewers with the minimum amount of information displayed. If the same stereo image-pair is sent to every viewer then the display only has to show two images irrespective of the number of viewers. If N viewers are to see motion parallax then 2N images must be displayed in order to give each viewer their own personal perspective of the scene. This is economical in display usage terms as redundant images are not directed into the viewing field into positions where no viewer's eye is present.

The advantage of head tracking over other display types is even greater when both vertical and horizontal parallax is to be displayed. When the images are rendered in accordance with viewer position there is no additional display resolution requirement for both par-allaxes, but for a holographic or light field display vertical parallax will require additional pixels in the order of a factor of the square of the number required for horizontal-only parallax.

Head tracking operates by producing regions in the viewing field, referred to as exit pupils, where a particular image is seen across the complete area of the screen and directing these pupils eyes under the control of the tracker. These are extended vertically to allow vertical head movement (Figure 10). where the regions with a diamond-shaped section produced by the ATTEST display extend above and below the viewers.

An exit pupil pair in the region of a viewer's head is shown in Figure 7(a). In the case of a fixed exit pupil display such as the Sharp RD3D Actius notebook mentioned in Section 5.2.3 the amount of viewer movement is limited, however if the pupils are steered by the output of a head tracker the viewer can freely move around as if 3D glasses were being worn, but without the inconvenience of wearing glasses. Figure 7(b) shows several viewers being served by multiple exit pupil pairs.

Early Head Tracked Displays

Before the European Union-funded head tracked 3D display research commenced in 2002 there had already been considerable activity in this area. In the following survey, the approximate chronological order of the work is indicated by the year at the start of section which shows when the papers cited were published.

1985: A very early head tracked display was that of Alfred Schwartz (Schwartz, 1985). CRT projection modules project the left and right images on to the Fresnel lens that acts as a field lens with respect to the image information. This lens focusing real images of the projector lenses in the viewing field, and then extending these images vertically forms exit pupils. The pupils are steered by moving

Figure 7. Exit pupils

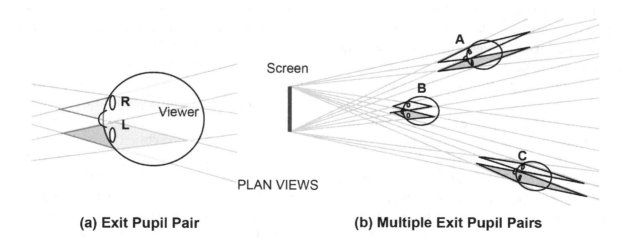

(a) Exit Pupil Pair **(b) Multiple Exit Pupil Pairs**

the Fresnel lens laterally with a servo-motor controlled by the head tracker output.

1989: Lenticular screens can be used in various ways in order to produce exit pupils. The easiest method is simply to place a lenticular screen, with its lenses running vertically, in front of an LCD. If the pitch of the lenses is slightly less than double the LCD horizontal pitch, left and right exit pupils are formed when left and right images are displayed on alternate pixel columns. The exit pupils can be steered by effectively changing the relative lateral position between the lenticular screen and the LCD. Simply swapping the images on the pixel rows can alter the relative positions between the pixel columns. This method is used in a display developed by the Nippon Telegraph and Telephone Corporation (NTT) (Ichinose et al, 1989). It is historically interesting as it the first use of a flat panel display for head tracked stereo that the authors have found. When the head tracker detects the centre of an eye-pair exit pupils are formed by a vertically-aligned lenticular screen such that each eye receives its appropriate image. The left and right images are formed on alternate pixel columns. As the head changes position laterally the left and right images are swapped when necessary so that the eyes continue to receive the correct images. As a series of left and right pupils is produced across the viewing field, considerable lateral movement is enabled. Although this method is a vast improvement over fixed exit pupil displays, degraded images will still be seen when eyes are located near the exit pupil boundaries. Another drawback of this method is that the LCD must be used in the portrait orientation.

In retro-reflecting screen displays, the projector illuminates the screen from the front, and the light is reflected back in the same vertical plane. In the early display of NTT (Tetsutani et al, 1989), projectors in an array are switched on selectively under the control of a head tracker in order to allow a single viewer a degree of lateral freedom of movement. The retro-reflecting screen consists of a vertically orientated lenticular sheet. This has a diffuser on its flat back face, which is in the focal plane of the lenses. Side lobes are also produced where this diffused light exits through adjacent lenses, but these are located outside the viewing region.

1993: In the Dimension Technologies Inc. (DTI) display a lenticular sheet is used to form a series of vertical illumination lines on a diffusing layer located behind the LCD (Eichenlaub, 1993). These lines are images of narrow vertical sources situated behind the diffuser with the exit pupils produced with a parallax barrier. The head-tracking version of the display has three sets of light sources so that three sets of overlapping exit pupils are formed. As the viewer's head moves, the appropriate lamps are lit so that the eyes always fall completely within a left and a right exit pupil.

1994: Parallax can be used in various ways in a head tracking display. In a method proposed by Hattori (Hattoti et al, 1994), a parallax barrier is used to block the light from a pair of projectors that form exit pupils by utilising a large convex mirror.

Exit pupils can be formed in the viewing field by using a pair of lenticular screens with a diffusing layer between them. If the lenses of the screens run vertically and are on the outer surface, and the diffusing layer is located between the two inner flat surfaces, the assembly acts as a one-dimensional version of the Gabor superlens (Hembd et al, 1997). The action of the diffuser is to produce a vertical exit pupil, therefore giving the viewer more freedom of vertical movement. In an early paper relating to this method, the use of lenticular screens with different pitches was proposed (Tetsutani et al, 1994). If the screen on the emergent light side of the screen has a slightly larger pitch than the other, the arrangement has the ability to magnify the movement of the projectors.

1996: Exit pupils can also be produced by holographic optical elements (HOE). The method used by Reality Vision has the advantage that a single HOE performs both the exit pupil formation and the image multiplexing (Trayner & Orr, 1996). The HOE produces the left exit pupil after the light has passed through the LCD pixels showing the left image and the right exit pupil after it has passed through the right image pixels. A vertical plane is formed in front of the screen where a left image can be seen to the left, and vice versa. When this plane lies between the viewer's eyes, stereo is seen. When both eyes are outside of the central zone, either a left image, or a right image only is seen. The stereoscopic viewing zone can be made to follow the viewer's head position by moving the light source.

A variant of this principle was developed by an Australian company, Xenotech (Harmon, 1996). In this display, light from two projectors is directed on to the screen via a semi-silvered mirror and two folding mirrors. This enables the projectors to be located behind the screen. The screen options are glass bead or prismatic. The prismatic screen consists of an array of corner cube elements where reflection takes place on three orthogonal surfaces. This type of display can provide screen sizes of 50" or more.

1997: In order to retain full resolution Sharp Laboratories of Europe (SLE) developed a display incorporating two LCDs whose images are combined with a semi-silvered mirror (Figure 8(a)). Lenses produce the exit pupils (referred to by Sharp as 'windows') and the light source consists of an array of cold cathode fluorescent tubes with outputs guided by Perspex wedges. This display is termed 'macro-optic' by Sharp as the focusing element is of a size comparable to the LCD.

Figure 8. Sharp and Seaphone head tracked displays

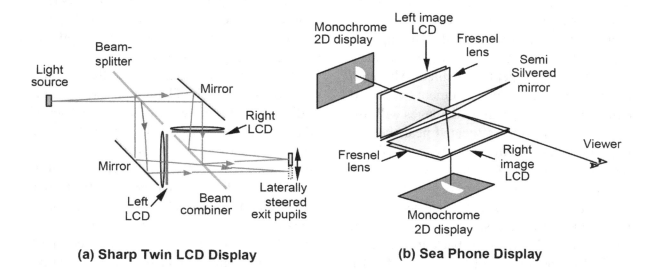

(a) Sharp Twin LCD Display   **(b) Sea Phone Display**

Novel optics was developed by Sharp where a novel LCD mask structure called PIXCON provides contiguous light across the viewing field (Woodgate et al, 1997). The pitch of the lenticular screen is set to provide exit pupils whose width is around 2/3 the average interocular separation. This arrangement reduced the observed intensity fluctuation at the exit pupil boundaries.

1998: Steering exit pupils in the Z direction can be achieved by using tapered lenticular optics in conjunction with an infrared retro-reflective head tracker (Street, 1998). The methods described so far steer the exit pupils by operating on the light emerging from the display. Lenticular screens can also be employed to operate on the light entering the LCD pixels.

There are several methods of producing a head tracking display; amongst these are prismatic arrays, macro-lens arrays and Holographic Optical Elements (HOEs). Dresden University has developed a display that replaces a parallax barrier with a prism array (Schwerdtner et al, 1998). Tracking is achieved by switching the images on the RGB sub-pixels. This enables the display to be used in the normal landscape orientation. In order to prevent crosstalk, four sub-pixels are allocated to each image pixel. The prism is designed to present all of two sub-pixels, for instance the green and blue, and parts of two red pixels. This arrangement enables the pixel positions to be shifted in increments of one sub-pixel pitch, but lowers the horizontal resolution by a factor of 3/8. The prism pitch is slightly less than 8/3 of the pixel pitch. The display is a compact method of providing a single viewer with stereo, with no moving parts required. Also, the prism array allows greater light throughput than a parallax barrier.

A system using a large mirror to produce exit pupils was developed by Hitachi (Arimoto et al, 1998). Although moving projectors could be used in this configuration, the Hitachi display increases the viewing zone width by employing an array of four projectors, therefore making the display a multi-view type.

1999: In the parallax display of New York University (NYU) a dynamic parallax barrier is located 100 millimetres in front of the display (Perlin, 1999). The barrier is a fast SLM array that consists of a one-dimensional array of elements that can block the light in selected vertical strips. The appearance of the barrier transmission structure is eliminated by rapidly cycling the barrier through three different positional phases for every frame. The image on the screen is changed at the same rate in order to present the appropriate image to each eye. The fast frame rate required by the image is obtained by using Texas Instruments' Digital Light Processor (DLP) that can handle sequential RGB. DLP is a projection system that incorporates a Digital Micro-mirror Device (DMD) and enables monochrome images to be presented at several times the normal frame rate. A ferroelectric shutter in front of the projector lens blocks the light in the period the barrier is switching. The barrier is a custom-built pi-cell liquid crystal screen. An IBM BlueEyes tracker, that uses infrared reflection from the viewer's retina, controls the barrier. The display can only support one viewer.

2000: A twin-screen display that uses Fresnel lenses is described in information from Sea Phone, a Japanese company (Hattori, 2000a). This uses a semi-silvered mirror to combine the images from two LCDs as shown in Figure 8(b). The principle of operation is very similar to that of Sharp, with the exception that the illumination is provided from two two-dimensional monochrome displays, as opposed to the single linear array and beam-splitter. The monochrome displays show images of each side of the head, as the figure illustrates. This enables exit pupils to be formed around the regions of the respective viewer's eyes. These are captured with an infrared camera that uses the same optics as the display in order to eliminate parallax error. An infrared reflecting mirror is mounted at 45° to the vertical in the light path between the left-image LCD and its monochrome display; this is not shown in the figure.

Another method of 3D display is also described in same Sea Phone report (Hattori, 2000b). Like the MIT display, it uses a micro-polariser array to separate the illumination to the left and right pixel rows and a Fresnel lens to form the exit pupils. In this case, however, the light is provided from linear arrays of white LEDs that have polarizers in front of them.

2001: The use of polarised light to perform a similar function to the DTI and Sharp displays was proposed (Tsai et al, 2001). In this display, the effective 'light lines' for the right exit pupils are vertical strips of one polarization on a 'micro tracking device', and the 'light lines' for the left exit pupils are strips with orthogonal polarization. A head tracker controls the positions of these regions. A lenticular screen that forms the exit pupils is located in front of this. A micro-polariser array with its elements running horizontally allows light from the right light line to pass only through the LCD pixel rows displaying the right image, and vice versa. It is doubtful whether this approach was very effective due to the high resolution required for the 'micro-tracking device' in order for the pupils to be directed in sufficiently small increments.

The Korea Institute of Science and Technology (KIST) developed a system that projects the images of two 6.5" LCD panels on to a 20" diagonal Fresnel lens (Son et al, 2001). Exit pupil steering is achieved with the use of an SLM that selectively blocks the light in two separate paths, one path for the right exit pupil and one for the left.

In the Varrier™ display of the University of Illinois, a physical parallax barrier with vertical apertures is located in front of the screen (Sandlin et al, 2001). The complete system can be considered as comprising two parts, these are: the real part, consisting of the display, a barrier and the viewer's eyes and the virtual part consisting of the image on the screen, a virtual barrier and a virtual projection point for each eye. In this system, the only object that is effectively fixed is the barrier. As the eyes of the viewer move the image on the screen is controlled to direct a left and a right image to the appropriate eyes by the output of a head tracker. The image simulates the effect of two virtual projection points behind the screen passing through a virtual fixed barrier that is effectively just behind the screen. The system has the disadvantages that the horizontal resolution is reduced and only one viewer can be served.

2002: The Massachusetts Institute of Technology (MIT) Media Lab built a display that uses a micro-polariser array to spatially separate the left and right images (Benton, 2002). The patent relating to the display describes the use of a two-dimensional SLM that rotates the polarization by 90° in the region that produces an exit pupil in the right-eye area of the viewing zone. Head tracking is achieved simply by imaging the right side of the viewer's head with reflection from an infrared source.

EUROPEAN HEAD TRACKED DISPLAY RESEARCH

Between 2002 and 2011, the European Union (EU) funded three projects where head tracked 3D displays have developed; the time scale of these is shown in Figure 9. The first of these was ATTEST (Advanced Three-Dimensional Television System Technologies) that ran from 2002 to 2004. ATTEST had the objective of proving a novel concept for a 3D TV broadcast chain with essential requirements being the backwards compatibility with existing 2D broadcast and flexibility to support a wide range of different 2D and 3D displays. This was achieved by providing depth as an enhancement layer on top of a regular 2D transmission. The project addressed the complete 3D-TV broadcast chain: 3D content generation (novel 3D camera and 2D-to-3D conversion of existing content), 3D video coding complying with 2D digital broadcast

and streaming internet standards and developing single and multi-viewer 3D displays for visualisation. The project was led by Philips and comprised seven partners including De Montfort University (DMU), Fraunhofer HHI and the Technical University of Eindhoven (TUE). Within the project the role of DMU's imaging and Research Displays Group (IDRG) was to produce a proof-of-concept head tracking multi-viewer autostereoscopic display.

In June 2006, the MUTED (Multi-user Television Display) project commenced. This was led by the IDRG and had seven partners including Sharp Laboratories of Europe (SLE), Fraunhofer HHI, TUE and Light Blue Optics. Its aim was to develop a practical 3D television system that could serve multiple viewers who do not have to wear special glasses. As more immediate niche market application was envisaged for the display an additional partner developing medical systems, Biotronics3D, was enlisted. In this

Figure 9. European head tracked display research

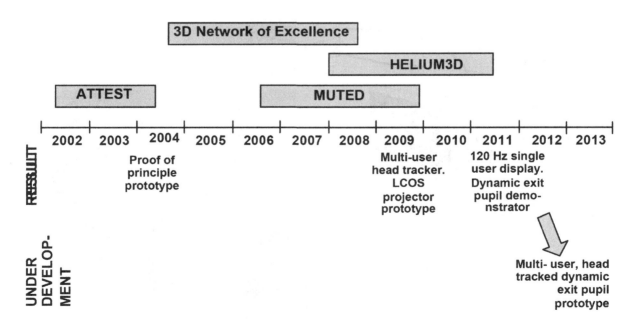

project an advanced version of the ATTEST display was developed. The project finished in October 2009.

The third project was HELIUM3D (High-efficiency Laser-based Multi-user Multi-modal 3D Display) that ran from January 2009 until June 2011.This was also led by the IDRG and the other partners were: Philips Consumer Lifestyle, Barco, University College London, Fraunhofer Heinrich-Hertz-Institut, Eindhoven University of Technology, Koç University and Nanjing University. In this project a display that uses laser illumination for its ability to be precisely controlled was developed. .Although the project has finished the work is ongoing and the latest developments are reported in Section 5.8.4.

ATTEST: MULTI-USER DISPLAY

In ATTEST, a 3D display was developed that was targeted specifically at the television market and capable of supplying 3D to several viewers who do not have to wear special glasses and who can move freely over a room-sized area. The display consists of a single liquid crystal display that presents the same stereo pair to every viewer by employing spatial multiplexing. This presents a stereo pair on alternate pixel rows, with the conventional backlight replaced by steering optics controlled by the output of a head position tracker. Illumination is achieved using arrays of coaxial optical elements in conjunction with high-density white Light Emitting Diode (LED) arrays.

Figure 10 is a schematic diagram of the display also showing the head tracker and viewers. In the figure the head tracker is shown

Figure 10. Schematic diagram of the display

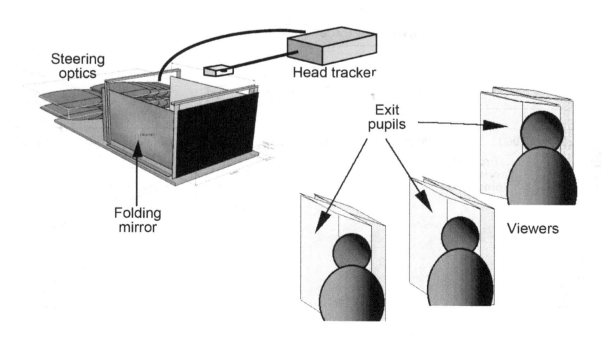

as using a camera but as the development of a non-intrusive tracker was not within the remit of the project a Polhemus electro-magnetic tracker that required the wearing of a pickup on the head was used.

At an early stage of the project, a simplified version incorporating a Fresnel lens was built. The illumination source is a pair of halogen lamps that illuminate the back of the screen via folding mirrors that keep the housing to a reasonable size. In front of the screen assembly is a vertical diffuser that produces elongated real images of the lamp filaments in the viewing field; this enables vertical movement of the viewer. The display does not employ tracking so the viewer has limited lateral movement.

In addition to the Fresnel lens and vertical diffuser, the screen incorporates an LCD with its normal backlight removed. Spatial image multiplexing is achieved with a lenticular screen having horizontally aligned lenses. The operation of the lenticular screen is described later in this section. It requires the two illumination sources to be at different vertical positions so that one source is focused on the odd numbered pixel rows and the other source on the even rows. The sources are also displaced laterally so that two exit pupils are formed side-by-side. This display is similar to the MIT display described at the end of Section 5.2.1.

The purpose of this prototype was to demonstrate the principle of spatial multiplexing; it is not particularly useful as a display as aberrations in the lens restrict the size of the viewing region. Also it is difficult to move the exit pupils in the Z direction as this would involve physically moving the lamps also in the Z direction.

In order to overcome the problems of aberrations encountered in the Fresnel lens the ATTEST display uses an optical array.

In this case the exit pupil is formed from a series of intersecting collimated beams that originate from the series of illumination sources. In order to steer the exit pupil laterally the illumination sources move in the opposite direction to the exit pupil. If the pupil moves towards the array the spacing of the sources increases and vice versa. In this way the exit pupil can be steered over a large area but with the illumination sources all lying in a single plane as opposed to a single source moving over a large area in the X and Z directions behind a lens. It there is more than one viewer, additional exit pupils are formed by introducing additional illumination sources. The number of exit pupils is only limited by the number of viewers that can physically fit into the viewing field.

In order to overcome off-axis aberrations an optical configuration referred to as *coaxial* is used where the illumination and refracting surfaces are cylindrical and are centred on a common axis (Figure 11 (a)). For a refractive index of 1.5 the radius of the illumination surface is double that of the refracting surface. The width of the output beam is limited by the aperture that is also centred on the common axis. The width of the output beam is approximately 1.5 times the width of the aperture. Light is contained within the upper and lower surfaces of the optical element by total internal reflection.

The element width is larger than the beam width and in order to provide a contiguous light source across the array the elements must be arranged in a 'staircase' configuration as in Figure 11(a). The illumination source must be curved and this is achieved by 256 white LEDs (Figure 11(b)). These are mounted on 16-element sub-assemblies that have their own drivers, heat sinks and linear microlens arrays located in front of the LEDs to condense the light output. The LED array is situated at

Figure 11. ATTEST optical array and illumination source

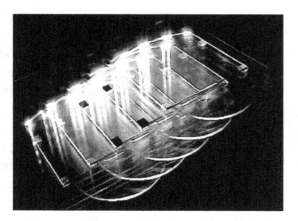

(a) Optical Array Section

(b) 256-elelemt White LED Array

the back of the optical element so it is edge lit. The light is contained within the element until it reaches the front refracting surface where it is diffused vertically.

Spatial multiplexing is achieved by placing a lenticular screen behind the LCD so that its pixels are in the focal plane of the lenses as shown in Figure 12(a). Two sets of exit pupils must be formed; one set for the left eyes and the other for the right eyes. This requires the use of two arrays that are separated vertically. In order to reduce costs the lenticular screen and LCD are the same as those used in the Fraunhofer HHI Free2C display that is described later; in this display the lenticular screen is in front of the LCD as it is used to produce exit pupils in front of the screen. Also, in the HHI display the lenticular screen lenses are vertically aligned and the LCD is in the portrait orientation. Allowance must be made for parallax and as the array is at a finite distance from the screen the lenticular screen is slightly less than double the LCD vertical pitch.

The complete display is shown in Figure 12(b). As the arrays are around 800 millimetres behind the screen two side mirrors are used to effectively increase the width of the array by producing reflected inverted virtual array sections either side of the directly-viewed central section. The mirrors are surface-silvered to prevent double reflections. Although there is a vertical diffuser mounted in front of the LCD the axes of the rays exiting the screen diverge and the screen brightness is enhanced by placing a cylindrical convex lens in front of it. As a custom-built Fresnel lens is expensive a large liquid–filled lens was built for the purpose.

The ATTEST display was produced as a proof-of-concept and the performance was not expected to be exceptional. User trials were carried out by partner TUE and the following points were concluded and solutions offered:

- **Brightness:** This was in the region of around 10 nits. The high étendue (product of source area and solid angle) of the LEDs in relation to their pitch did

Figure 12. ATTEST spatial multiplexing and display set-up

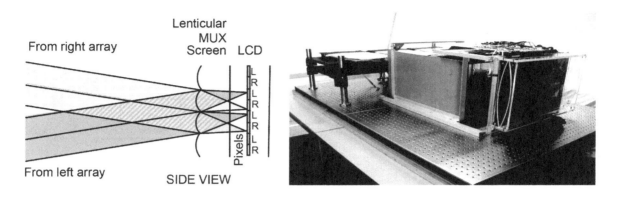

(a) Spatial Multiplexing   **(b) ATTEST Display Set-up**

not provide very effective condensing and there was considerable light loss. Redesign of the illumination source is necessary.

- **Vertical banding in the image:** This is also referred to as mura and is caused by variations in the colour and output of the LEDs. Care was taken to reduce vertical striations by selecting LEDs from a single batch and by using soft apertures in the optical elements. Again, this indicates that the illumination source required modification.

- **Crosstalk:** This was high, being in the region of 15%. The principal cause of this was the LCD where a fine structure in the sub-pixels produced excessive diffraction. There are two ways to combat this; rotate the LCD through 180° or select a panel that has a different sub-pixel structure.

FREE2C SINGLE USER DISPLAY

In addition to the multi-user display developed by DMU a single viewer display, called Free2C, was built at Fraunhofer HHI. In this display an exit pupil pair is produced in front of the LCD with a vertically-aligned lenticular screen. Unlike the DMU prototype the single viewer display retains the backlight supplied with the LCD. The viewer position is determined by processing the output from a pair of cameras mounted above the screen. The exit pupils are steered by moving the lenticular screen in both the X and Z direction (Figure 13a) with the use of voice coil actuators. This enables viewer movement over the large region shown in Figure 13(b).

The display uses a non-contact non-intrusive video based system that provides a near to real-time high-precision single-person 3D video head tracker. The fully automated tracker employs an appearance-based method for initial head detection (requiring no calibration) and a modified adaptive block-matching technique for head and eye location measurements after head location. This approach compares the current image with eye patterns of various sizes that are stored during initialization. Tracking results (shown as locating squares on the eyes) for three different viewers with three different scene backgrounds and illumination conditions are shown in

Figure 13. Fraunhofer HHI single user display

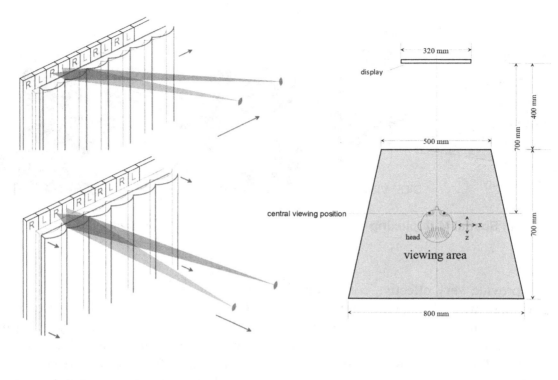

(a) Tracked Optical System　　　　**(b) Tracking Region**

Figure 14.The tracking algorithm also works for viewers wearing glasses.

For automatic initialization the tracker finds the viewer's eye positions by either looking for simultaneous blinking of the two eyes, or by pattern fitting face candidates in an edge representation of the current video frame by applying a predefined set of rules. These face candidates are finally verified by one of two possible neural nets. After initial detection the eye patterns that refer to the open eyes of the viewer are stored as a preliminary reference. Irrespective of the initialization method applied, the initial reference eye patterns are scaled using an affine transformation to correspond to six different camera distances.

MUTED: HOLOGRAPHIC PROJECTOR DISPLAY

The multi-user display developed in the MUTED project used ATTEST as a basis with the principal difference being in the illumination arrangement. The ATTEST display employed 5120 white LEDs that had variations in both colour and brightness as mentioned in the

Figure 14. Fraunhofer HHI head tracker

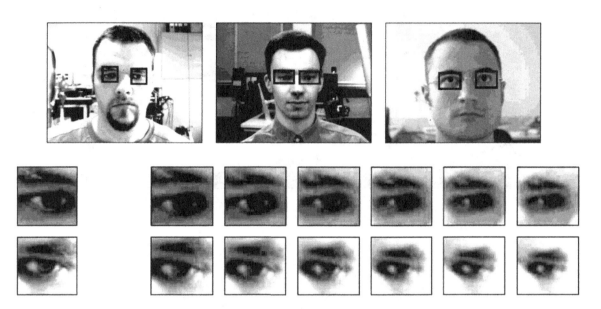

previous section. The natural solution to this appeared to be with the use of some form of projector producing a two-dimensional image as opposed to the effective two-dimensional matrix formed from 20 off 256 linear arrays. A typical pattern required for the production on one exit pupil is shown in Figure 15. This is the pattern for only one viewer but even if there are several viewers it can be seen that the pattern is still sparse. As most of the light is blocked with a conventional projector it was decided that a holographic projector would be used. This concentrates the light from the complete wavefront into the pattern as the image is produced by interference.

With a projector the illumination matrix lies on a flat plane as opposed to a series of curved regions as in ATTEST. The array element now has a different shape (Figure 16a) where the lens is not circular but has a profile optimised with the use of ZEMAX software. The optical arrays have the same 'staircase'

configuration as ATTEST and are arranged in two sets of 49 elements, one above the other as in Figure 16(b).

Figure 17(a) is a schematic diagram of the MUTED display set-up. The illumination source is a high throw angle holographic projector supplied by MUTED partner Light Blue Optics. This is located below the screen and points toward the back of the display. The light entering the back of the arrays must be collimated and this is achieved with the large parabolic field mirror. This is potentially an expensive component and an inexpensive solution was achieved with the use of a flexible surface-silvered plastic mirror whose shape is obtained by mounting it in a frame having parabolic retaining grooves.

The LCD used in the screen assembly is a 20" 4:3 panel that has clear sub-pixels that have no microstructure within the sub-pixel. A custom-built lenticular screen was obtained whose pitch was matched to the LCD in order

Figure 15. MUTED illumination pattern

Figure 16. MUTED optical array

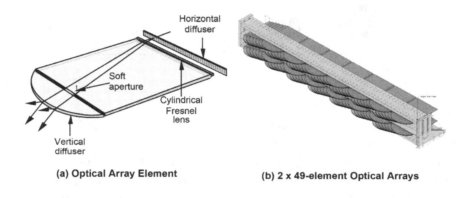

(a) Optical Array Element   (b) 2 x 49-element Optical Arrays

Figure 17. MUTED display set-up and holographic projector

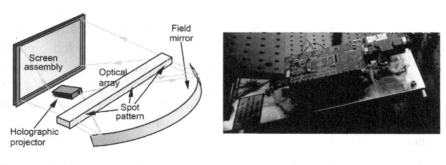

(a) MUTED Schematic Diagram   (b) LBO Holographic Projector

for the array to be located 450 millimetres behind the screen. Another advantage of the particular display panel chosen was that it does not have an anti-glare surface on the front diffuser so that light passing through is not scattered. In the MUTED display the cylindrical brightness enhancement lens incorporated is a custom-built Fresnel lens. In the holographic projector pictured in Figure 17(b), a computer-generated hologram is illuminated with coherent light. This has the advantage that the complete wavefront is used and the light delivered is controlled by diffraction rather than simply being blocked. Unlike conventional projection, this device utilizes a laser light source, and a phase-modulating LCOS microdisplay on which a hologram pattern, rather than the desired image, is displayed. The hologram patterns are computer-generated such that, when the LCOS is illuminated by coherent laser light, the light interferes with itself in a complex manner through the physical process of diffraction resulting in the formation of a large, high-quality projected image. However, since the phase- modulating SLM devices are binary, a conjugate (mirror) image is produced, thus reducing the efficiency by half. Higher orders produced by the SLM pixellation effectively limit the diffraction efficiency to approximately 40%. A Hologram produced for each location in the viewing field that is split up into approximately 5 millimetre X increments and 20 millimetre Y increments. This gives around 12,000 locations and rather than perform the calculations in real time 12,000 holograms are stored on a Look-Up Table (LUT).

As the light exiting the projector is polarised light throughput is maximised by making all the optical components in the light path non-birefringent and matching the polarisation with a half wave plate. The materials used in the path include: Pokalon, polymethylmethacrylate (PMMA), BK7 glass and Triacetylcellulose (TAC).

The single viewer head tracker developed by Fraunhofer HHI for their Free2C display was adapted for multi-user applications. This implementation is divided into fully automated initial face detection and subsequent feature tracking. The initial face detection is based on a decision cascade of Haar basis functions (Viola & Jones, 2001). By combining these simple functions it is possible to construct a classifier which is also able to discriminate classes with more complex distributions with sufficient accuracy. An essential advantage against other methods is the high speed in the detection process. The decision cascade is preliminary learned by boosting (Freund & Schapire, 2001) on a large face dataset and a very large dataset of a complementary non-face class.

After the initial face detection has been done specific facial feature points are detected by several image processing methods. One method used is the calculation of the radial symmetry to detect pupils (Loy, 2003). After a set of facial features has been successfully detected the information is used to track these features in a computationally inexpensive and fast tracking process. The tracking is achieved by employing enhanced adaptive block matching methods. For that purpose tracking features (image elements around a facial feature with properties that make this element simple and reliable to track) for the specific facial feature are selected and tracked. The combination of the tracking results of these tracking features increases the accuracy of the facial feature up to sub-pixel accuracy. The use of calibrated stereo camera pairs provides spatial position data with sufficient accuracy in all dimensions.

Figure 18. MUTED LCOS projector prototype

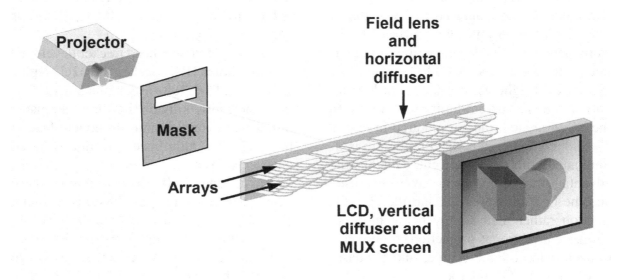

A fundamental problem in video-based tracking technology is the limitation of the camera optics. A limited depth of focus and a constant focal length restrict the possible detection area to unsuitable dimensions. A single optic of a system that tracks an object with an adequate accuracy provides a limited variation of the tracking distance. For that reason the approach uses multiple cameras, a camera system with long focal length to observe far areas and a short focal length system with large opening angle to observe near areas.

The basic design of the tracking system supports the use of multiple, almost independent stereo camera pairs. These camera pairs observe a sub-area of the overall tracking area. The detection and tracking results are sent to a data manager which collects the data of all camera pairs, evaluates the data and merges all information to globally defined instances of the tracked individuals. As a result, a person is constantly tracked even though he moves from one sub-area to another.

In evaluation trials, the head tracker was capable of tracking four viewers between 1000 millimetres and 2500 millimetres from the screen with an opening angle of 50° and consequently it performed to the required specification. The display however was extremely dim with an output of less than 5 nits. Whilst the light throughput was maximised, it was not possible to increase the laser powers above 300 mW for each of the red, green and blue as above this power they are not single mode and so are unable to function with the projector.

LCOS PROJECTOR DISPLAY

In order to overcome the brightness problems with the laser projector, a version using a Liquid Crystal On Silicon (LCOS) conventional lamp projector running at 60 Hz has been built (Figure 18). At present this produces a single pair of exit pupils 1000 millimetres directly in front of the screen whose positions are fixed. The display does however provide images that are sufficiently bright for use for comparative evaluation purposes.

As the output of this projector is elliptically polarised and also the polarisation is different for each of the primary colours there was nothing to gain by using non-birefringent optics. For this reason a Plexiglass array is used in this version. The configuration of this prototype is similar to the holographic projector-based prototype with two exceptions; the laser projector is replaced by a Canon 60 Hz LCOS conventional projector and the parabolic field mirror replaced by with a large one-metre focal length Fresnel lens. Originally it was anticipated that the faceted structure of a Fresnel lens would create fringing artefacts but subsequent investigation showed this not to be the case. Conventional projection also allows the pattern to be a series of rectangles, as opposed to the spots from the laser projector that have a Gaussian profile. This enables greater control over the exit pupil intensity profile.

Head tracking has not been applied to the prototype but the optical performance can be determined without this. Although relatively low at 25 nits, the screen brightness is sufficient to enable it to be viewed under reasonably bright ambient lighting conditions.

HELIUM3D

Principle of Operation

The HELIUM3D display operates by forming exit pupils so that a stereo image pair can be directed to each viewer. The display can operate in two modes. In the first mode a single image pair is formed the same pair can be directed to the left and right eyes of all uses. In this mode the display acts in a similar manner to a conventional glasses display, with the exception that the glasses are not necessary. Images are produced sequentially so that they must have a frame rate of 120Hz in order to eliminate flicker. If higher frame rates can be obtained this would enable more than one viewer to see their own dedicated images. For example a 240 Hz frame rate allows four images to be presented every 1/60th second so that two viewers can see two separate image pairs; this would enable motion parallax and other interesting modes of operation to be obtained.

Figure 19 is a simplified schematic diagram of the display. It is essentially a projection display where images are formed in a light engine and transferred to a viewing screen via a relay lens system that contains an SLM. The light engine forms a horizontally scanned image on L_2. In the figure L_2 is a field lens that concentrates the light from the light engine projection lens (L_1) on to the second projection lens L_3. A horizontal diffuser spreads the real image of L_1 across the complete width of L_3. L_3 is adjacent to a linear SLM and relays the image on L_2 on to the screen and. This SLM controls the light input to the front screen but its image is not seen as it is in the Fourier transform plane of L_3. A real image of the SLM is formed in the viewing field and light beams from the screen are directed to the images of the apertures; exit pupils are formed at the intersections of the beams.

The exit pupils are created dynamically; in MUTED the pupils are formed simultaneously so that the complete width of the screen is illuminated at any one time and in HELIUM3D an image column scans the screen horizontally with the directions of the light emerging from the column controlled by an SLM as in Figure 20. This shows light being diverted to two viewers' eyes by opening two transmitting apertures in the SLM.

Figure 19. HELIUM3D schematic diagram

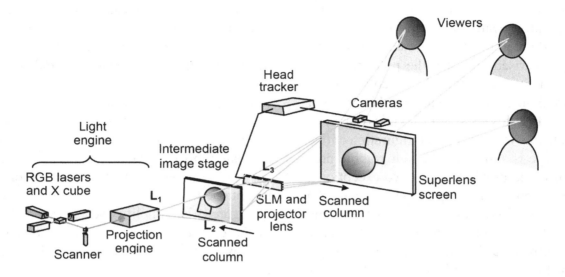

The front screen assembly includes a Gabor superlens that comprises two sets of microlens arrays as shown in Figure 21(a). This has different imaging properties to conventional lenses as input and output ray angles remain on one side of the normal to the lens surface and image distances become less as the object distance is reduced (Figure 21b).

The purpose of the superlens screen in the display is to effectively magnify and focus an image of the SLM into the viewing field so that this fills its complete width. Magnification is achieved with the use of a superlens having different focal lengths. Each superlens element behaves as a telescope where the first lens acts as the 'objective 'and the final lens as the 'eyepiece'. In order to prevent light

Figure 20. HELIUM3D exit pupil formation

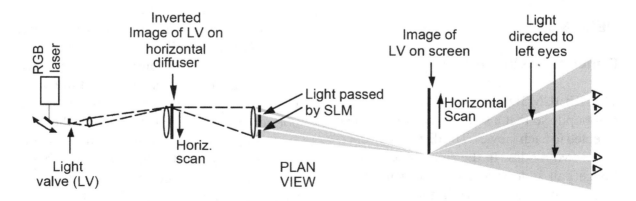

Figure 21. HELIUM3D Gabor Superlens

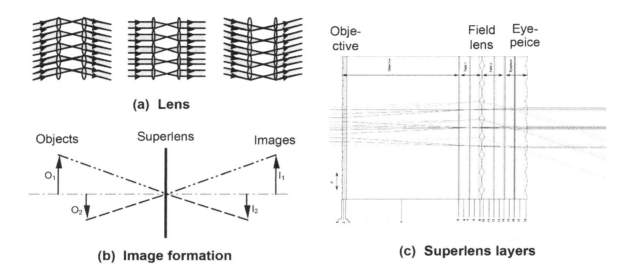

(a) Lens

(b) Image formation

(c) Superlens layers

passing into adjacent elements a field lens is located in the focal plane of the 'objective' as shown in Figure 21(c). The lens surfaces are only curved in one direction so that a complete lens array is a lenticular screen having vertically aligned lenses. Angular magnification takes place only in the horizontal direction and its value is equal to the ratio of the objective focal length to the eyepiece focal length.

With reference to Figure 21(b) it can be seen that the SLM will form a real image in the viewing field, the position of which is determined by the focal lengths of collimating lenses in the screen assembly. The position of this image is referred to as the conjugate plane and when an eye is located in this plane the position of the transmitting region in the SLM does not change position over the duration of a scan. When however the eye is away from the conjugate plane the transmitting region must traverse the SLM during the scan with the distance traversed being proportional to the distance of the eye from the conjugate plane.

In Figure 22, the case for an eye closer than the conjugate plane is shown. As the effective position of the source must be closer than the SLM in this case, a point C must be formed where the light always passes through over the scan period. In order to achieve this it can be seen that the transmitting region in the SLM must move from point A to point B. If the eye is further than the conjugate plane the point C must lie behind the SLM so that it is a virtual intersection point.

The original intention was to use a linear diffractive light valve in the display as this is capable of running at a high frame rate. It is also designed to operate in the horizontally scanned image column mode with illumination from a laser source. Unfortunately, these are not readily available now and the HELIUM3D prototype uses an analogue LCOS light engine with a scanned laser illumination source. A DLP light engine cannot be used

Figure 22. HELIUM3D dynamic exit pupil formation

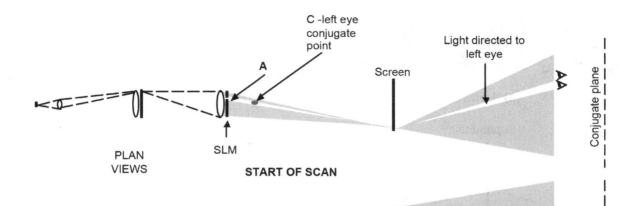

as grey scale is obtained by Pulse Width Modulation (PWM) and this requires a constant illumination on each pixel which scanned Illumination does not provide; the illumination beam is a vertical fan of rays from a combined RGB laser source that is concentrated into a narrow beam and scanned horizontally.

Prototype 1

Display Prototype 1 (Figure 23) is a simplified version built half way through the project; its functionality is limited as full specification components were not available at that time of its construction. The three key components of the display are the light engine, the SLM and the Gabor superlens front screen. It was the SLM in particular that was the principal cause of delay due to its fast response time requirement.

As HELIUM3D is a projection display images are produces by a light engine. At the start of the project 120 Hz light engines were not available and Prototype 1 incorporated a 60 Hz engine obtained from an off-the-shelf projector. Images are presented time sequentially so this version provided 30 Hz per eye.

In order to produce dynamically-formed exit pupils an SLM with a response time in the order of 150 µs is necessary. This requires the use of ferroelectric liquid crystal (FLC) technology and expertise in this area was difficult to find. For this reason the delivery of the SLM was delayed and Prototype 1 incorporates a 120 Hz LCD to perform this function. This allows exit pupil control in the lateral direction only and not in the Z direction.

The design and construction of the Gabor superlens also proved to be challenging and the design of a lens with sufficiently well-collimated output took longer than originally

Figure 23. HELIUM3D Prototype 1

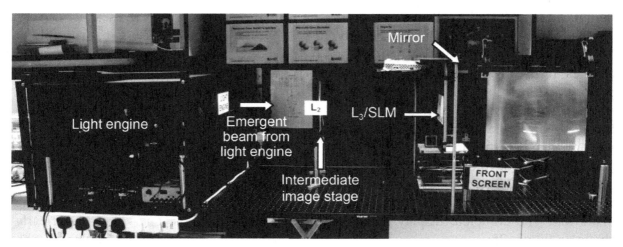

anticipated. Prototype 1 incorporated a super-lens made from available lenticular screens that give an angular magnification of two.

Even with limited functionality Prototype 1 did allow the basic principles to be demonstrated and enabled limited user trials and objective measurements to be carried out.

Prototype 2

The light engine developed at Koc University for Prototype 2 comprises an RGB Laser, beam-shaping optics, scanning mirror, liquid crystal, LCOS unit and projection optics. These were successfully integrated with the transfer screen and head tracker unit to form Prototype 2. The illumination source is three lasers with wavelengths of 640 nm, 532 nm and 473 nm whose outputs are combined in free space with the use of an X- cube. The lasers are multi-emitter laser diodes, as opposed to single emitter lasers used in Prototype 1. The diameter of a single beam coming from the lasers is typically 0.08 millimetres and has a maximum diameter of 0.15 millimetres, with the typical divergence angle of 10mrad. The maximum optical power of the red laser is

typically 4 W and it has an absolute maximum rating of 4.4 W; the red laser is of a different design that is specially produced for laser display applications.

Fibre coupling is used to provide a better speckle contrast ratio and better homogenization (uniformity). An objective is placed at both ends of the fibre to provide coupling with high efficiency and a collimated output beam. The outputs of the multi-emitter lasers are homogenized with the use of a cylindrical microlens array. In the light path after this, two anamorphic lenses are used to provide the necessary beam size and shape.

Vibrating the fibre is an effective way for providing temporal averaging in order to reduce the speckle contrast and increase the homogenization but without any loss in laser power. The fibre is vibrated using a DC motor spinning at about 1,000-2,000 rpm, which is effective when scanner is not in motion. However, to achieve effective temporal averaging and speckle reduction during scanning requires much faster vibration speeds that are in the kilohertz range.

When the light engine was first developed there was not a commercially available 120 Hz light engine. A 120 Hz light engine had to be built using two 60 Hz LCoS engines (Figure 24a) and displaying their outputs sequentially. This is achieved by mounting two projection engines from off-the-shelf Canon SX800 projectors and combining their outputs in free space (Figure 24b). Illumination is supplied by dividing a single sweep from the scanner into two channels that are separated with an arrangement of mirrors. The resolution of the projectors is 1024 x 768 but this is reduced to around 700 x 700 in order to overcome distortion. The complete display set-up is shown in Figure 25.

As the fast SLM was not available in time for it to be incorporated into Prototype 2 an interim solution was provided with the use of 3D shutter glasses synchronised the images produced by the projection engines (Figure 26). This is achieved with the use of an infra-red link whose signal is produced by an 8-bit

PIC12F675 micro-controller whose input is from the same source as the scanner driver.

With the existing set-up, it is possible to provide a stereo image pair to a viewer at a fixed position. This gives fixed viewing regions that allow only limited lateral movement and +/-100 millimetres motion in the Z direction.

Although the fast SLM is not in place, pupil tracking is integrated into the system thus giving content changes in accordance with the motion of the viewer in both the Z and Y directions. The viewer can clearly observe the behaviour of the pupil tracker by moving freely in the vertical axis or moving ±100 millimetres from the optimum viewing distance.

There are various ways in which the content can change; for example, a viewer can see a view of a 3D object from different directions. It is also possible to control the content with special motion of the hand, for

Figure 24. HELIUM3D Prototype 2 light engine

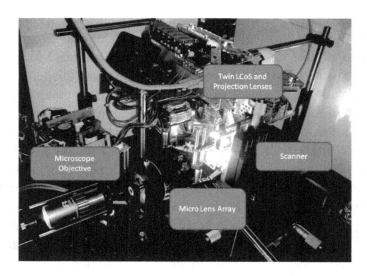

(a) Light Engine

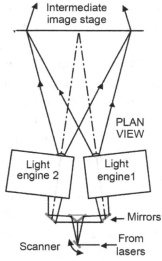

(b) 120 Hz Projection Engine

Figure 25. HELIUM3D Prototype 2

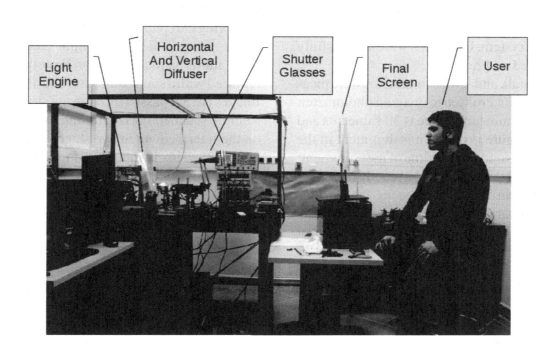

Figure 26. HELIUM3D Prototype 2 SLM

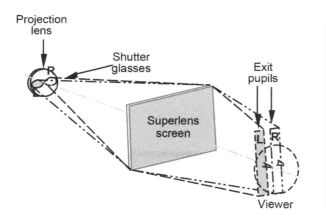

(a) Exit Pupil Formation

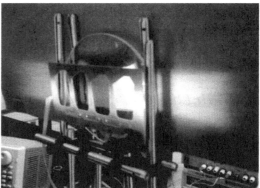

(b) Shutter Glasses Set-up

example a viewer can turn a virtual page by his/her hand. All these properties are made possible by integrating the pupil and the hand tracking system. User trials were successfully completed and reported.

Crosstalk and screen luminance were measured using a ProMetric imaging photometer. The maximum luminance is 30 value nits and this low figure is due to misalignment in the RGB laser source at the time of measurement; it estimated that it this can be readily increased to more than 200 nits.

The separation of the images is shown in Figure 27 which is a series of photographs of a matrix of displayed 'L's and 'R's taken at 30 millimetre intervals. At -30 and 30 millimetres which correspond to the locations of the eyes, the images are completely covered by the appropriate letter. The measured crosstalk value was 1.6%.

DYNAMIC EXIT PUPIL DEMONSTRATOR

Due to delays in obtaining the SLM and its driver a demonstrator to show its function was built after the official end of the project in June 2011. This does not include a light engine and the performance is established by observing exit pupils projected on to a screen in the viewing field. The light engine and RGB illumination source of Prototype 2 is replaced with a single blue laser and a scanner. Images are not passed through the system and a single scanned horizontal line is produced on the intermediate image stage as shown in Figure 28. This is transferred to the SLM as another horizontal line where the control of the exit light directions from the front screen is carried out. The light emerging from the SLM is diffused in the vertical direction only so that a scanned vertical line is produced on the front screen. Although this diffuser has no effect on the horizontal direction of the light it does enable viewers located at the exit pupil positions to observe illumination over the complete height of the screen.

The SLM is a linear 128-element FLC array comprising four tiled 32-element cells and is located in position D in Figure 29(a). These are connected to the driver board via four 30 millimetre long ribbon cables to the voltage level shifters (in region C). The board is an FLC driver from GarField Microelectronics Ltd, a company based in the UK; it

Figure 27. HELIUM3D Prototype 2 image separation

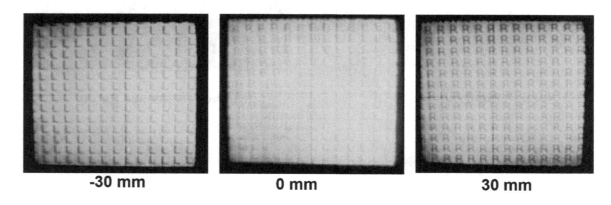

-30 mm   **0 mm**   **30 mm**

Figure 28. Dynamic exit pupil demonstrator

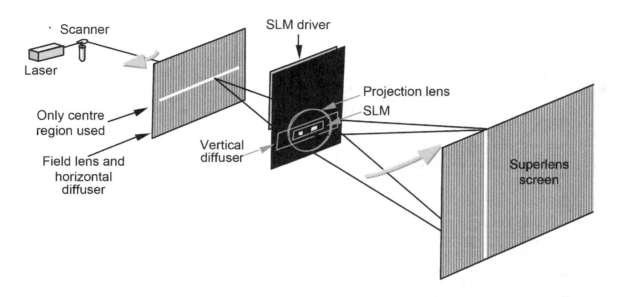

incorporates a field programmable gate array (FPGA) that has been programmed to accept the HELIUM3D display parameters and the head position outputs from the HHI head tracker. A summary of the functionality of the driver board is listed below, and a block diagram is shown in Figure 29(b).

- Synchronisation with source video frame-rate.
- Drive scanning mirror in the light engine section of the display and keep it in syncronisation with the rest of the system.
- Accept eye coordinate data from external PC connected to the head tracker unit.
- Generate SLM pixel control data from eye coordinate data.
- Drive the SLM using ± 10 volt signals generated from SLM pixel control data
- Calculate and control DC balancing of the SLM.

The demonstrator currently operates in three modes, these are:

- Move a single exit pupil exit pupil in the X and Z directions under the control of a mouse.
- Pre-program the motion up to four independently moving exit pupils – Figure 30 shows three exit pupils projected on to a screen.
- Control the lateral position of a single exit pupil with the use of a simple infrared tracker.

FUTURE WORK

As it is not certain whether or not a fast light engine will be available in the near future it is envisaged that the commercialised version of the HELIUM3D display will be in the form of a steerable backlight for a 120 Hz LCD (or faster if it becomes available), as opposed to its current rear projected form.

Figure 29. HELIUM3D SLM and driver

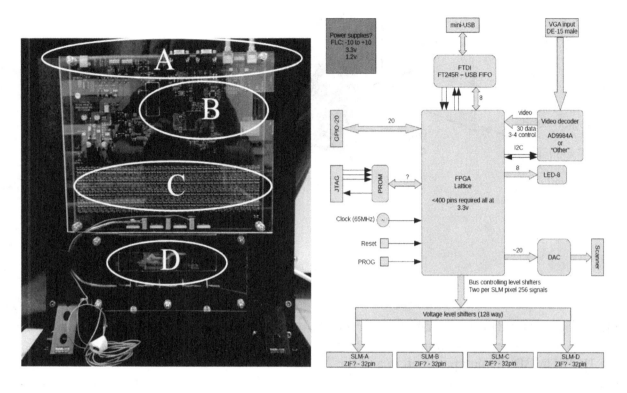

(a) SLM and Driver Board   **(b) SLM Driver Schematic Diagram**

Figure 30. HELIUM3D dynamic exit pupil demonstrator

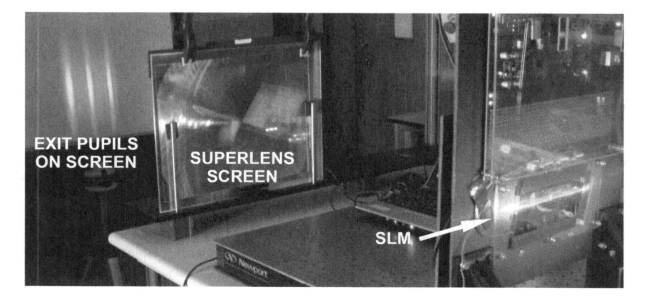

The demonstrator is currently being converted into a display by placing a 120 Hz Alienware LCD panel in front of it and synchronising its frame rate to the SLM and scanner. As the demonstrator has no intermediate image stage the total volume occupied by the light path and the hardware between the scanner and the SLM is small and could be folded as illustrated in Figure 31 This would occupy only a small volume behind the screen. Further development is required in order to reduce the volume of the light path between the SLM and the front screen. Investigation into how this would be possible has been carried out within the project where one of the deliverables covers some of the issues associated with free-space and piped methods.

CONCLUSION

Other Display Types

The survey of display types in Section 5.2 can be used to give an indication of the advantages and disadvantages of head tracked displays over other approaches. The consensus of opinion is that users will not find glasses displays as acceptable as autostereoscopic displays and it is fairly certain that 3D television will be glasses-free within a few years.

Volumetric displays have the advantage of having no A/C conflict and having full parallax, however they generally suffer from image transparency and their hardware tends to be complex and expensive. The anisotro-

Figure 31. HELIUM3D hang-on-the-wall version

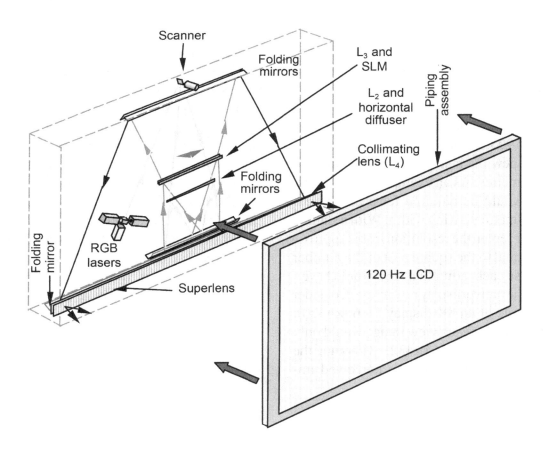

pic displays under development have the potential to provide images having surfaces that are opaque and capable of having a non-Lambertian distribution. The cost of this is greater hardware complexity, considerably more information being displayed and generally, loss of vertical parallax.

Multi-view displays provide a potentially inexpensive display as the hardware at its simplest comprises only a flat panel display and a view-directing screen that is either a lenticular screen or a parallax barrier. Although there is a trade-off between resolution and the number of views, the use of a slanted view-directing screen distributes the resolution loss between the horizontal and vertical directions. If the number of views is low, say 9 as in the Philips Wow display, the depth of field is limited due to the overlapping views causing points in the image in space, equivalent to the voxels in a volumetric display, to be elongated in the Z direction.

A multi-view display with a large number of views is referred to as Super Multi-View (SMV) and If the criteria for the appearance of continuous motion parallax given by various sources are adhered to (Pastoor, 1992; St Hilaire, 1993; Kajiki, 1997) it would appear that the view pitch should be in the order of 2.5 millimetres. As an example, 100 parallax images would be required for a continuous 250 millimetre viewing width. With a slanted lenticular screen the resolution loss is approximately equal to the square root of the number of views so each individual view would have a tenth of the native display resolution. Consider the example of an 8K display (7860×4320); the resolution of each view image would only be in the order of 768×432. Although the addition of 3D enhances the perceived im-

age quality, it is not clear that this resolution reduction would be justified.

An interesting recent development is the Toshiba 55" autostereoscopic display. The native resolution of the display panel is 4K (3840×2160) which enables a series of nine 1160×720 images to be formed. The addition of head tracking permits the viewing zones to be located in the optimum position for the best viewing condition for several viewers. Toshiba states that this can be up to nine viewers but best performance is obtained with four or fewer. In addition to the resolution loss, the other disadvantage of this display is that viewers have to be close to a fixed distance from the display.

If very high resolution displays do eventually become available at reasonable cost they might possibly enable an alternative to head tracking as the mass consumer 3D display of the future.

Light field displays require the presentation of a large amount of information, for example the Holografika display requires around 100m pixels in order to give an angular resolution that provides a good depth of field. This involves the display of a large amount of redundant information where light rays are sent out into the viewing field where no eyes are present to receive those rays. If the cost per effective pixel drops sufficiently, this could be a viable approach in a similar way to the SMV display.

Most proposed light field methods do not have vertical motion parallax so that there will still be rivalry in the vertical direction; the eyes will be attempting to focus at two different distances at the same time possibly giving a similar effect to astigmatism. This is an area that would be a useful subject of future human factors investigations.

The research described in Section 5.2.5 indicates that although holographic displays are possibly some years away, the work on means of reducing the amount of information that must be displayed, for example by showing discrete view panoramagrams as opposed to continuous motion parallax, or developing new display media such as photorefractive polymers, show that the research is moving in the right direction.

It is becoming clear that capture is unlikely to ever be purely holographic. The richness of a naturally-lit will not be recorded by illuminating it with laser illumination and also it would not be practicable to do so.

Other Head Tracked Displays

The SeeReal approach was first announced in 2007. In September 2009 they announced the production of an 8" holographic demonstrator with 35 micron pixels forming a phase hologram and that a 24" product was planned. It does not appear at the time of writing (2012) that a product is actually available yet.

Microsoft Applied Sciences Group has announced a system where the 'Wedge' waveguide technology developed at Cambridge University is adapted to provide steering of light beams from an array of LEDs (Travis et al, 2010). The display has a thin form factor of around fifty millimetres. Images are produced view-sequentially so that the focussing of the rays exiting the screen changes position each time a new view is displayed. The display in operation can be seen via the link in reference (Microsoft, 2010).

Another means of directing light to viewers' eyes is described in a patent assigned to Apple (Krah, 2010). This uses a corrugated reflecting screen where the angle of the exiting beam is a function of its X position. The text of the patent mentions that head tracking can be applied to this display. Proposed interactive applications for the display, including gesture and presence detection, can be obtained on the Patently Apple blog (Purcher, 2011).

Summary

The aim of the European head tracked display research is to produce the next generation of 3D display, in particular one that is suitable for the enormous 3D television market. This must be ready to become a product within the next five years. Head tracking was chosen for the basis of the display as this appears to be the method most likely to be supported by currently available enabling technology. At present the limitation is set by the amount of information that be displayed at a cost that is within the budget for a mass-market product, however if very high resolution panels do become available at reasonable cost it might be the case that the extra overhead of tracking multiple viewers cannot not justified.

Another issue regarding available technology is that the head tracked displays require a transmissive display; that is an LCD, as opposed to an emissive display such as OLED (organic light emitting diode). If these do become the mainstream display technology this could possibly impact on the adoption of head tracked displays.

REFERENCES

Arai, J., Kawai, H., & Okano, F. (2006). Microlens arrays for integral imaging system. *Applied Optics*, *45*(36), 9066–9078. doi:10.1364/AO.45.009066 PMID:17151745

Arimoto, A., Ooshima, T., Tani, T., & Kaneko, Y. (1998). Wide viewing area glassless stereoscopic display using multiple projectors. [SPIE.]. *Proceedings of the Society for Photo-Instrumentation Engineers*, *3295*, 186–191. doi:10.1117/12.307163

Baird, J. L. (1945). *Improvements in television* (UK Patent 573,008). London: UK Patent Office.

Balasubramonian, K. et al. (1982). On the merits of bicircular polarisation for stereo colour TV. *IEEE Transactions on Consumer Electronics*, 28.

Baloch, T. (2001). *Method and apparatus for displaying three-dimensional images* (United States Patent No. 6,201,565 B1). Washington, DC: US Patent Office.

Barabas, J., Jolly, S., Smalley, D. E., & Bove, V. M. (2011). Diffraction specific coherent panoramagrams of real scenes. In *Proceedings of the SPIE*, (vol. 7957). SPIE.

Benton, S. A. (2002). *Autostereoscopic display system* (United States Patent No. 6,351,280). Washington, DC: US Patent Office.

Blanche, P.-A. et al. (2010). Holographic three-dimensional telepresence using large-area photorefractive polymer. *Nature*, *468*(7320), 80–83. doi:10.1038/nature09521 PMID:21048763

Blundell, B. G., & Schwartz, A. J. (2000). *Volumetric three-dimensional display systems*. New York: Wiley- IEEE Press.

Boev, A., Raunio, K., Gotchev, A., & Egiazarian, K. (2008). GPU-based algorithms for optimized visualization and crosstalk mitigation on a multiview display. In *Proceedings of SPIE-IS&T Electronic Imaging* (vol. 6803). Retrieved 29 January 2012 from http://144.206.159.178/FT/CONF/16408309/16408328.pdf

Bradshaw, G. (1953, July 25). We can have 3-D television. *Picture Post*.

Brown, D. (2009). *Images across space*. London: Middlesex University Press.

Collender, R. (1986). 3-D television, movies and computer graphics without glasses. *IEEE Transactions on Consumer Electronics*, *32*(1), 56–61. doi:10.1109/TCE.1986.290119

Cossairt, O., Napoli, J., Hill, S. L., Dorval, R. K., & Favalora, G. E. (2007). Occlusion-capable multiview volumetric three-dimensional display. *Applied Optics*, *46*(8). doi:10.1364/AO.46.001244 PMID:17318244

Dimenco. (2012). *Products – Displays 3D stopping power – 52″ professional 3D display*. Retrieved 29 January 2012 from http://www.dimenco.eu/displays/

Eichenlaub, J. B. (1990, May). Stereo display without glasses. *Advanced Imaging (Woodbury, N.Y.)*.

Eichenlaub, J. B. (1993). Developments in autostereoscopic technology at Dimension Technologies Inc. [SPIE.]. *Proceedings of SPIE Stereoscopic Displays and Applications*, *1915*, 177–186.

Favalora, G. E. (2009). Progress in volumetric three-dimensional displays and their applications. *Opt Soc Am*. Retrieved 29 Jan 2012 from http://www.greggandjenny.com/gregg/Favalora_OSA_FiO_2009.pdf

Freund, Y., & Schapire, R. E. (1999). A short introduction to boosting. *Journal of Japanese Society for Artificial Intelligence, 14*(5), 771–780.

Graham, C. H. (1965). *Visual space perception in vision and visual perception.* New York: Wiley.

Harman, P. (1996). Autostereoscopic display system. *Proceedings of the Society for Photo-Instrumentation Engineers, 2653,* 56–64. doi:10.1117/12.237458

Hattori, T. (2000a). *Sea phone 3D display 1-8.* Retrieved 28 January 2012 from http://home.att.net/~SeaPhone/3display.htm

Hattoti, T. (2000b). *Sea phone 3D display: 7-9.* Retrieved 28 January 2012 from http://home.att.net/~SeaPhone/3display.htm

Hattoti, T., Sakuma, S., Katayama, K., Omori, S., Hayashi, M., & Midori, Y. (1994). Stereoscopic liquid crystal display 1 (general description). *Proceedings of the Society for Photo-Instrumentation Engineers, 2177,* 146–147.

Haussler, R., Schwerdtner, A., & Leister, N. (2008). Large holographic displays as an alternative to stereoscopic displays. *Proceedings of the SPIE, 6803.*

Hembd, C., Stevens, R., & Hutley, M. (1997). Imaging properties of the Gabor superlens. *European Optical Society Topical Meetings Digest Series, 13,* 101–104.

Hitachi. (2010). *Hitachi shows 10 glasses-free 3D display.* Retrieved 24 January 2012 from www.3d-display-info.com/hitachi-shows-10-glasses-free-3d-display

Ichinose, S., Tetsutani, N., & Ishibashi, M. (1989). Full-color stereoscopic video pickup and display technique without special glasses. *Proceedings of the SID, 3014,* 319-323.

Ijzerman, W. (2005). Design of 2d/3d switchable displays. *Proc of the SID, 36*(1), 98-101.

Jones, A., McDowall, I., Yamada, H., Bolas, M., & Debevec, P. (2007). Rendering for an interactive 360° light field display. *Siggraph 2007 Emerging Technologies.* Retrieved 29 January 2012 from http://gl.ict.usc.edu/Research/3DDisplay/

Jorke, H., & Fritz, M. (2012). *Infitec – A new stereoscopic visualization tool by wavelength multiplexing.* Retrieved 25 January 2012 from http://jumbovision.com.au/files/Infitec_White_Paper.pdf

Kajiki, Y. (1997). Hologram-like video images by 45-view stereoscopic display. *Proceedings of the Society for Photo-Instrumentation Engineers, 3012,* 154–166. doi:10.1117/12.274452

Kanolt, C. W. (1918). *United States patent 1260682.* Washington, DC: US Patent Office.

Krah, C. (2010). *Three-dimensional display system* (US Patent 7,843,449). Washington, DC: US Patent Office. Retrieved 4 February 2012 from www.freepatentsonline/7843339.pdf

Lambooij, M., IJsselsteijn, W., Fortuin, M., & Heynderickx, I. (2009). Visual discomfort and visual fatigue of stereoscopic displays: a review. *The Journal of Imaging Science and Technology, 53*(3). doi:10.2352/J.ImagingSci.Technol.2009.53.3.030201

Lippmann, M. G. (1908). Epreuves reversibles donnant la sensation du relief. *Journal of Physics, 4,* 821–825.

Lipton, L. (1988). Method and system employing a push-pull liquid crystal modulator. US patent 4792850.

Loy, G. (2003). *Computer vision to see people: A basis for enhanced human computer interaction*. (PhD thesis). Australian National University, Canberra, Australia.

McCarthy, S. (2010). *Glossary for video and perceptual quality of stereoscopic video*. Retrieved 24th January 2012 from http://www.3dathome.org/files/ST1-01-01_Glossary.pdf

McCormick, M., Davies, N., & Chowanietz, E. G. (1992). Restricted parallax images for 3D TV. *IEE Colloq Stereoscopic Television, 173*, 3/1-3/4.

Microsoft. (2010). *The wedge - Seeing smart displays through a new lens*. Retrieved 4 February 2012 from http://www.microsoft.com/appliedsciences/content/projects/wedge.aspx

Moller, C., & Travis, A. (2004). Flat panel time multiplexed autostereoscopic display using an optical wedge waveguide. In *Proceedings of the 11th Int Display Workshops,* (pp. 1443-1446). Niigata, Japan: Academic Press.

Mora, B., Maciejewski, R., & Chen, M. (2008). Visualization and computer graphics on isotropically emissive volumetric displays. *IEEE Computer Society*. Retrieved 28 Jan 2012 from https://engineering.purdue.edu/purpl/level2/papers/Mora_LF.pdf

Okui, M., Arai, J., & Okano, F. (2007). *New integral imaging technique uses projector*. Retrieved 28 January 2012 from http://spie.org/x15277.xml?ArticleID=x15277

Otsuka, R., Hoshino, T., & Horry, Y. (2006). Transpost: 360 deg-viewable three-dimensional display system. *Proceedings of the IEEE, 94*(3). doi:10.1109/JPROC.2006.870700

Pastoor, S. (1992). Human factors of 3DTV: An overview of current research at Heinrich-Hertz-Institut Berlin. *IEE Colloquium 'Stereoscopic Television', 173*, 1/3.

Perlin, K. (1999). *A sisplayer and a method for displaying* (International Publication No WO 99/38334).

Purcher, J. (2011). *Apple wins a surprise 3D display and imaging patent stunner*. Retrieved 4 February 2012 from http://www.patentlyapple.com/patently-apple/2011/09/whoa-apple-wins-a-3d-display-imaging-system-patent-stunner.html

Sandlin, D. J., Margolis, T., Dawe, G., Leigh, J., & DeFanti, T. A. (2001). Varrier autostereographic display: Stereoscopic displays and virtual reality systems VIII. *Proceedings of the Society for Photo-Instrumentation Engineers, 4297*, 204–211. doi:10.1117/12.430818

Schwartz, A. (1985). Head tracking stereoscopic display. In *Proceedings of IEEE International Display Research Conference,* (pp. 141-144). IEEE.

Schwerdtner, A., & Heidrich, H. (1998). The Dresden 3D display (D4D). *Proceedings of the Society for Photo-Instrumentation Engineers, 3295*, 203–210. doi:10.1117/12.307165

Sexton, I., & Crawford, D. (1989). Parallax barrier 3DTV. *Proceedings of the SPIE Three-Dimensional Visualization and Display Technologies, 1083*, 84–94. doi:10.1117/12.952875

Son, J. Y., & Shestak, S. A. KIM, S.-S., & Choi, Y. J. (2001). A desktop autostereoscopic display with head tracking capability. *Proceedings of the SPIE, 4297,* 160-164.

Soneira, R. M. (2012). *3D TV display technology shoot-out.* Retrieved 27 January 2012 from http://www.displaymate.com/3D_TV_Shoot-Out_1.htm

St Hilaire, P. (1994). Modulation transfer function and optimum sampling of holographic stereograms. *Applied Optics, 33*(5), 768–774. doi:10.1364/AO.33.000768 PMID:20862073

Street, G. S. B. (1998). *Autostereoscopic image display adjustable for observer location and distance* (United States Patent 5,712,732). Washington, DC: US Patent Office.

Sullivan, A. (2004). DepthCube solid-state 3D volumetric display. *Proceedings of the SPIE, 5291.*

Tanaka, K., & Aoki, S. (2006). A method for the real-time construction of a full parallax light field. *Proceedings of the Society for Photo-Instrumentation Engineers, 6055.*

Tetsutani, N., Ichinose, S., & Ishibashi, M. (1989). 3D-TV projection display system with head tracking. *Japan Display, 89,* 56–59.

Tetsutani, N., Omura, K., & Kishino, F. (1994). A study on a stereoscopic display system employing eye-position tracking for multi-viewers. *Proceedings of the Society for Photo-Instrumentation Engineers, 2177,* 135–142. doi:10.1117/12.173868

Tilton, H. B. (1987). *The 3-D oscilloscope - A practical manual and guide.* Upper Saddle River, NJ: Prentice Hall Inc.

Traub, A. C. (1967). Stereoscopic display using rapid varifocal Mmrror oscillations. *Applied Optics, 6*(6), 1085–1087. doi:10.1364/AO.6.001085 PMID:20062129

Travis, A., Emerton, N., Large, T., Bathiche, S., & Rihn, B. (2010). Backlight for view-sequential autostereo 3D. *SID 2010 Digest,* 215–217.

Trayner, D., & Orr, E. (1996). Autostereoscopic display using holographic optical elements. *Proceedings of the Society for Photo-Instrumentation Engineers, 2653,* 65–74. doi:10.1117/12.237459

Tsai, C. H., Lee, K., Hseuh, W. J., & Lee, C. K. (2001). Flat panel autostereoscopic display. *Proceedings of the Society for Photo-Instrumentation Engineers, 4297,* 165–174. doi:10.1117/12.430815

Viola, P., & Jones, M. (2001). Rapid object detection using a boosted cascade of simple features. In *Proceedings of IEEE Conference on Computer Vision and Pattern Recognition.* IEEE.

Woodgate, G. J., Ezra, D., Harrold, J., Holliman, N. S., Jones, G. R., & Moseley, R. R. (1997). Observer tracking autostereoscopic 3D display systems. *Proceedings of the Society for Photo-Instrumentation Engineers, 3012,* 187–198. doi:10.1117/12.274457

Zworykin, V. K. (1938). *Television system* (US Patent no 2,107,464). Washington, DC: US Patent Office.

Chapter 6
Integral 3-D Imaging Techniques

Hans I. Bjelkhagen
Centre for Modern Optics, UK

ABSTRACT

In 1891 the optical physicist Gabriel Lippmann developed a method of reproducing colour in photography without dyes, instead using pure light from the solar spectrum. Later study took his interest into the research of three-dimensional imaging via a method of integral photography in which a fly's eye lens array is used to record images in complete three-dimensional fidelity. Other noteworthy workers in the field such as Ives, Burckhart & Doherty, Bonnet and Montabello followed up the principle, but today Lippmann is acknowledged as being a founding father of the micro-lens technique for three-dimensional imaging. Advances in micro-lens production has led to the easy availability of lenticular print and consumer electronic companies are eager to develop 3-D TV system that incorporates much Lippmann theory. This chapter offers a brief history of Gabriel Lippmann and his subsequent legacy.

INTRODUCTION

Few photographers today are familiar with the name Gabriel Lippmann (1845-1921). Even fewer have seen a Lippmann colour photograph. But in 1908 Lippmann was awarded the Nobel Prize in Physics for his invention of interference photography, a colour technique

exploiting the phenomenon of optical standing waves (Lippmann, 1891). This represents the only time this prestigious award has been made for a photographic invention. However, the process proved difficult and impractical: exposures ran into minutes, the image was difficult to view, and it was almost impossible to copy. Fewer than 500 original examples of this arcane process still exist. Attempts at commercialization floundered, and by the

DOI: 10.4018/978-1-4666-4932-3.ch006

time Lippmann was awarded his Nobel Prize the more practical Autochrome plates had become available. Despite the difficulties, Lippmann's photography remains to this day the only process of true colour photography. Lippmann photographs possess exquisite beauty, both technically and aesthetically. The ultra-finegrain plates essential to the medium display the highest photographic resolution so far achieved. In addition the encoding of colour as a diffraction grating of pure silver offers unrivalled longevity. Lippmann's papers (1891, 1894, 1902) describe the technique in more detail, including modern applications. At the time of Lippmann's invention, stereophotography was very popular. Special viewing devices called *stereoscopes* were used to view the pairs of images. Lippmann himself made a few remarkably realistic colour stereophotographs using his new colour photography technique. His new 3-D autostereoscopic technique, now known as *integral imaging* (*I/I*), employs a microlens array in the recording and display of the images.

INTEGRAL PHOTOGRAPHY

After Lippmann had perfected his colour process, he began work on techniques for recording 3-D photographs that could be viewed without any optical viewing devices. In 1908 he presented his *integral photography* (I/P) in which an array of small closely spaced spherical lenses (a *fly's-eye array*) is used to photograph a scene, recording images of the scene as it appears from many different horizontal and vertical locations (Lippmann, 1908a & b) (see Figure 1). The recorded microlens photograph is then reversal processed, generating a positive image. When this is viewed mounted behind a similar array of lenses, a

Figure 1. Fly's-eye lens array

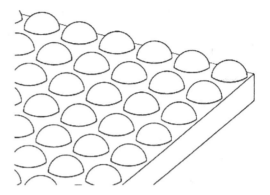

single integrated image composed of small portions of all the images is seen by each eye. The photograph presents different images to the left and right eye respectively, thus creating the 3-D effect, the position of each eye determining which parts of the small images it sees. The overall effect is that the visual geometry of the original scene is reconstructed so that the limits of the array seem to be the edges of a window through which the scene appears, realistically exhibiting both vertical and horizontal parallax and perspective shift with any change in the position of the observer. Thus all the microimages are combined into a single 3-D image.

Integral photography is based on a principle known as the *lens sampling effect*. To achieve this effect, the thickness of the lens array sheet is chosen so that incoming light rays focus on the opposite (flat) side of the sheet, at which plane the microimages are recorded on a single film sheet. It is possible to utilise arrays of pinhole apertures in an opaque plate instead of the microlenses. A pinhole behaves like a lens, but depends on the rectilinear propagation of light rather than on focusing by refraction. The radius of the pinholes is selected to provide the best compromise between the effects of geometrical spreading and diffraction. However, when the pinhole array is used as a viewing screen, the radius needs to be much larger; otherwise the image will be too dark to view with normal illumination.

RECORDING

The recording principle is shown in Figure 2. As can be seen in Figure 2, the integral image gives a reversed-perspective or *pseudoscopic* image when viewed. By using a two-step recording process the reversed image can be reversed a second time, and thus turned into an *orthoscopic* image. Much research has been devoted to solving this problem. The simplest and most obvious method would be simply to re-photograph the image using a similar optical arrangement. However, this

copying process results in an unacceptable deterioration in image quality.

Researchers continuing to work on the process included Ives (1933), who substituted a cylindrical lens array for Lippmann's integral lens array. Such a *lenticular sheet* consists of a linear array of thick plano-convex cylindrical lenses, known individually as *lenticules*. The lens sheet is transparent, and the rear face (which is in the focal plane) is flat. With this system only horizontal parallax was possible. Burckhardt and Doherty (1969) solved the pseudoscopic problem by a new and simple recording method. The integral photographs were recorded on beaded high resolution photographic plates, i.e. with a layer of glass balls affixed to the emulsion surface. These had a refractive index of 1.92 and acted as miniature lenses. Kodak 649 GH plates were chosen. The size of the

Figure 2. Recording an integral photograph

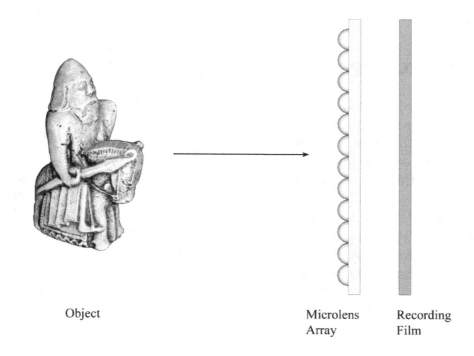

Object   Microlens   Recording
                          Array   Film

balls gave a resolution corresponding to a printer's 230-270 dpi (dots per inch). The high refractive index of the glass balls meant that their focal length was equal to their diameter, and thus they formed an image on their back surface. The beaded plate was exposed in the central image plane of a real pseudoscopic reconstruction of another integral photograph. After reversal development and upon illumination from the rear, the beaded plate displayed an orthoscopic three-dimensional image of the original object, with a depth identical to that of the pseudoscopic image (Figure 3).

A number of researchers advanced the process of integral photography in the 1960s and 1970s, including in particular de Montebello (1977), who produced hundreds of 3-D images using his *Integram* system. There were also important advances in the technique by Bonnet (1942), who developed a number of patented 3-D camera designs and imaging techniques. Two examples of Bonnet's images are shown in Figure 4 & 5. Notice how the lens array affects the image quality in the enlarged detail photo in Figure 5.

IMAGE RESOLUTION

The main problem with integral photography is the poor image resolution. This was appreciated by Burckhardt (1967), who investigated its limitations. He found that for conventional objects it was comparable to the resolution of a TV picture in the USA (which at that time was poor). He also computed the number of resolvable spots needed to record an integral photograph, and showed that this number was proportional to the fourth power of the linear resolution in the image display.

In another paper on image resolution by Hoshino et al. (1998), the authors considered the spatial frequency measured at the viewpoint, which the authors called viewing spatial frequency. Its maximum value defined the resolution of an I/P. In estimating this, the authors derived the optimum width of the lens aperture. They also calculated the width of the viewing area of an I/P, and discussed the practical display capability of an I/P. To simplify the mathematical model they assumed a one-dimensional I/P. The resolution of images

Figure 3. Viewing an integral photograph with an inverted-perspective image

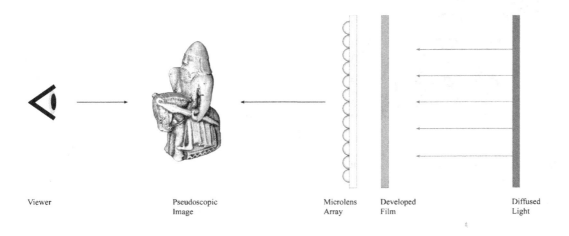

Viewer   Pseudoscopic Image   Microlens Array   Developed Film   Diffused Light

Figure 4. Example of Bonnet images (Photographs by the author, 1991)

Figure 5. Example of Bonnet images (Photographs by the author, 1991)

produced close to the exit pupil, they found, was determined only by the Nyquist frequency of the sampling at the exit pupil (in this case, the aperture of the lens). Thus this (spatial) frequency can be called the 2-D resolution of an I/P image. As long as the depth factor

is greater than unity, degradation of image resolution distant from the exit pupil does not occur. The authors explained that their lens array could produce images at a great distance with the same resolution as conventional 2-D displays provided aberration-free lenses were used; in practice this represented a serious limitation. The difficulty of displaying the element image with respect to the pixel pitch also remains.

Lenticular Images in Print

In the 1960s, as large corporations recognized its advertising potential, the lenticular technique showed rapid progress. It was less difficult than the integral technique for printing 3-D images because the latter required very accurate register between the printed image and the microlens array. Not until digital photography and computer registration were introduced did it become feasible to mass-produce integral images. The first mass-produced lenticular image appeared on 25 February 1964, when Look Magazine featured the 'first ink-printed postcard-sized "parallax panoramagram"'. This was a black-and-white still life of a bust of Thomas Edison, surrounded by some of his more famous inventions, and required a special camera tracked in a programmed arc, to record the image, which is illustrated in Figure 6. The manufacturing process involved printing the image using a 300-line offset press and a special technique for coating and embossing a thin layer of plastic on the image at high speed. Within the magazine readers found a stiff postcard-size insert, on which was the black-and-white 3-D picture. The image, known as an *Xograph*, was processed by Eastman Kodak in Tennessee. The Harris-Intertype Corp. built the equipment needed to add the plastic lens coat. Over seven million copies were sold. Look

Magazine followed with a colour lenticular on 7 April 1964 (Figure 7).

The parallax panoramagram process is laborious, costly and slow, and was not adaptable to high-speed printing. In addition to recording the photograph – which took two full days of work – five additional weeks were required to engrave the photograph, print 7 million copies on a sheet-fed offset press and then pour on and precisely shape the clear plastic film that covered the picture. Another early example (also labelled 'Xograph' is shown in Figure 8) and was published in Venture.

Venture believed that 3-D photography on the covers, and later on inside pages, would dramatically enhance the excitement of the world of travel. These types of 3-D images were used for only a short time and did not justify Venture's hopes. This was echoed by the experience of National Geographic, who in the 1980s placed embossed holograms on three of their covers, but thereafter abandoned the practice. The use of 3-D images on journal covers has seldom been more than a novelty feature. Lenticular images need their plastic raster plate on the cover: this is difficult to

Figure 7. Look Magazine's first colour lenticular image- 7 April 1964

Figure 8. Venture's colour lenticular image-The Venture Xograph image

Figure 6. Look Magazine's first b/w lenticular image- 25 February 1964

add, and reduces the image resolution. And holograms need to be viewed under a spotlight.

In the USSR, Dudnikov (1970) was aware of holography and of Yuri Denisyuk's early work in the medium. Like Denisyuk, Dudnikov became interested in Lippmann integral photography, and wanted to improve the imaging technique as an alternative to holography. The advantage would be that coherent light was not necessary to record integral images. Another important advantage was the ability to record live subjects. Dudnikov *et al.* (1980) describes the recording of integral photographic portraits.

INTEGRAL PRINTING

The interest in lenticular images began to wane not long after their appearance in the 1960s, and by the 1980s only a few manufacturers were left. More recently, owing to the availability of computer graphics, and to the introduction of new and improved methods of manufacturing lenticular sheets, interest has revived. This has opened up the process to creative computer artists. For the first time, any ambitious printer who wishes to produce lenticular 3-D images now has the means to do so. Today, the quality of lenticular sheets and microlens arrays has improved considerably. The combination of new manufacturing technology with better imaging and print resolution has resulted in new applications of the integral 3-D technique. Modern autostereoscopic techniques are based on the ability to manufacture high-quality lenticular sheets and microlens arrays. *Integral Printing* (I/P) is a technology of 3-D printing that can record perspective image data in both horizontal and vertical directions. However, the technique requires advanced imaging processes involving very precise registration of the microlens arrays with respect to the printing dots. For this reason I/P has not yet been widely adopted by the printing industry.

Grapac Japan Co., Inc. (Tokyo) is a company that develops new graphic design techniques to create static 3-D graphics. It has developed a modern version of 3-D printing technique. In spite of recent major progress in the field of animated 3-D images, research into static 3-D graphic expression is still in its early stages. *Honeycomb Array Lens Sheet* (*HALS*) is a micro-array lens sheet manufactured by the company. By printing special patterns on the underside of HALS lens sheet, 3-D effects can be obtained. The 3-D effect of HALS can be seen with the unaided eye over a full 360°. Fabrication of a HALS is carried out on the surface of a base material such as PP or PET. There are two types produced, namely the *HALS Honeycomb Series*, where the microlenses are arranged in a honeycomb array of 0.25 – 0.30 mm thick and suitable for lamination with other materials, and the *HALS Square Series*, where the microlenses are arranged in a square array 0.40 – 0.50 mm thick, suitable for posters and POP displays. For both these types microlenses are formed on the upper side, creating surface dimples of no more than a few micrometres. The underside is flat. The HALS pattern is printed on the smooth underside, and the 3-D image is seen when the sheet is viewed from the other side (i.e. through the lens surface). Four-colour printing can be carried out on the front side (lens surface). Figure 9 illustrates the HALS principle, and Figure 10 & 11 show enlarged images of HALS micro-lenses. Examples of Graphic patterns on different packages is featured in Figure 12. Figure 13 shows a HALS security card in which the violet image floats above the card surface, while Figure 14 shows examples of HALS labelled merchandise.

Another company, Rolling Optics AB (R/O), located in Sweden, has introduced an I/P technique for creating 3-D effects in a similar way to the Japanese products. The materials were developed at the Ångström

Figure 9. The HALS label principle

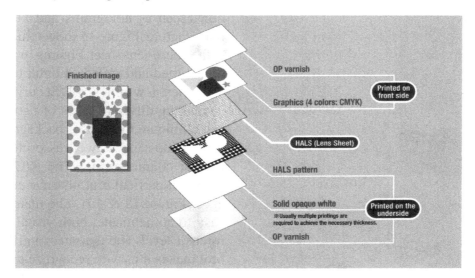

Figure 10. Transmitted light pattern

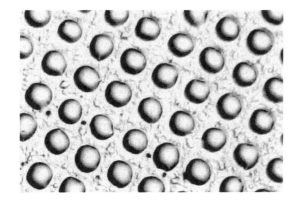

Figure 11. Reflected light pattern

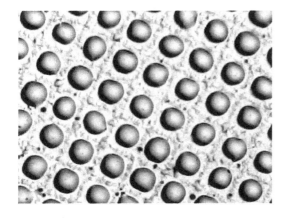

Figure 12. Graphic pattern

Figure 13. A HALS security card in which the violet image is floating above the surface of the card

Figure 14. The use of graphic 3-D patterns on product packages

laboratory at Uppsala University. The 3-D label was developed as a result of research on optical materials and microstructures, and on the scattering of light in thin foils. Rolling Optics was formed in 2004, with a team consisting of a rare mix of scientific and industrial expertise.

R/O optical films can display both printed and covert graphics, providing a double level of security. The materials are resistant to counterfeiting, tampering, distortion, heat and moisture. The material is flexible and fully transparent. In addition, the material is printable on both sides, and can be combined with any type of packaging material. The process employs high-speed rotation die-cutting and flexo-printing in register. The 3-D labels are available on rolls or in customised sheeting sizes.

By creating unique patterns in the plastic film, light is reflected in such a way that different images reach the left and right eye respectively. According to the company the impression of depth surpasses that of conventional security holograms, and it is cheaper to manufacture. Figure 15 shows enlarged views of the micro-lens sheet. Figures 16 and 17 show transmitted and reflected light, respectively. With their new optical 3-D Security Seal, R/O is creating different types of patterns which can eliminate the risk of package and document tampering as well as protecting fully against counterfeiting. The R/O labels are aimed in particular at pharmaceutical packages (Figure 18 & 19). The material is very easy to authenticate, both visually and on a forensic level. The sharpness and depth of the seal makes it easy to recognize and virtually impossible to copy. An opened seal breaks and cannot be resealed. The first R/O label product was a 4 cm diameter circular 3-D label for the Swedish hair care line *Grazette* of Sweden (Figure 20 &Figure 21). Both the Japanese and the Swedish companies typically use patterns floating behind or in front of the label, with printed text on its surface.

In the USA 3M has developed similar products containing 3-D integral images on a film containing microlenses (Pingfan Wu *et al.*, 2009), employing a pulse laser beam

Figure 15. Microscopic views of the R/O micro-lens sheet

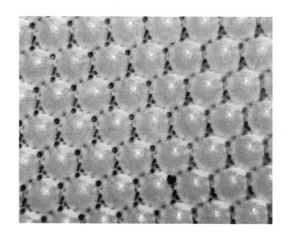

Figure 16. Transmitted light

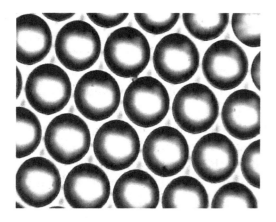

Figure 17. Reflected light

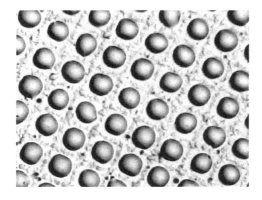

Figure 18. An R/O 3-D security seal and sample pharmaceutical box

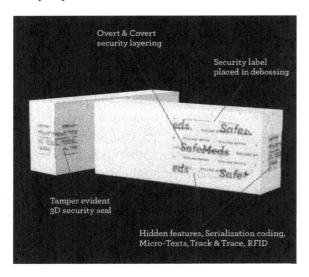

Figure 19. An R/O 3-D security seal

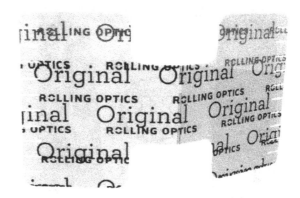

to write the image point by point. The laser pulse is focused to a spot by a writing lens, and is imaged in turn by each of an array of microlenses. In viewing, the recorded micro-images on the rear surface of the lens array are re-imaged by the lenses, and reconstruct the original spot. Scanning and overlapping sequential focused laser spots gives continuous control over the height and depth of the image. The two types of micro-optic film work in a similar manner, and allow a variety of lighting opportunities. In viewing, the recorded micro-images on the rear surface of the lens array are re-imaged by the lenses and reconstruct the original spots. 3M has several patents, the first US patent having been issued in 2001 (Florczak *et al.*, 2001).

Large-Format Integral Images

More recently, large-format integral images have become a reality. Central to this development is the use of the powerful 3-D graphic computers now available. The German company *KaraSpacE* is developing large-format integral images which can appear in front of or behind a flat projection wall. This wall can be applied in any dimension from 0.5×0.5

Figure 20. The Grazette hair shampoo tube with its 3-D label

Figure 21. The Grazette 3-D label

m up to several square metres. The imaging quality and the 3-D effect are exceptionally good. *KaraSpacE* is intended as a medium for presenting large-scale advertisements. The image information is stored on a microfilm layer within the projection wall. The picture is a static 3-D image, but when the viewpoint alters the changing image perspective gives an overwhelming impression of depth. In order to create the 3-D image, a computer is used to optically assemble the image scene from all angles. It then exposes a photographic film inside the image wall, pixel by pixel. The resulting 3-D imagery conveys an impression of realism matched only by museum-quality holograms. The integral image allows differential focusing of the eyes on foreground and background elements, something not possible with lenticular or barrier methods. Animation is also possible with this system. Computer-generated integral images can create high-quality orthoscopic 3-D images, avoiding the pseudoscopic image difficulty, though the restricted image depth and viewing angle are still a problem. The technology for production of lens arrays is now very advanced, though the costs are still high. The picture modules together with the photo layers are the core of the *KaraSpacE* product line. The photolayer is laid on the back of the picture module and stores the 2-D image information separately for each pixel. With the appropriate backlight system the modules project millions of individual images into the eyes of the viewer. The modules are built as a sandwich of optical plastics lens plates which can possess different resolutions, view alignments and visibility ranges to fit different applications.

The *KaraSpacE* software is an important component of the technology. The software *KaraSpacE Developer 1.0* combines all aspects of integral image creation into one extensive software package. The basis for the system is the *Autodesk Maya* 3-D graphics development system. After finishing the image composition work, everything is finally saved, and the exposure work begins by preparing the exposure hardware. During the exposing process the software controls the exposure machine and calculates the correct perspective image in high quality. After inaugurating the exposure process, the computer calculates the perspective image for each pixel and shows it on an LCD monitor inside the exposure machine. This machine takes a photo sheet from the magazines in a drum and lays a special exposure image module on it. Each pixel is now moved to the optical system with the LCD monitor and is exposed with its individual image. This process is repeated for every pixel, every photo layer and every drum. The exposure machine mechanically codes the magazines inside the drums so that an exposing protocol printout can later show exactly what order the photo layers have to be placed in. The processing of the exposed film is carried out in HOPE processing machines. Being a R&D company, *KaraSpacE* has not yet commercialised the technology, but is continuing the development process. The company is seeking additional funding in order to be able to bring its 3-D images to market.

Create3D, located in England, is another R&D company involved in large-format integral images. The company is connected with the 3-D Imaging Technologies Group at De Montfort University in Leicester, where the original development work was carried out. The integral images are similar to the German products. The Create3D images are available up to 2.5×1.25 m. It is possible to produce animated images extracted from video footage. When processed using patented software the final 3-D image is a single print and can be produced in poster form with special display light boxes.

INTEGERAL IMAGING

When based on Lippmann's original technique, only static 3-D images could be recorded. Over the years experimental images have been produced by a large variety of other methods exhibiting 3-D, animation and other effects. These have been based on lenticular, barrier, or (in some cases) integral techniques. Since the 1990s research into integral photographic technology has experienced a renaissance, and today, with new high-resolution electronic recording and display devices, this technique has become much improved.

When digital imaging was first introduced it was not possible to record and display moving 3-D images, but the situation has changed. Today we tend to use the term *Integral Imaging* (I/I) rather than integral photography (I/P) to describe 3-D display techniques based on micro-lens arrays.

FURTHER DEVELOPMENTS

Creating 3-D integral images by digitally interlacing many computer generated two-dimensional views was first demonstrated in 1978 by Igarashi *et al.* (1978) of Japan. In 1990 Okano *et al.* (1997) described a real-time recording method for the display of a three-dimensional image based on integral photography. A real image of the object was created by each micro-lens behind the lens array. These real images were then recorded by an HDTV camera positioned behind the array to produce an I/I image in the form of a television signal. This image could be shown on a display for auto-stereoscopic observation through either a pinhole sheet or a lens array. As explained earlier, this primary image would be pseudoscopic. It was re-inverted longitudinally by an electronic concave–convex converter within the system. The converted images were displayed on an LCD panel. To obtain acceptable image resolution a television camera with an extremely large number of pixels was needed in order to display 3-D images of HDTV quality. The principle of electronic integral imaging is shown in Figure 22.

There are three main research groups involved in the development of integral TV and video imaging systems. The Japanese group Okano, Arai *et al.* (2006a &b) are working on the recording and display of moving electronic integral images using a LCD panel. They have presented a method of converting the pseudoscopic image into an orthoscopic one by using an array of three convex lenses. They have developed an integral 3-D television that uses a 2000-line video system that can record and display 3-D colour moving images in real time. The new system uses

160×18 elemental images, six times as many as in previous work. They have confirmed that the new system is superior to the previous one in terms of the resolution of the reconstructed image and the viewing area. This experimental system has yet to achieve the resolution of current broadcasting receivers. Okui *et al.* (2006) described a way to display three-dimensional images by I/I using an ordinary projector. The authors explained a method that uses a large-aperture converging lens and a proposed method using two sets of lens arrays. Based on this new principle, both front and rear projection would be possible. Only a limited viewing area can be formed on the optical screen by this method, which improves the brightness of the screen image. The projector itself does not need an additional optical system. The authors reported on a method of displaying 3-D images by I/I using an ordinary projector designed for 2-D projection. They showed that an optical setup with a converging lens could be used to produce a single bright viewing area. A two-lens

Figure 22. Principle of integral imaging with electronic pseudo/orthoscopic conversion

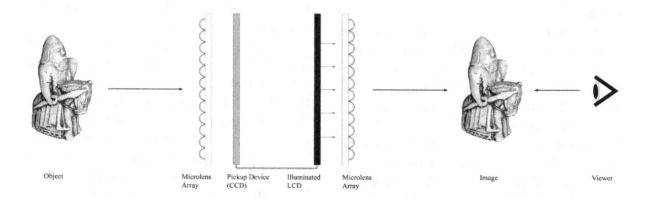

Object   Microlens Array   Pickup Device (CCD)   Illuminated LCD   Microlens Array   Image   Viewer

array arrangement showing an improved viewing area was also described.

In collaboration with universities in Korea and Spain, workers in the USA have also contributed to electronic I/I development. Jang and Javidi (2002) introduce synthetic aperture integral imaging, in which an effectively enlarged viewing aperture is obtained by movement of a small I/I system. This system also increases the viewing resolution.

By the introduction of double-device systems, I/I can be enhanced in image depth and viewing angle or image size. Lee *et al.* (2003) confirmed that double-device I/I systems can pick up and display images at two separate image planes. Another method of improving the viewing angle was by using lenslet arrays with a low fill factor, as described in a further paper by Jang and Javidi (2003) However, the viewing resolution is lowered because the spatial sampling rate is reduced. However, it was shown that both viewing resolution and viewing angle can be improved by adopting a moving array lenslet technique.

Digital techniques for formation of real undistorted orthoscopic integral images by direct pickup do exist. However they are confined to symmetric image record/display systems. Navarro *et al.* (2010) introduced a more general algorithm which allows pseudoscopic/orthoscopic transformation with full control over the display parameters so that a set of synthetic elemental images that suits the characteristics of the I/I monitor can be generated, also permitting control over the depth and size of the reconstructed 3-D scene.

Another I/I research group is the *3D & Imaging Technologies Group* at De Montfort University, UK. McCormick, Davies *et al.* (1988) describe an optical system which produces 3-D images exhibiting continuous parallax, with no flipping or 'cardboarding'.

Both integral and lenticular images can be recorded on photographic film without recourse to a specialized environment. Standard photographic processing and enlarging techniques are used to obtain hard copy. Measurements on commercially available retrodirective screens and a new retro-screen arrangement show significant advantages. A retro-directive and a transmission version of the system have been constructed. A further paper by Davies *et al.* (1994) concentrates on the image transfer system using micro-lens arrays. A systematic evaluation of the spatial resolution capabilities of micro-lens optical combinations and the synthesis of a large-aperture objective lens by a segmented lens form is reported. It is shown that by employing an optical combination comprising macro-lens arrays in conjunction with micro-lens focusing screens, a suitable transmission element retaining the required angular and lateral resolution can be constructed. The group is working on the possibility of developing 3-D TV systems based on I/I, a task originally presented by McCormick (1995). One requirement is to create continuous parallax in pixellated integral 3-D image displays, covered in a paper by Brewin *et al.* (1995). A possible integral 3-D TV system was outlined. The novel application of existing compression schemes were described; this indicated that the bandwidth requirements for I/I 3-D TV are the same as for HDTV. The aim was to bring 3-D imaging to the new high-resolution computer monitors and TV. The method of integral photography turned out to be practicable for 3-D TV, and a few such monitors have been developed. An HDTV monitor with 1920×1080 pixels can generate an integral image of 480×270 pixels, where each pixel consists of 4×4 pixels inside (Figure 23), though still with a limited image resolution.

Figure 23. 3-D HDTV LCD Monitor using a lenticular method

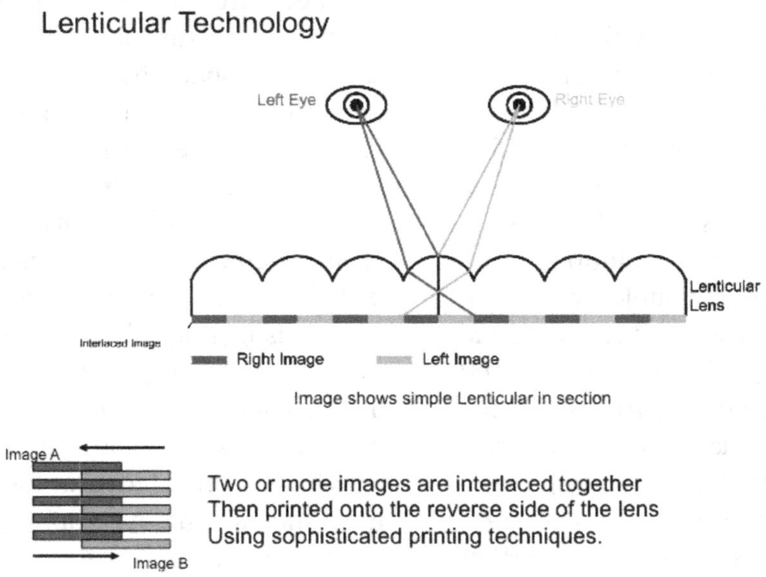
Lenticular Technology
Left Eye
Right Eye
Lenticular Lens
Interlaced Image
Right Image
Left Image
Image shows simple Lenticular in section
Image A
Image B
Two or more images are interlaced together
Then printed onto the reverse side of the lens
Using sophisticated printing techniques.

PUBLIC DISPLAYS

In 2010 Toshiba presented a 21-inch autostereoscopic high-definition display for use with next-generation 3-D monitors. The new product is actually based on a lenticular technique, but with overlapping images to avoid the image-jumping problem seen in many lenticular systems. Toshiba employs what it refers to as an 'integral imaging system' to reproduce a real object as a 3-D image that can be viewed without glasses over a wide range of viewing angles, employing a multi-parallax design that allows images to change depending on the viewer's position. This system is superior to other systems based on conventional lenticular screens.

In a recent paper Kim *et al.* (2010) review I/I techniques in displays. According the authors, 3-D display hardware technique based on I/I has been vigorously researched as a promising way for realising 3-D displays. It can display real-time 3-D movies with full colour and parallax using a lens or pinhole array and a commercial 2-D display device. Although none of the 3-D display techniques has yet satisfactorily spanned the possible3-D display market, I/I with its many advantages can be a promising candidate. According to the authors, I/I methods will not be limited to academic research but will certainly be commercialized some time in the future. A recent publication is a book by Javidi and Okano (2009) in which state-of-the-art 3-D display and visualisation technologies are presented and analysed.

SUMMARY AND CONCLUSION

The main advantages of integral imaging are:

- Full vertical and horizontal parallax.
- Quasi-continuous viewpoints.
- Full colour real-time images.
- No need for special viewing eyewear.
- Wide viewing angle for observation.

The main disadvantages are:

- Limited image resolution.
- Limited viewing angle.

When an image is recorded the person recording the image needs to decide on which plane to focus, as the lens array can only focus on a single plane. The intent to have an ideal 3-D image with objects reaching far in and out of the plane does not at present seem feasible. Integral 3-D images have not yet achieved significant commercial success, mainly owing to the manufacturing problems of microlens arrays. These optical elements are an important factor in the image quality. (The same problem also affects lenticular images.) The microlens optical element needs to be attached to the photograph or printed image. In 3-D display systems, the microlens array is also included in the display system, which further affects the image resolution. The problem of obtaining the highest possible image resolution in eyewear-free 3-D display systems based on some form of optical elements is difficult to solve from a purely practical aspect. Possibly holographically fabricated screens should be considered. Holographic optical elements (HOEs) can certainly be designed and used as projection screens in autostereoscopic display systems. It seems likely that in the future, 3-D display systems will be based on holographic principles combined with ultrafast computers and electronic ultrahigh-resolution displays.

REFERENCES

Arai, J., Kawai, H., & Okano, F. (2006). Microlens arrays for integral imaging systems. *Applied Optics*, *45*, 9066–9078. doi:10.1364/AO.45.009066 PMID:17151745

Arai, J., Okui, M., Yamashita, T., & Okano, F. (2006). Integral three-dimensional television using a 2000–scanning-line video system. *Applied Optics*, *45*, 1704–1712. doi:10.1364/AO.45.001704 PMID:16572684

Bjelkhagen, H. I. (1999). Lippmann photography: Reviving an early colour process. *History of Photography*, *23*(3), 274–280.

Bonnet, M. (1942). *La photographie en relief. Centre de documentation*. Maison de la Rhimie.

Brewin, M., Forman, M., & Davies, N. (1995). Electronic capture and display of full parallax 3D images. *Proceedings of the Society for Photo-Instrumentation Engineers*, *2409*, 118–124. doi:10.1117/12.205851

Burckhardt, C. B. (1967). Optimum parameters and resolution limitation of integral photography. *Journal of the Optical Society of America*, *58*, 71–76. doi:10.1364/JOSA.58.000071

Burckhardt, C. B., & Doherty, E. T. (1969). Beaded plate recording of integral photographs. *Applied Optics*, *8*, 2329–2331. doi:10.1364/AO.8.002329 PMID:20076020

Davies, N., McCormick, M., & Brewin, M. (1994). *Design and analysis of an image transfer system*. Unpublished.

Davies, N., McCormick, M., & Yang, L. (1988). Three-dimensional imaging systems: A new development. *Applied Optics*, *27*, 4520–4528. doi:10.1364/AO.27.004520 PMID:20539602

de Montebello, R. L. (1977). Wide-angle integral photography – The integram system using microlens arrays. *Optical Engineering (Redondo Beach, Calif.)*, *33*, 3624–3633.

Dudnikov, Y. A. (1970). Autostereoscopy & integral photography. *Optics Technology*, *37*, 422–426.

Dudnikov, Y. A., Rozhkov, B. K., & Antipova, E. N. (1980). Obtaining a portrait of a person by the integral photography method. *Sov. J. Opt. Tech.*, *47*(9).

Florczak., et al. (2001). *Sheeting with composite image that floats* (US Patent 6,288,842). Washington, DC: US Patent Office.

Hoshino, H., Okano, F., Isono, H., & Yuyama, I. (1998). Analysis of resolution limitation of integral photography. *Journal of the Optical Society of America. A, Optics, Image Science, and Vision*, *15*, 2059–2065. doi:10.1364/JOSAA.15.002059

Igarashi, Y., Murata, H., & Ueda, M. (1978). 3-D display system using a computer generated integral photograph. *Journal of Applied Physics*, *17*, 1683–1684. doi:10.1143/JJAP.17.1683

Ives, H. E. (1933). Optical properties of Lippmann lenticulated sheet. *Journal of the Optical Society of America*, *21*, 171–176. doi:10.1364/JOSA.21.000171

Jang, J.-S., & Javidi, B. (2002). Three-dimensional synthetic aperture integral imaging. *Optics Letters*, *27*, 1144–1146. doi:10.1364/OL.27.001144 PMID:18026388

Javidi, B., Okano, F., & Son, J.-Y. (2009). *Three-dimensional imaging, visualization, and display*. Berlin: Springer–Verlag. doi:10.1007/978-0-387-79335-1

Kim, Y., Hong, K., & Lee, B. (2010). Recent researches based on integral imaging display method. *3D Research, 1*, 17–27

Lee, B., Min, S.-W., & Javidi, B. (2002). Theoretical analysis for three-dimensional integral imaging systems with double devices. *Applied Optics, 41*, 4856–4865. doi:10.1364/AO.41.004856 PMID:12197653

Lippmann, M. G. (1891). La photographie des couleurs. *Comptes Rendus Hebdomadaires des Séances de l'Académie des Sciences, 112*, 274–275.

Lippmann, M. G. (1894). Sur la théorie de la photographie des couleurs simples et composées par la méthode interférentielle. *Journal of Physics, 3*(3), 97–107.

Lippmann, M. G. (1902). La photographie des couleurs [deuxième note]. *Comptes Rendus Hebdomadaires des Séances de l'Académie des Sciences, 114*, 961–962.

Lippmann, M. G. (1908). Épreuves réversibles donnant la sensation du relief. *Journal de Physique (Paris), 7*, 821–825.

Lippmann, M. G. (1908). Épreuves réversibles: Photographies integrals. *Comptes Rendus Hebdomadaires des Séances de l'Académie des Sciences, 146*, 446–451.

McCormick, M. (1995). Integral 3D image for broadcast. In *Proceedings of the Second International Display Workshop*. ITE.

Navarro, H., Martínez-Cuenca, R., Saavedra, G., Martínez-Corral, M., & Javidi, B. (2010). 3D integral imaging display by smart pseudoscopic-to-orthoscopic conversion (SPOC). *Optics Express, 18*, 573–583. doi:10.1364/OE.18.025573 PMID:21164903

Okano, F., Hoshino, H., Arai, J., & Yuyama, I. (1997). Real-time pickup method for a three-dimensional image based on integral photography. *Applied Optics, 36*, 1598–1603. doi:10.1364/AO.36.001598 PMID:18250841

Okui, M., Arai, J., Nojiri, Y., & Okano, F. (2006). Optical screen for direct projection of integral imaging. *Applied Optics, 45*, 9132–9139. doi:10.1364/AO.45.009132 PMID:17151752

Wu, P., Dunn, D. S., Smithson, R. L., & Rhyner, S. J. (2009). Development of integral images. In *Proceedings of the Frontiers in Optics (FiO), 3-D Capturing, Visualization and Displays*. Stockholm, Sweden: Rolling Optics AB.

Chapter 7
Principles of Holography:
Wavefront Reconstruction and Holographic Theory

Martin Richardson
De Montfort University, UK

ABSTRACT

The discovery of diffraction and interference led eventually to the holographic principle, the recording and reconstruction of the shape of a wavefront. Transmission and reflection holograms are detailed in this chapter, along with the principles of rainbow holograms and holographic stereograms and their applications. Digital holography is described in the form of multiplexed images. The psychological and philosophical implications of the holographic image are discussed with some examples from the field of creative art.

BACKGROUND

Modern holography straddles the grounds of imaging science and artistic endeavor. Still enveloping the first moments of excitement in the mid-sixties as three dimensional images where for the first time captured and reconstructed using laser beams. It fed into the popular imagination with visions of how, in a technical and scientific advanced future, animated holograms would walk amongst us. Most of us have seen small holograms on our bankcards, and peppers ghosts and other visual effects that are called holograms without actually being so. Still relatively few have had the opportunity to see large scale holograms or the latest developments in digital holography. The proposed exhibition

DOI: 10.4018/978-1-4666-4932-3.ch007

illustrates contemporary up to the very most recent technical and creative developments in the field. It also demonstrates the constant inter disciplinary discourse between the practitioners and theorists in the fields of advanced technology, design and art.

In recent years holograms have revolutionised security where documents are employed to authenticate identity. Today most passports, ID cards, credit cards, travel cards and high-value including bank notes all incorperate holograms and are refered to as Optical Verifacation Devices (OVDs). The reason for this is a hologram has the ability to create an easily recognisable three-dimensional image and importantly very difficult to counterfeit thus reduceing the threat from organised criminal activity.

Apart from being a security device holography has the potential to become the ultimate 3-D format within the next few decades as the holographic image can be called a true replica indistinguishable from reality. But there is much more to holographic imagery than mere replication. Holography had its beginnings in 1948. Denis Gabor, a Hungarian-born physicist working for the electrical company British Thompson Huston based in Rugby, UK, who was atempting to improve the resolution of electron microscope images and hit upon the idea of recording the actual radiated wavefront emanating from the object. As a beam of electrons could not at the time be made coherent he used green-filtered light light from a mercury vapour lamp. In a seminal paper describing his findings (Gabor 1949) he explained how light of a single frequency carried all the information describing the object contained in the light wavefront and soon after named his invention holography, the word *holography* comes from the Greek words ὅλος (*hólos*; "whole") and γραφή (*grafē*; "writing" or "drawing"). This, he argued, could be recorded on a photographic emulsion. He succeeded in achieving this after a fashion, though his mercury lamp produced a band of wavelengths nowhere near as narrow as he would have desired. As a result of this large bandwidth his earliest images had to be very small two-dimensional transparencies no larger than a pinhead, and even these were blurred, and distorted by an unwanted complementary image directly in front. But although his experiment was crude and unconvincing his theory was sound, and he was eventually to receive the Nobel Prize for Physics in 1971 for his ingenuity, though others had needed to find methods of making his ideas workable. Today holograms appear everywhere, on bank cards, passes and any document that needs security protection. Holography also finds applications in the decorative trade, and in engineering, where it is used in nondestructive testing in the form of holographic interferometry.

THE PERCEPTION OF DEPTH

A binocular stereoscopic image, either in a viewer or projected on a screen, can reconstruct only a single viewpoint. Moving one's viewing position does not change the perspective: it merely foreshortens the image. A photographic image is necessarily two-dimensional. Even a stereoscopic pair of photographs records only a pair of 2-D views as seen from the position of the viewer's right and left eyes. Because of the way our perceptive mechanism works the two slightly different images combine to form a three-dimensional impression. The effect is severely restricted compared with that of a holographic image, because traditional photography can record only the time-averaged light intensity The relation between the phases of the combined reflected light waves incident on the sensitive material, which are the part that contains information about the depth of the subject matter, are not recorded. But a hologram does record this information.

But how does it work? When we 'see' an object, what is happening is that the light waves reflected from the object (or emitted by it) have a shape that is unique to that object, and our eyes intercept part of it. We call this shape a *wavefront*. When the illumination is by a laser the light waves all have the same wavelength, and are all in phase (in step, like a radio wave). Of course, being made up of light, the wavefronts are travelling at a great speed, but by the optical trick due to Gabor's insight, a hologram actually records their shape, coded within a photographic emulsion. This wavefront can be recalled and sent on its way by simply illuminating the hologram with laser light, so that when we examine the hologram our eyes intercept the duplicate of the wavefront that was caught by the emulsion. So we can see the image of the object from every viewpoint: it appears three-dimensional. In scientific terms it is said to have full parallax, i.e., both vertical and horizontal perspective.

In 1801 the English physicist Thomas Young was the first to show interference and diffraction phenomena, in an experiment involving a light beam from a point source, which he passed through a pair of narrow slits onto a screen. He saw that the illumination on the screen was not a uniform patch of light but was an array of much smaller light and dark patches. He interpreted this pattern as the consequence of interference between the two waves emanating from their respective slits. This phenomenon could not easily be incorporated into Newton's particle theory of light. Newton had tried to explain optical phenomena such as refraction and reflection in terms of classical mechanics, but Young showed with his experiment that this model could not account for his 'fringes'.

At that time the only monochromatic light source available (sodium) produced a wide band of wavelengths with little correlation between them. This light is now classed as *incoherent*, and white-light sources can produce Young's interference effect only when heavily filtered, as Gabor's mercury arc source was. Today, though, we have access to the well-behaved beams emitted by lasers, which contain only a very narrow band of wavelengths, all in the same phase. This light is said to be *coherent*, and it can produce bright interference fringes over a considerable distance. For a given laser, the distance over which these interference effects can be observed is known as the *coherence length* of the beam, and this may be anything from a few centimetres to many metres, depending on the type of laser (see Figure 1).

The German lens maker Joseph von Fraunhofer took Young's experiment further, by greatly multiplying the number of slits, which

Figure 1. Incoherent and coherent light

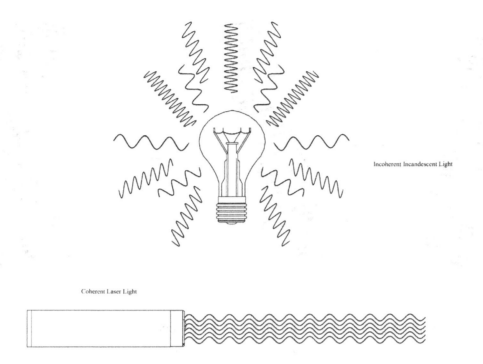

Incoherent Incandescent Light

Coherent Laser Light

he ruled in glass, thus producing the first diffraction grating. When illuminated by coherent light such a grating produces a much brighter, uniform array of spots. This is the clue to holography. If we take a laser beam and divide it into two parts using a partial mirror (*beamsplitter*), then divert one of the beams back across the other, they will form an interference pattern in this space similar to the one Fraunhofer saw (Figure 2). Although the light itself is travelling at enormous speed, the interference pattern is stationary, and if we position a photographic film or plate within this space it will record it. We will have actually produced a diffraction grating photographically, by interference.

The exciting thing about this grating is that if you now illuminate it with just one beam

it will reproduce both beams! And this is the basis of the holographic principle. You see, if one of the beams isn't just a simple beam, but has been reflected off an object first, there will still be a stationary interference pattern, and you can still record it on film; but the pattern will be a coded version of the shape of the wavefront, and if you now illuminate the developed film with the original undisturbed beam it will reconstruct the other beam exactly as it was when it left the object and arrived at the film. If you look through the film illuminated in this way you will appear to be seeing the object itself, because the light beam re-created by the hologram is identical with the one that was reflected off the object. This is illustrated in Figure 3.

Figure 2. Interference effect of two crossed laser beams

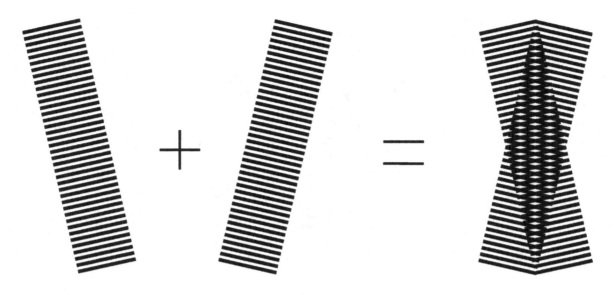

In practice the photographic emulsion has to have a very high resolution, as the interference pattern detail is 1/2000 of a millimetre or less in size; and this means that the emulsion will also be very slow, so that exposures have to be several seconds. During this time the subject matter and the laser have to be stationary within about a tenth of a wavelength (0.00005 mm), so only still-lifes of rigid objects are possible unless a pulse laser is used. These give a flash of laser light that is only a few nanoseconds (billionths of a second) in duration.

TRANSMISSION AND REFLECTION HOLOGRAMS

The two basic types of hologram, invented independently in the 1960s by Emmett Leith and Yuri Denisyuk, are known respectively as *laser transmission holograms* and *white-light reflection holograms*. These qualifications are given because the former requires to be reconstructed (replayed) by laser light, whereas the latter may be replayed using white light. Both methods are made by two laser beams derived from the same source, one of which (the *reference beam*) is undisturbed and falls directly on the emulsion, whereas the other (the *object* beam)only reaches the emulsion after it has been reflected from (or transmitted by) the object. The difference between the two types of hologram is that in a transmission hologram both beams are incident on the emulsion from the same side, whereas in a reflection hologram the beams are incident on it from opposite sides. In the reflection hologram the interference pattern is in the form of layers of varying transmittance (or refractive index) which form interference planes ('Bragg planes') approximately one half-wavelength apart and roughly parallel to the emulsion surface; when processed, this type of hologram can produce its image using a beam of white light because constructive interference occurs for only the wavelength

Figure 3. Principle of making a hologram and reconstructing (replaying) the image

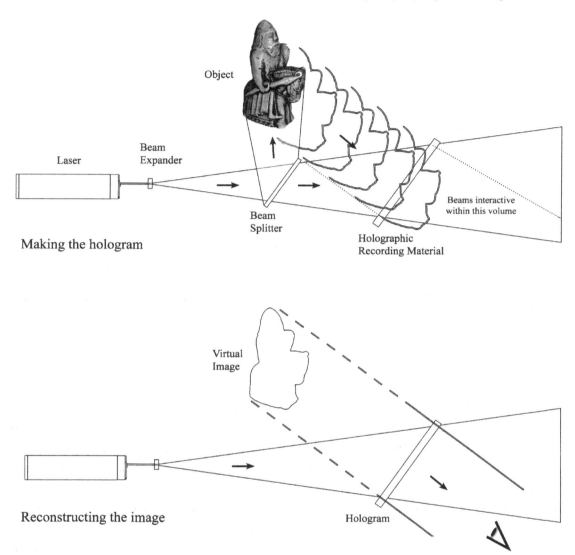

that created it: other wavelengths simply pass through. In a transmission hologram the interference pattern is in the form of planes roughly perpendicular to the emulsion surface, and in this case the Bragg effect is not powerful enough to select out a single narrow waveband. So a transmission hologram must have its image replayed using light from the original laser (or a quasi-monochromatic source); otherwise all that can be seen is a smudgy spectrum of colour. Figure 4 shows the beam configurations for making and viewing the two types of hologram.

One type of reflection hologram is particularly easy to make, as a single beam can do duty for both reference and object beams.

Figure 4. Making and replaying the two basic types of hologram

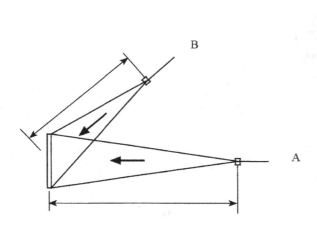

Recording a transmission hologram
Object beam (a) and reference beam
(b) both illuminate the light sensitive
material from the same side.

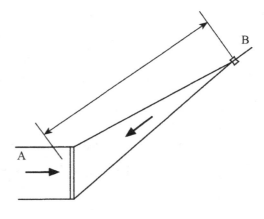

Recording a reflection hologram
Object beam (a) and reference beam
(b) are directed toward the light sensitive
material from opposite sides.

The object is positioned behind the film, and the reference beam passes through it and is reflected back from the surface of the object, forming the object beam. As this was Denisyuk's original format, this type of hologram is named after him (Figure 5).

WHITE-LIGHT TRANSMISSION HOLOGRAMS

In the 1970s Stephen Benton discovered a way of making large transmission holograms with great depth of image that were viewable by white light. From their brightly coloured appearance they became known as *rainbow holograms*. They are made using the transfer principle. If a transmission hologram is illuminated by the replay beam reversed in direction (i.e. from the other side of the film), a real image is formed in front of the hologram. This can be used as an object to make a second or transfer hologram. As the image formed in this way is pseudoscopic (has its perspective inverted) the final image also has to be flipped for viewing. (This technique is standard for making display reflection holograms too.) Benton's insight was to make the transfer with the 'master' hologram masked down to a narrow slit. The result, when examined by laser light, was an image with full horizontal parallax but little or no vertical parallax, much like viewing the image through a narrow letter box. But when illuminated with white light, the hologram acted like a diffraction grating in the vertical plane, spreading the wavelengths into a spectrum, so that when the image was viewed from directly ahead its hue was green, but from lower down it was blue or violet, and from high up it appeared in yellow or red. The

Figure 5. Denisyuk's arrangement for a reflection hologram

fact that there was no vertical parallax in such an image was usually unimportant in normal viewing as one's eyes are set horizontally and vertical parallax is ignored in normal visual situations. Figure 6 shows the schematic setup for a rainbow hologram.

Rainbow holograms can have a much greater image depth than reflection holograms, and are very bright. They are thus the main type used in large displays. Since the entire spectrum can be used in replay it is possible, by combining the images from more than one master, to produce white or multicoloured results.

HOLOGRAPHIC STEREOGRAMS

A holographic stereogram is assembled from a series of 2-D photographs made using a moving camera in a similar manner to a photographic stereogram, and giving an image with horizontal parallax in the same way. It uses the slit transfer principle as in the rainbow hologram, but this time the slit is vertical, and narrow 'slit' holograms are produced adjacent to one another on a film that is moved behind the slit one step for each exposure. The resulting master can be transferred in the usual way to either reflection or

Figure 6. Schematic arrangement for a rainbow hologram

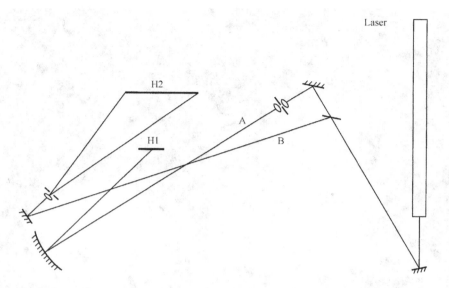

Recording a rainbow transmission hologram
Object beam (a) illuminates the master hologram (H1) with a slit of laser light. The reference beam

rainbow format, with the resulting set of images in the plane of the film. At each viewing point the viewer sees only one of the images, a different one for each eye, the effect being similar to that of a lenticular photographic stereogram, but viewable over a much larger angle, and by several viewers at the same time. The principle is shown in Figure 7.

Holographic stereograms are autostereoscopic, that is, they can be viewed without glasses or any other form of optical device, and – an important point often ignored – they present an image with three-dimensional depth even to those who are unable to obtain binocular fusion using stereoscopes. Like conventional rainbow holograms, they can be multicoloured. Using a carefully calculated setup a rainbow stereogram can even produce images in natural colour from colour originals

by the superposition of red, green and blue images from separation prints. Figure 8 shows the camera system in operation, and Figure 9 is a 'still' taken from a double-image portrait stereogram by the author.

PULSE LASER HOLOGRAPHY

The very first lasers delivered their light in short pulses of somewhat low energy. As the technology developed, pulse lasers became more powerful, and their pulse duration shorter, until today the types suitable for holography can produce flashes of coherent light of 10 or more joules (J) of energy and a duration as short as 5 nanoseconds (ns). This amounts to a power of several gigawatts (GW) (a watt is a joule per second), so such pulses

Figure 7. Principle of the holographic stereogram

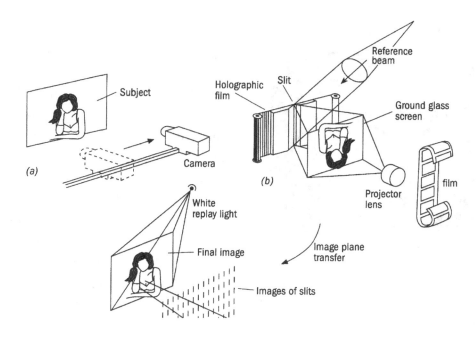

Figure 8. Recording a set of portrait exposures for a holographic stereogram

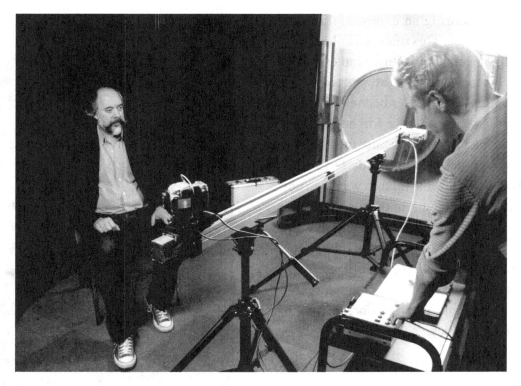

Figure 9. Psychedelic Amy, a double-exposure stereogram by the author with the originals made using the setup of Figure 7

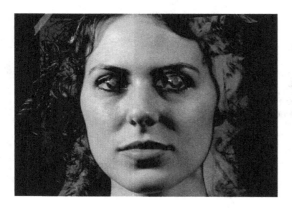

have to be treated with caution. For holography of stationary subjects a continuous-wave (CW) laser is generally used, as it is readily controllable and much cheaper, but the subject has to remain stationary within a tenth of wavelength during an exposure that may be several seconds. This rules out holography of moving objects, portraiture, and even live flowers. When sufficiently attenuated by diverging optics a pulse laser bean is no longer dangerous, and is suitable for holographic portraiture; it can even record fast-moving subject matter such as flying birds. Originally the only type of laser available was the ruby laser, with a wavelength in the far red. When used for portraiture this resulted in waxy complexions and a 'dead' look (Figure 10). More recently, green and even blue pulse lasers have become available, making for more natural results, and even colour portraiture using three laser beams.

DIGITAL HOLOGRAPHY

Holographic stereograms made using a digital camera or computer imagery for the primary images are sometimes referred to carelessly as 'digital holograms'. This, of course, is a misnomer. The original may have been made digitally, but the hologram will have been made using conventional exposing and processing methods. In its accurate definition, digital holography is a method of forming the holographic interference image itself by digital means. It is possible to draw the interference patterns generated by simple objects in a computer program and to transfer these, shrunk to the required dimensions, onto a suitable material to form a surface hologram

Figure 10. Holographic portrait of Martin Scorsese, 1998, by M. Richardson

that can be mass-produced using embossing methods. However, this technique requires a very large amount of memory space even for a simple wire-frame model, and at present has few commercial possibilities. A different view of digital holography is the making of holograms on LCD arrays. The big problem about this is that at present the smallest pixel size is at least an order of magnitude too coarse to record the fringe pattern of a conventional off-axis hologram, and the only way to make the interference fringes broad enough to record on digital material is to use Fourier-transform or in-line (Gabor) optical layouts and to eliminate the spurious image by means of computer software.

One type of computer-generated hologram that has considerable commercial application is fabricated using the dot matrix principle. Each dot is produced by a pair of tightly-focused laser beams, and so is in itself a trivial hologram, i.e. it carries no image but merely reflects light as a tiny diffraction grating. The angles and intensities of the beams are varied from one dot to the next under the control of a computer program, and produce a surface relief image that can be used to make an embossing shim. The result is not a true hologram, though it produces its image by holographic means.

EVALUATING HOLOGRAPHY'S IMPACT ON AUDIENCES

The philospher Roland Baths once asked of Photography 'Who is the author of the moving image; the artist who created the sculpture, the photographer that captured it or the par-

ticipating, moving, audience?' (Baths, 1982). This view of authorship may also be applied to holography. It suggests the postmodernist philosophy of co-creation. Analysis of three-dimensional spatial imaging compels us to make decisions about 'real-time' and 'real-space', for our instincts are as exact in their recognition of the unfamiliar as the familiar, but far less exact when a three-dimensional image gives no clues as to its origins being holographic ie; how can we distinguish it from the real thing? We primarily recognize the object rather than the hologram, particularly if we are not familiar with the medium. We generally continue to employ the criteria of recognition, i.e. the comparison between it, ourselves and our spatial surroundings, our orientation to it and its orientation to us, even though it is an illusion that we are assessing and not tactile information. In this sense holography may be used as a cognitive tool that offers a new insight into the way we perceive and evaluate our surroundings (see Figure 11).

The success of an artistic hologram depends less on this realisation (conscious or unconscious), or, indeed, on the craftsmanship incorporated in it, than on the level of mystery that the image sustains through its visual power. Holography's realism makes it the ideal medium for the reproduction of sculptural artwork, as in Figure 12.

A 3-D holographic image seems to possess a special kind of quasi-visible reality, somewhat like the force fields so popular in science fiction. The term 'Personal distance' has been used by the psychologist Heidegger (1969) to designate the distance consistently separating individual members of a group of people.

Figure 11. The viewer is so involved with the image that she attempts to touch it as it hangs in the air

Figure 12. Hologram of bronze figure, by Inaki Beguiristan

It might be thought of as a small protective sphere that we maintain between others and ourselves. A separation of some 45– 150 cm is generally considered a 'normal' distance. This is also relevant to the viewing of a holographic image. If such an image is examined from too close or too distant a viewpoint, some distortion of perspective may occur, just as it does with a photograph taken with a wide-angle or a long-focus lens, though in a holographic image the effect is more obvious. The typical distance at which a viewer will observe a holographic image tends in practice to be determined largely by the size of the image, but it is a subjective matter, and at a distance that would correspond to touching or holding the subject matter as imaged, any visual distortion is not apparent. But especially in holographic portraits a close approach may result in an uneasy feeling of intruding into someone else's space. 'Keeping at arm's length' is a way of expressing this optimum personal distance. It usually extends from a point just outside the point where two people can just touch fingers. This can perhaps be thought of as the limit of physical domination: beyond this distance a person cannot easily get their hands on somebody else, so to speak.

APPLICATIONS FOR HOLOGRAPHIC TECHNOLOGY

When we see a comparatively unfamiliar image our instinct is to rationalise it, to think of it as a real object; trying to square this impression with the knowledge that it is merely a technological trick is unsettling, at least until we get used to viewing these images (the same effect was true of early cinema, which exploited the audience's naïveté with images of oncoming railway trains, bolting horses and so on). The MIT Media lab is even now experimenting with *haptic images* – images

you can actually *feel*. There is undoubtedly a wide potential market for display holography with its unique and overwhelming realism, haptic or not. Figure 13 shows suggested marketing areas, some of which are already thriving, and others where a few entrepreneurs have adopted the ideas with enthusiasm.

HOLOGRAPHIC DATA STORAGE

Holographic data storage, as a concept, has been around for a long time. It is clear that the resolution capabilities of a surface hologram such as an embossed image are of the order of one thousand bits per millimetre, and even that is in a horizontal direction only. The amount of information that can theoretically be stored in a volume hologram with its three-dimensional capability is staggering. In a volume (thick) hologram the reconstruction angle is critical: several images can be stored within a square degree, and any image, with its gigabytes of information, can be recovered

Figure 13. Overview of potential market areas for holograms

POTENTIAL MARKETS FOR HOLOGRAPHIC MATERIALS

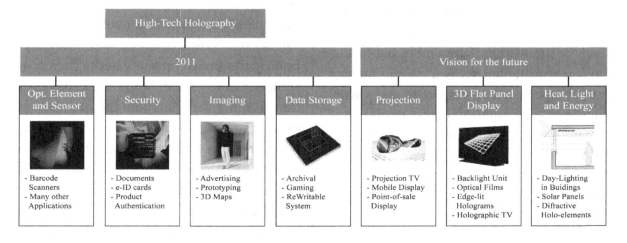

simply by varying the angle of incidence of the replay beam. Already by 1990 research workers had managed to code as many as 2000 2-D images into a single cubic centimetre of emulsion, and that was less than a thousandth of what is theoretically possible. Holographic data storage research went into decline as other methods of digital information storage appeared within computer technology, but it now seems to be entering a stage of revival.

CONCLUSION: SOME PHILOSOPHICAL CONSIDERATIONS

The experience of viewing a hologram is like looking at the original object through a window – 'a window with a memory' (Abramson, 1981). Each point on the hologram codes an entire image from that viewpoint. If we break up a first-generation hologram we will still be able to see the entire image through any piece, however small, though the smaller it is the weaker and more restricted the image will be. Some physiologists have used the holographic principle as a model to describe the way permanent memories are stored in the brain, since damage to small areas of it does not eradicate whole specific memories or parts of memories, but just weakens the whole memory bank. However, this model is not a *theory* of the working of memory but just a convenient metaphor, as is a current model for the Universe in holographic terms, which includes an accounting for some of the missing dimensions essential to string theory. There is of course no suggestion that the Universe *is* a hologram.

Holography has long been given a misleading face by science-fiction films and literature.

As a result many members of the public believe that any sort of three-dimensional image must be some kind of hologram. True holographic movies and video are not only possible but have been repeatedly demonstrated; however, the technical problems are difficult to overcome, and research continues worldwide. There are, of course, several methods of displaying true three-dimensional images (i.e. not-binocular stereo). These include generation within a locally-created atmospheric fog, rapidly rotating screens, dynamic micromirrors and the production of aerial plasmas at the intersection points of several computer guided laser beams (Blundell & Schwarz, 2000). Most of these are directed towards advertising displays and special effects in theme parks, etc.

Are holograms simply recordings of reality or are they creative artefacts demanding interpretation and possessing self-expression, or even capable of deception? In the nineteenth century the same question was posed with reference to photography. Dismissed initially by artists as merely a mechanical recording medium, photography became in the hands of such craftsmen as Henri Cartier-Bresson and Bill Brandt a subtle artistic tool capable of the most delicate expression; others such as Man Ray, who were themselves practising artists, combined their photography with surrealism. Skilful framing and manipulation of the image, and a careful choice of subject matter and lighting, are capable of transforming the banal into the dramatic or even sinister. Techniques of image manipulations have always been possible – think of the Cottingley Fairies, or the 'vanishing commissars' of the Soviet photographs progressively doctored during Stalin's wicked reign. Such montages and image eliminations required much darkroom expertise, not to mention some nifty finger work with scissors and paste; though in these days of digital manipulation such work requires only a few keystrokes. My own experience with three-dimensional portraiture via pulse laser holography or the holographic stereogram, and in particular with the portraiture of celebrities, suggests that the ultra-realistic holographic 'digital' portrait necessarily reflects the personality of both the subject and the craftsman in the pursuit of reality. Other technologies such as artificial intelligence and robotics may contribute towards the same goal.

In the early 1960s new ways for artists using holography began to evolve, and since then a body of work under the name of *Conceptual Art* has grown up. For workers such as the American artist Bruce Nyman, what began to matter was not so much the physical work involved as that of the ideas their creations embodied; in this respect holography provided an excellent vehicle for expression. Conceptual Art was treated by some of its practitioners as a political tool, embodying a stand against exploitative art galleries: to them art was not as object that could be bought and sold. In the hologram any practical use for the subject matter is removed, and so it becomes itself impotent. This attitude has once again become relevant in photography. An increased analytical and critical attitude towards society focuses on communication in visual terms. Photographs taken by self styled artists are often used to form visual essays that are displayed in art galleries and published in books. Conceptual art shuns the idea of art as a commodity.

Since the early days of creative holography, artists (including the author) have attempted to bridge the gap between optical imaging and solid reality by including real objects in their creations. One of the earliest examples of this type in holographic art was by the American artist Rick Silbermann around 1979 (Figure 14). 'The Meeting' is a conjunction of a broken wineglass occupying the same space as a holographic image of its former intact self. projected over its original, only the original has been broken. Silbermann's illusion makes a visual statement about our assumptions of reality. Having accepted the content as 'possible', the viewer is led by the veracity of the details and of the space they occupy to accept the illusion as reality. Some viewers find this image disturbing; others find it mind-expanding. With the advent of full-colour holographic imaging and the continuing development of hybrid artistic creations using photography, holography, graphics and animation, there is still plenty of scope for artistic expansion of this fascinating visual medium.

Figure 14. The Meeting (Rick Silbermann)

REFERENCES

Baths, R. (1982). Death of the author. In *A Barthes reader*. New York: Hill and Wang.

Blundell, B., & Schwarz, A. (2000). *Volumetric three-dimensional displays systems*. New York: Wiley-Interscience.

Gabor, D. (1949). Microscopy by reconstructed wavefronts. *Proceedings of the Royal Society*, *197*(1051), 454–487. doi:10.1098/rspa.1949.0075

Heidegger, M. (1969). *Identity and difference* (J. Stambaugh, Trans.). New York: Harper & Row.

ADDITIONAL READING

Benton, S. A. (1977). White light transmission/reflection holography. In Marom, E. (Ed.)(1977). Applications of Holography and Optical Data Processing, pp. 401-9. Oxford: Pergamon Press.

Collier. R, J. (1971). *Optical Holography*. New York: Academic Press.

Denisyuk, Yu. N. (1962). On the reflection of optical properties of an object in a wave field of light scattered by it. *Doklady Akademii Nauk SSSR*, *144*(6), 1275–1278.

Graube, A. (1974). Advances in bleaching methods for photographically recorded holograms. *Applied Optics*, *13*, 2942–2946. doi:10.1364/AO.13.002942 PMID:20134813

Leith, E. N., & Upatnieks, J. (1962). Reconstructed wavefronts and communication theory. *Journal of the Optical Society of America*, *52*(10), 1123–1130. doi:10.1364/JOSA.52.001123

Phillips, N. J., & Porter, D. (1976). An advance in the processing of holograms. *Journal of Physics. E, Scientific Instruments*, 631. doi:10.1088/0022-3735/9/8/011

Saxby, G. (2004). *Practical Holography* (3rd ed.). Boca Raton, FL: CRC Press.

Toal, V. (2012). *Introduction to Holography*. Boca Raton, FL: CRC Press.

Upatnieks, J., & Leonard, C. (1969). Diffraction efficiency of bleached photographically recorded intereference patterns. *Applied Optics*, *8*, 85–89. doi:10.1364/AO.8.000085 PMID:20072177

Van Renesse. R, L. (2005). Optical Document Security. New York: Artech House.

Chapter 8
The Reinvention of Holography

Frank C. Fan
AFC Technology Co., Ltd., China

C. C. Jiang
AFC Technology Co., Ltd., China

Sam Choi
AFC Technology Co., Ltd., China

ABSTRACT

In this chapter, the authors examine the concept of 3-D displays with sampled spatial spectra in discrete hoxels, reducing the redundancy present in digital holography, which prints each individual hogel. The underlying theory deals with the principles of 3-D imaging using the quantum model and 4-D (x,y,z,t) Fourier transforms. The image information is treated as a single spatial spectrum and its 3-D information recovered by digital vector treatment through hoxels. This enable the achievement of real-time communication in 3-D.

INTRODUCTION

What is light? The question has existed for over three millennia, and there have been many theories over the years. There is still no full answer, though there are several models, each effective in its own way at predicting the behaviour of light. One of the most advanced models suggests that light consists of *photons*, stable, chargeless, massless fundamental particles that exist only at a speed c (2.997 m s^{-1}in empty space) with different energies and states of polarisation. Unlike ordinary objects, photons cannot be seen directly; what is known of them comes from observation of the result of their creation or annihilation. A photon is observed by detecting the effect it has on its surroundings, and it has an observable effect only when it either comes into or goes out of existence. An object illuminated by the Sun, for example, is perceived as a result of the quantised energy distribution generated by the different states of the photons emitted. The *geometric optics* model treats the photon stream as light rays; it explains the phenomenon of imaging through a pinhole.

DOI: 10.4018/978-1-4666-4932-3.ch008

Newton's theory of lenses was instrumental in bringing about the birth of photography, the first practical medium to permanently record the visual experience. *Wave optics*, on the other hand, treats light as an electromagnetic disturbance propagated through space as a transverse wave motion, and accounts for the spatial distribution of the interference fringes that are the basis of holography. *Lippmann colour photography* also employs the model of interference planes recorded directly within a layer of light-sensitive material. However, we can only record the distribution of light energy in space directly by means of *hoxels* (holographic pixels), which are a visualisation of this quantized energy distribution. Our goal is to record and reconstruct our visual experience in three dimensions.

In this chapter, we make use of four-dimensional Fourier transform integrals of the *wave function f(x,y,z,t)* of this complex energy distribution so as to create a new description of light with its two kinds of spectrum, *temporal* and *spatial*. We expand the concept of *spatial frequency* in present-day imaging systems to the *spatial spectrum* and the recording and reconstruction of a complex wavefront in conventional holography. Applications based on this new description include holographic video. We also propose a model for holographic information technology (HIT). Its content could be construed as a kind of re-invention of holography with new possible applications, by discarding the interference concept for wavefront recording and reconstruction that is at present employed in conventional holographic theory.

THE BEGINNINGS OF HOLOGRAPHY

The images that light can form represent important ways to convey information to our consciousness. The two-dimensional (2-D) image has until recently been the main way of recording the kind of information human vision gives us. For most of the nineteenth century and the first quarter of the twentieth, before the invention of television, still photography and, later, motion pictures were the only method of producing images that truly represented the visual appearance of the world. The development of integrated semiconductor circuits in the middle of the twentieth century heralded the introduction of computer science, and with the advent of the laser, optical fibres and the Internet, it could now be asserted that the existence of our present-day society depends almost entirely on the successful application of information technology. One of its most important applications is the transfer of 2-D imagery via the modulation of 1-D electronic signals to illustrate the 3-D nature of the world we are living in – even 4-D, if we include the dimension of time. To our conscious perception Nature herself appears in three dimensions and in real time. But the 2-D information that is still standard for conveying visual information can show us only a single perspective in an image, restricting our perception of the 3-D world we live in. For many years people have asked whether it might be possible to record – and recover – the total; information present in a 3-D image.

Since the beginning of the nineteenth century, scientists have endeavoured to find a practical way to solve the problem of recording three-dimensional images. Restricted by conventional concepts of the imaging process, they have treated 2-D imagery as the standard model for human perception (Okoshi, 1976), regarding the stereopsis and dynamic parallax which form our subjective perception of three dimensional space as little more than add-on information, mere subject matter for a Victorian toy. Nevertheless, there have been successful technologies, beginning with the viewing of stereo pairs of 2-D images through stereoscopes (Norling, 1953) and continuing with parallax barrier technologies (Ives, 1903), to integral photography (Lippmann, 1908) and the lenticular stereogram (Burckhardt & Doherty, 1969). Although stereoscopic cinema and the successful production of lenticular stereograms in the middle of the twentieth century represented a breakthrough, we are some way from realizing the ultimate 3-D display. The inevitable shortcoming of present technologies is that some of the image quality is sacrificed in order to achieve the three-dimensional illusion, requiring some complicated techniques and often special eyewear.

The holographic principle was first proposed by Dennis Gabor (1948, 1949) who was chiefly interested in improving the image quality of electron microscopy. However, his inspiration for recording and reconstruction of actual wavefronts spurred scientists to develop his ideas in practical ways, introducing a new branch of optics: holography and optical information processing. This new concept had a profound influence on the modern development of information technology. Gabor's insight eventually earned him the Nobel Prize in Physics in 1971. Further progress demanded a highly coherent light source, and with the invention of the laser in 1962, American scientists Emmett Leith and

Juris Upatnieks (1962, 1963, 1964) were able to develop off-axis transmission holography, which created the first true 3-D images by interference of two coherent light beams In the Soviet Union Yuri Denisyuk (1962, 1963, 1965), inspired by the work of Gabriel Lippmann (1891), another Nobel Prize winner, developed reflection holography. These two inventions brought about the birth of a completely new visual medium distinct from photography, capable of recording an image of an object in true 3-D.

By the end of the 1960s creative artists worldwide were researching techniques of display holography both as a commercial prospect or simply as a hobby (Richardson, 2006). In contrast to the many scientists concerned with military applications, their research was wholly peaceful, and laid the foundations of modern holography. The most important developments included pulse laser holography, which could freeze motion for holographic imaging. Tung Jeong (1967) recorded 3-D information directly in cylindrical format. Huff and Fusek (1980), and Cross and Cross (1992) developed the principle of holographic stereograms. An important step forward had been provided by the so-called rainbow hologram, invented in 1969 by Stephen Benton (1969), which was later incorporated into embossing technology, introducing what has now become a worldwide industry in its role as a guardian of the security of documents and as an anti-counterfeiting measure for brand merchandise. In addition to realizing 3-D displays, holography was found to have applications with unique advantages in nondestructive testing (Stetson & Powell, 1966), holographic optical elements and data storage (Hiselink *et al.*, 2004), all of which are now regularly applied within their own areas. Compared with the rapid growth of other modern information technologies using microelectronics technology, and with the information revolution that spawned such enterprises as the Internet, the progress of applied holography was slow; but it was steady.

In the early 1990s, with the help of new types of Spatial Light Modulator (SLM), the research team at the Massachusetts Institute of Technology headed by Benton, who devoted his life's work to the development of practical technologies for display holography, developed the *ultragram* (Klug *et* al., 1992), which was a display hologram with true colour, full parallax, large viewing angle and large format, eventually to employ entirely digital techniques. This proposed technology was intended to simplify the generation of computer generated holograms (Lohmann & Paris, 1967) because more efficient holographic recording methods could do away with complicated calculations and encoding as well as simplifying the image recovery process. Today, full-parallax digital holographic stereograms are still in an early stage of development (Zebra Imaging, 2008) and the ability to print a perfect display hologram by this means is still severely restricted. Benton died in 2003, but left behind many promising ideas for the evolution of 3-D imaging in time to come.

THE PRESENT SITUATION

Since the invention of holography there have been many suggestions concerning the future development of the technology, but most of these have been discarded owing to theoretical shortcomings or technological problems. However, recent advances in video technology and high-resolution SLMs, 3-D have brought the possibilities of 3-D videos in real time within our grasp (St Hilaire *et al.*, 1992, Huebschmann *et al.*, 2003, Slinger *et al.*, 2004). So far these techniques, actual or proposed, depend, as do cine and TV images, on the physiological phenomenon of *persistence of vision*, which gives the impression of movement on viewing a rapid succession of projected images; but the result is still not a truly continuously moving holographic image. Bjelkhagen & Kostuk (2008) have shown the full capabilities of holography to record 3-D images in true colour, and Tay *et al.* (2008) have predicted the future of holographic video, but the bandwidth necessary for the transmission of such a huge quantity of information is likely to limit the rate of further advances for some time, in spite of 'Moore's Law' (which attempts to quantify the rate of miniaturisation of electronic circuitry).

In this chapter we treat light as a quantized energy probability distribution. By means of four-dimensional (4-D) Fourier transform integrals of the wave function or *probability amplitude*, we propose a new description of light in terms of its *temporal* and *spatial spectrum*. We expand the concept of spatial frequency in the present imaging system into its own spatial spectrum, and the reconstruction of the complex wavefront (i.e. both amplitude and phase) in holography in terms of the recovery of the spatial spectrum of the image. As a demonstration of our theory, we show the experimental results of our real-time holographic display instrument as a rudimentary version of the holographic video of the future. Our work to date has been published recently (Shang *et al.*, 2000, Fan *et al.*, 2010), and was demonstrated at the 8th International Symposium on Display Holography, 2009. This chapter represents a detailed description of our ideas.

In a recent book edited by Caulfield (2004), Stephen Benton said of holography: 'One cannot converse about holography for long in any environment without soon finding a one or two degrees-of-separation link with either Emmett Leith or Yuri Denisyuk, followed by fond approbations.' He also said: 'Holography has come to take on two meanings in our culture. Firstly, it means wavefront reconstruction by interference and diffraction/reflection. More widely, it has come to mean the ultimate 3-D imaging method of the future, and it stands as an optimistic hope for the progress of our science and technology relating to everyday life. Hopefully, future reinventions will be as productive and useful as holography's evolution has been.' We are happy to be involved in this reinvention.

The Fourier transform principle is well known as a fundamental theory in information processing. It has been successfully adopted in the IT industries, in 1-D for signal communication, 2-D for image processing, and 3-D for colour recognition. In this chapter, we propose to develop a 4-D Fourier transform of the wave function (or probability amplitude) $f(x,y,z,t)$ of the quantized energy probability distribution $I(x,y,z,t)$ to form a new description of light with its two kinds of spectra, temporal and spatial. We expand the concept of spatial frequency in present-day imaging systems as the natural spatial spectrum, and the reconstruction of a complicated wavefront (i.e. both amplitude and phase) in holography as leading to the recovery of spatial and temporal spectra of the original material. We also demonstrate a real-time holographic video display as an example of the application of this theory.

4-D Fourier Transform of the Wave Function

Light is composed of photons, which possess the energy of electromagnetic radiation expressed by the formula $\varepsilon = h\upsilon$, where ε represents energy, h is Planck's constant $(6.626 \times 10^{-34}$ J s), υ is the frequency of a photon of wavelength λ $(= c/\upsilon)$, c being the speed of light in empty space $(\approx 2.996 \times 10^8$ m s^{-1}). The momentum of this corresponding photon is expressed by $p = hk$ where k is the propagation vector and $k = 1/\lambda$. This all fits in satisfactorily with special relativity, which relates the mass, energy, and momentum of a particle by the formula $\varepsilon = [(cp)^2 + (mc^2)^2]^{1/2}$. For a photon, $m = 0$ so that $\varepsilon = cp$ (see Figure 1).

Now let us consider a 3-dimensional object reflecting the light incident on it (or being itself a source of electromagnetic radiation). Suppose this object to be at the origin of 3-dimensional coordinates (x,y,z) within the limits Δx, Δy, Δz. Numerous photons with different states are emitted from this object at time t which form the corresponding quantized energy probability distribution $I(x,y,z,t)$ surrounding this object, to make it appear as shown in Figure 1.

The *wave function* or *probability amplitude* is defined as $f(x,y,z,t)$, which is a complex number, to form the *quantized energy probability distribution* as seen in Box 1.

If we rewrite (1), (2), (3) and (4) as vector expressions then the *wave function* or the *probability amplitude* of any quantized energy probability distribution by photons of this nature could be expressed by the follow-

Figure 1. Quantized energy probability distribution and wave function

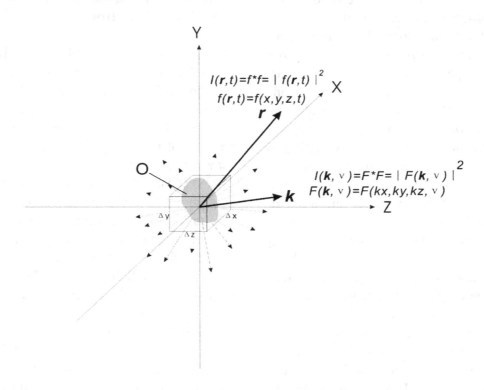

Box 1.

$I(x,y,z,t) = f^*f = |(x,y,z,t)|^2$ (1)

The 4-dimensional Fourier transform of this wave function $f(x,y,z,t)$ is defined as:

$$F(k_x, k_y, k_z, \nu) = \iiint\limits_{-\infty}^{+\infty}\int f(x, y, z, t) exp\,[i2\pi(k_x x + k_y y + k_z z + vt)]\, dxdydxdt \qquad (2)$$

And the inverse transform relation is:

$$f(x, y, z, t) = \iiint\limits_{-\infty}^{+\infty}\int F(k_x, k_y, k_z, \nu) exp\,[-i2\pi(k_x x - k_y y + k_z z + vt)]\, dk_z dk_y dk_z dv \qquad (3)$$

with its *quantized energy probability distribution* defined as:

$I(k_x k_y, k_z, \nu) = F^*F = |F(k_x k_y, k_z, \nu)|^2$ (4)

In which k_x, k_y and k_z are called *spatial frequencies* as the components of the photon propagation vector k with respect to the x, y, and z directions, where $k = 1/\lambda = \nu/c$; ν is the frequency of corresponding photons with the same energy state, where $\varepsilon = h\nu$ and $\lambda\nu = c$.

ing vector forms of 4-D Fourier transform integrals.

In the time-space domain:

$$I(\mathbf{r},t) = f * f = \big| |f(\mathbf{r},t)| \big|^2 \tag{5}$$

$$f(r,\nu) = \iint\limits_{-\infty}^{+\infty} F(k,\nu)\exp_0[-i2\pi(-\mathbf{k}\bullet\mathbf{r}+\nu t)]dkd\nu \tag{6}$$

In frequency-spectrum domain:

$$I(\mathbf{k},\nu) = F * F = \big| F(\mathbf{k},\nu) \big|^2 \tag{7}$$

$$F(k,\nu) = \iint\limits_{0}^{+\infty} (\mathbf{r},t)exp[i2\pi(-\mathbf{k}\bullet\mathbf{r}+\nu t)]drdt \tag{8}$$

in which $\mathbf{r}$ is the *position vector* of 3-D space where $r^2 = x^2 + y^2 + z^2$, $x = r \cos a$, $y = r \cos \beta$, $z = r \cos \gamma$, as shown in Figure 2; $\mathbf{k}$ is the photon propagation vector in the frequency coordinates where $k = 1/\lambda$, $k^2 = k^2_x + k^2_y + k^2_z$, $k_x = k \cos a$, $k_y = k \cos \beta$, $k_z = k \cos \gamma$, (Figure 3). (a,β,γ) is the space directional angle of a vector corresponding to the x,y,z-axes of a rectangular coordinates where $\cos^2 a + \cos^2 \beta + \cos^2 \gamma = 1$ and $\mathbf{k} \bullet \mathbf{r} = k_x x + k_y y + k_z z$. The minus symbol before $\mathbf{k} \bullet \mathbf{r}$ in Equation (8) means that the photon is emitted from the object rather than converged to form this object. Therefore the phase of any photons in the space is always behind the initial phase of the photons from the object.

The integral of a function with a vector variable from 0 to $+\infty$ means that the super-

Figure 2. Position vector $\mathbf{r}$ and propagation vector $\mathbf{a}$

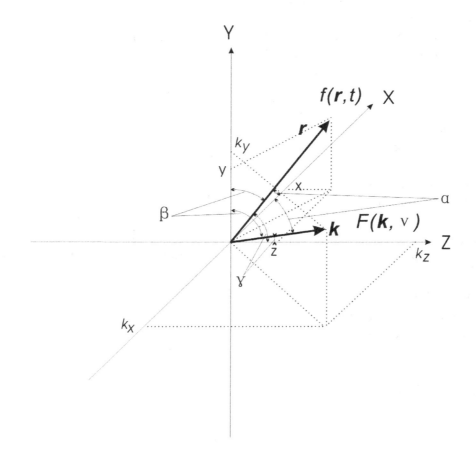

Figure 3. Integral surface with vector variables

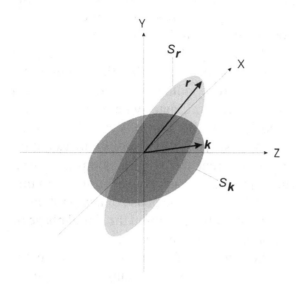

position of this function corresponds to the vector variables in the whole space in both amplitude and phase of this vector, i.e. along any enclosed surfaces encircling the zero point sweeping by the vector variables. The integral on the same direction vectors is replaced by other scalar variables t and ν in the time-space coordinates (x,y,z,t) or (r,t) and the corresponding *frequency-spectrum* coordinates (k_x, k_y, k_z, ν) or (k, ν) respectively to make the integral linear superposition of this 4-D function, as shown in Figure 3. Obviously, equations from (1) to (8) are a purely mathematical treatment of the complex function

$I(x,y,z,t)$ with four variables of the quantized energy probability distribution. However, these expressions have a real physical meaning. Although (1) to (4) are very wide, with their time and frequency integrals from -∞ to +∞, since the negative time and frequency have no meaning here, the vector expressions from (5) to (8) are more realistic in physics, the definite integrals being from 0 to +∞. Almost all useful concepts in modern

information technology can be derived from, or traced back to, these equations.

If we consider the object as a point source of radiation with different energy states of photons in all directions where $\Delta x = \Delta y = \Delta z = 0$ at the time $t = 0$ shown in Figure 4, we have

$$f(r,0) = [f(0,0)/\text{r}]\ \exp\ [i2\pi(kr)] \qquad (9)$$

which is the standard expression for the amplitudes generated by a typical spherical wavefront with wavelength $\lambda = 1/k$ in all directions of the time-space coordinates (x,y,z,t) or (r,t). It means that in all directions the wave function corresponds to a monochromatic plane wave with amplitude inversely proportional to that of the source $f(0,0)$.

At time t we have the wave function expressed as:

$$f(r,\ t) = [f(0,t)/r]\ \exp\ (i2\pi k\ r)\ \exp\ (-i2\pi\nu t)$$
$$(10)$$

$$= [f(0,\ t)/ct]\ \exp\ \{-i2\pi\nu t\ [1-(k\ r/\nu t)]\}$$

$$= f(0,\ t)/ct$$

$$= f'(t)$$

in which $r = ct$, $kr = (1/\lambda)(ct) = (c/\lambda)t = \nu t$ makes the function $f(r,t)$ change into $f'(t)$ with only one variable of t naturally. Then Equations (5), (6), (7) and (8) could be written in 1-dimensional form as

$$I\left(t\right) = f * f = \mid f\left(t\right) \mid^2 = \mid f\left(0t\right)/r\mid^2 \qquad (11)$$

$$f'\left(t\right) = \int_0^{+\infty} F'\left(\nu\right) exp\ (i2\pi\nu t) d\nu \qquad (12)$$

$$I(v) = F'* F' = \mid F(v)\mid^2 \qquad (13)$$

Figure 4. Wave function of point source radiation

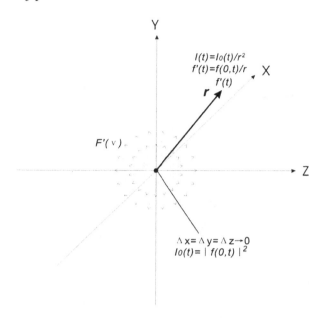

$$F'(\nu) = \int_0^{+\infty} f'(t) \ exp\,(i2\pi\nu t)dt \qquad (14)$$

These are standard 1-D Fourier transform pairs, here applied to the electromagnetic spectrum shown in Figure 5, in which the visible light band is very narrow. The conventional light spectrum in spectroscopy could be defined as the irradiance (average energy per unit area per unit time) $I(\nu)$ vs. the temporal frequency of the source. Figure 6 compares the energy distribution of sunlight compared with that of a tungsten filament lamp. The 'sunlight' distribution is the same whether we are looking at light from the sun, the moon, or a faint star.

When we consider a certain point in the space at the certain time where $x = x_0$, $y = y_0$,

Figure 5. Temporal spectral composition of electromagnetic radiation

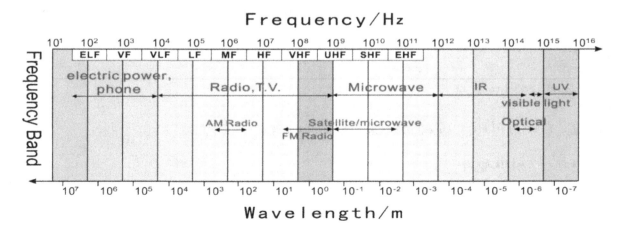

Figure 6. Comparison of temporal spectrum between sunlight and tungsten lamp

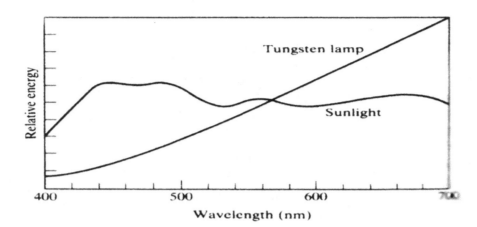

$z = z_0$ and $t = t_0$, equations (1) to (4) simplify, as seen in Box 2.

The case shown in Figure 7 resembles a pinhole imaging process where the wave function approximates to a Dirac delta function $\delta(x-x_0, y-y_0, z-z_0, t-t_0)$ at the pinhole location expressed by equation (15). Equation (16) is a typical expression of the Fourier transform of a delta function at the point (x_0, y_0, z_0) and the time t_0, where the displacement of the delta function from $(0,0,0,0)$ to (x_0, y_0, z_0, t_0) is transformed to the phase shift $\exp[-2\pi i(k_x x_0 + k_y y_0 + k_z z_0 + \nu t_0)]$. This means that the energy contributions at the point $(x_0, y_0, z_0) = r_0$ are the superposition of a series of monochromatic plane waves with different frequencies

ν and different propagation directions k at t_0 if we rewrite the equation (16) and (18) as a vector expression:

$$f\left(r_0, t_0\right) = \int \int_0^{+\infty} F\left(k\nu\right) [-i2\pi(k \bullet r_0 + \nu t_0)] dk d\nu \tag{19}$$

$$F\left(k, \nu\right) = f(r_0, t_0) \exp[-i2\pi (k \bullet r_0 + \nu t_0)] \tag{20}$$

At each point P on the object shown in Figure 7, only the photons along the two lines joining P and (x_0, y_0, z_0) could reach this point, and thus contribute energy to it. We may define all the photons propagating in the same direction as the *spatial spectrum* of the light. In Figure 8 a pinhole produces a 2-D image.

Box 2.

$I(x_0, y_0, z_0, t_0) = f^* f = \left\| f(x_0, y_0, z_0, t_0^2 \right\|$	(15)
$f\left(x_0, y_0, z_0, t_0\right) = \int \int \int \int_{-\infty}^{+\infty} F(k_x, k_y, k_z, \nu) exp\ [i2\pi(k_x x_0 + k_y y_0 + k_z z_0 + \nu t_0)] dk_x dk_y dk_z d\nu$	(16)
$I(k_x, k_y, k_z, \nu) = F^* F = \left\| F(k_x, k_y, k_z, \nu) \right\|^2$	(17)
$F\left(k_x, k_y, k_z, \nu\right) = f(x_0, y_0, z_0, t_0) \exp[i2\pi (k_x x_0 + k_y y_0 + k_z z_0 + \nu t_0)]$	(18)

Figure 7. Wave function at the point (x_0, y_0, z_0)

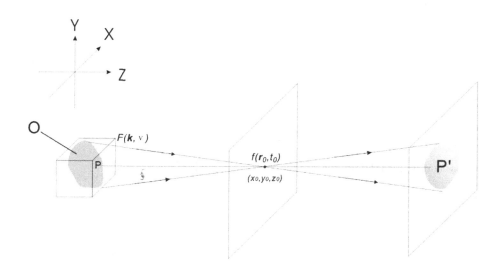

In Lippmann colour photography each point in the picture upon which light is incident is reflected from the rear of the light-sensitive layer, the incident and reflected wavefronts interfering to form fringes (Bragg planes) which when developed form an interference filter of the point source r_p and its mirror image r_p' with average wavelength $\lambda_a(P)$.

If the illumination is monochromatic so that $\nu = \nu_0$, equations (1), (2), (3) and (4) become simplified to the equations in Box 3.

The wavefronts illustrated in Figure 9 are a good representation of the object field in conventional laser holography. A TEM$_{00}$ laser beam is a close approximation to a single wavelength λ_0 where $k_0 = 1/\lambda_0$. In Figure 9, f

Figure 8. Principle of Lippmann colour photography

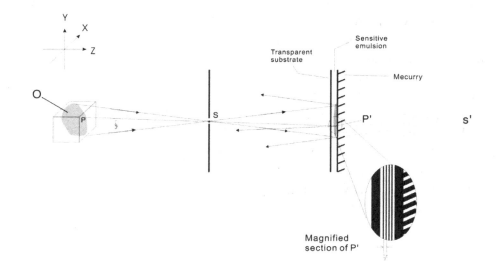

Box 3.

$$I(x,y,z) = f^*f = \left| f(x,y,z) \right|^2 \tag{21}$$

$$f(x,y,z) = \int\int\limits_{-\infty}^{+\infty}\int F\left(k_{0x}, k_{0y}, k_{0z}\right) \, exp \, [i2\pi(k_{0x}x+k_{0y}y+k_{0z}z)]dk_{0x} \, dk_{0y}dk_{0z} \tag{22}$$

$$I\left(k_{0x}, \; k_{0y}, \; k_{0z}\right) = \; F*F \; =\left| F\left(k_{0x}, k_{0y}k_{0z}\right) \right|^2 \tag{23}$$

$$F(k_{0x}, k_{0y}, k_{0z})_{-\infty} = \int\int\limits_{-\infty}^{+\infty}\int f\left(x,y,z\right) exp \, [i2\pi(k_{0x}x+k_{0y}y+k_{0z}z)]dxdydz \tag{24}$$

and the vector expression of (5), (6), (7) and (8) are simplified as:

$$I\left(\boldsymbol{r}\right) = \; f*f \; = \left| f\left(\boldsymbol{r}\right) \right|^2 \tag{25}$$

$$f\left(\boldsymbol{r}\right) = \int\limits_0^{+\infty} F\left(\boldsymbol{k}_0\right) \, exp \, (\text{-}i2\pi\boldsymbol{k}_0\bullet r)d\,\boldsymbol{k}_0 \tag{26}$$

$$I\left(k_0\right) = \; F*F \; =\left| F\left(k_0\right) \right|^2 \tag{27}$$

$$F\left(\boldsymbol{k}_0\right) = \int\limits_0^{+\infty} f\left(\boldsymbol{r}\right) \, exp \, (i2\pi\boldsymbol{k}_0\bullet r) \, d\,\boldsymbol{r} \tag{28}$$

$(\boldsymbol{r}_0)$ is the wave function of the photons incident on the object surface, where $I(\boldsymbol{r}_0)$ $= \left| f(\boldsymbol{r}_0) \right|^2$ is the quantised energy distribution on this object surface and independent of t, and $f(\boldsymbol{r}_0)$ is the *plural amplitude* of this object. The wave emitted by reflection from the object could be described as the *probability amplitude* of photons which are in the same phase as the photons on the surface of the object. In Figure 9 these so-called equal phase surfaces are expressed in terms of the out-of-plane irregularities of the object as integral multiples of λ_0. These equal phase surfaces (expressed by Equations 25–28) can be called the wavefront of the object. Each point on these surfaces has the real amplitude $fr_0/m\lambda_0$ and phase difference $2\pi m\lambda_0$ corresponding to the point on the studied object surface $\boldsymbol{r}_0$ where

$$f\left(\boldsymbol{r}_0\right) = \int\limits_0^{+\infty} F\left(\boldsymbol{k}_0\right) exp \, (i2\pi\boldsymbol{k}_0\bullet\boldsymbol{r}_0)d\,\boldsymbol{k}_0 \tag{29}$$

which is the superposition of a series of plane waves $F\left(\boldsymbol{k}_0\right) exp(2\pi i\mathbf{k}_0\bullet\boldsymbol{r}_0)$ at the position $\boldsymbol{r}_0$, and when propagation is along the normal to the surface of the object, Equation 29 can be written

$$f(\boldsymbol{r}_0) = F(k_{0n})exp(i2\pi\mathbf{k}_{0n}\bullet\boldsymbol{r}_0) \tag{30}$$

Equation 30 is only one component of series of wavefronts originating at the point $\boldsymbol{r}_0$ with the wavelength λ_0 expressed by Equation 29. It follows that

$$\left| f(\boldsymbol{r}_0) \right| = \left| F(\boldsymbol{k}_0) \right| = \left| F(\boldsymbol{k}_{0n}) \right|$$

Figure 9. Wavefronts and formation of hologram of object

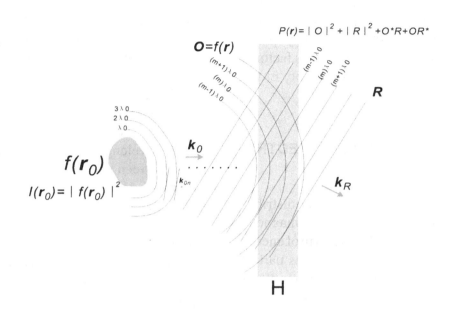

where k_{0n} is the propagation vector of the wavefront along the direction normal to the object surface at the point r_0. Although complicated to calculate in terms of equations 25–28, the wavefront of this object can be recorded holographically by causing the wavefront to interfere with another coherent beam with the same frequency ν_0, introduced as a plane wave R propagating in the direction of k_R as shown in Figure 9, where R is designated the *reference beam*. The intersections of a set of parallel planes with wavelength λ_0 and the wavefronts of the object O shown in Figure 9 and expressed by Equation 29 generate a complicated energy distribution P(r) composing the interference fringes of O and R as

$$P(r) = (O + R)(O + R)* = |O|^2 + |R|^2 + O*R + OR^*$$
(31)

which is called the hologram of the object O, when Equation 31 is recorded in a light-sensitive substance. This is possible because Equation 31

has a component that is independent of time and therefore represents a stationary energy distribution.

When the object is a 2-D quantized energy distribution I(x y) where $\nu = \nu_0$, equations 21–24 are written as

$$I(x, y) = f \ * f = \left|f(x, y)\right|^2 \qquad (32)$$

$$f(x,y) = \int\limits_{-\infty}^{+\infty} \int F(k_{0x}, k_{0y}) \ exp \ [-i2\pi(k_{0x}x + k_{0y}y)] dk_{0x} dk_{0y}$$
(33)

$$I(k_{0x}, k_{0y}) = F*F = |F(k_{0x}, k_{0y})|^2 \qquad (34)$$

$$F(k_0 k_{0y}) = \int \int\limits_{-\infty}^{+\infty} f(x, y) \ exp \ [i2\pi(k_{0x}x + k_{0y}y)] dx dy \qquad (35)$$

Equation 34 describes the 4f imaging system in terms of Fourier optics. Figure 10 is a depiction of the equal phase surfaces in a situation in which the image O′ is composed of different spatial frequencies (k_{0x}, k_{0y}) generated from the interference of the two wavefronts as a kind of grating. The recording medium should ideally have a linear response, but in practice the modulation transfer function (MTF) of the imaging system, and any spatial filtering, seldom do have one; so this factor is also relevant in describing the quality of the final image of such a system.

TEMPORAL AND SPATIAL SPECTRA OF LIGHT

The *temporal spectrum* of light is the distribution of the visible range of electromagnetic radiation, usually plotted in the form of energy versus temporal frequency *I(ν)*. It is part of the electromagnetic spectrum, which contains frequencies from ultralong-wave radio signals at around 50 kilohertz (50×10^3 Hz) to ultrashort X-radiation or gamma rays at around 100 exahertz (100×10^{18} Hz) corresponding to wavelengths approximately $3\times10^3 - 3\times10^{-12}$ m. Visible light lies in the range 390–790 terahertz (10^{15} Hz) but is usually described in terms of wavelength (380–760 nm in air).

Within the spectral range of visible light, the temporal frequency is related to the subjective appearance of colour (Table 1). Colour is, of course, not a property of light itself but a product of the perceptual mechanism of the brain. Early in the nineteenth century it was found that any colour can be synthesised from a mixture of no more than three different coloured lights – even lights consisting of single wavelengths. The three usually chosen are red, green and blue, these three frequencies being designated respectively ν_R, ν_G and ν_B for red, green and blue (these subscripts are sometimes given as ($\rho, \gamma \& \beta$); these frequencies correspond to values of ν in Equations 1 to 8.

The spatial spectrum of light is defined in terms of wavelength or its reciprocal spatial frequency (sometimes called *wave number*). It is formally described as the *luminance* (for electromagnetic radiation in general it is known as *irradiance*) in a specified spatial direction *I(k)*. Referring back to Equation 6, it can be seen that the wave function or probability amplitude *f(r,t)* of the quantized

Figure 10. Equal phase surfaces of 2-D quantized energy distribution and 4f imaging system

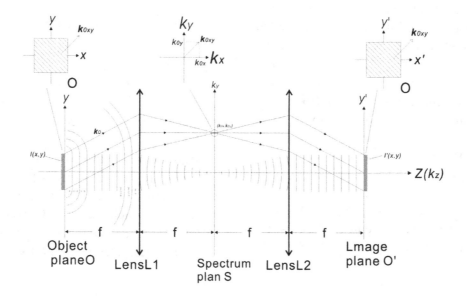

Table 1. Approximate frequency and wavelength ranges for various colour sensations

Color	λ_0 (nm)	ν (THz) *	Color	λ_0 (nm)	ν (THz) *
Red	780 – 622	384 – 482	Green	577 – 492	520 – 610
Orange	622 – 597	482 – 503	Blue	492 – 455	610 – 659
Yellow	597 – 577	503 – 520	Violet	455 – 390	659 – 769

* 1 terahertz (THz) $= 10^{12}$ Hz, 1 nanometer (nm) $= 10^{-9}$ m.

energy probability distribution $I(\mathbf{r},t)$ is the superposition of a series of plane waves

$F(\mathbf{k},\nu)\exp[2\pi i(-\mathbf{k}\bullet\mathbf{r}+\nu t)]$ generated at the object surface with irradiance $I(\mathbf{r}_p) = \left| f(\mathbf{r}_p) \right|^2$ where $f(\mathbf{r}_p)$ is so called the *plural amplitude* of the wavefront, and each point on the wavefront has a wavelength $\lambda_a(P)$ at this point.

We can express the spatial spectrum of light according to one of three different models, namely geometrical, wave and quantum optics. In geometrical optics, we define the spatial spectrum of the studied object as a perspective projected through a pinhole, as shown in Figure 11. This pinhole can be thought of as a point source containing all the information about the object from that viewpoint. From this point source the light spreads out in accordance with the law of rectilinear propagation, forming an image on a flat screen. If the pinhole is enlarged and a lens installed in the space it will form an image that appears sharp at a distance determined by the focal length of the lens and its distance from the object. Although an image formed in this manner is actually 3-dimensional, the image as seen on a screen is only 2-dimensional.

In wave optics, the spatial spectrum of the object is defined as a series of monochromatic plane waves of different frequencies along any given direction. Figures 10 and 32–35 give an account of the image forming process in terms of Fourier theory, but these deal only with a single wavelength in which the *spatial frequency* (k_{0x}, k_{0y}) is part of the corresponding *spatial spectrum*.

In quantum optics, the spatial spectrum of the studied object is defined as the quantized energy distribution formed by the various energy states of photons travelling in the same direction, expressed by projecting the

Figure 11. Pinhole imaging

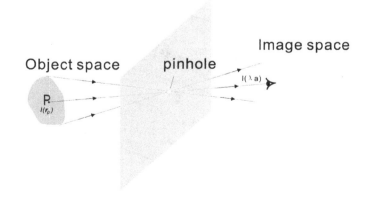

irradiance onto the surface of the object by parallel projective lines. It could be thought as the effective description of the property of photon momentum *p* with various photon energy states ε.

HOXELS AND THE SPATIAL SPECTRUM

Equations 5–8 give us two kinds of basic description of the spatial spectrum.

Equations 5 and 6 give us a description in which the quantized energy probability distribution $I(r,t)$ of each point is superposed by plane waves $F(k,\nu) \exp[2\pi i(-k \bullet r + \nu t)]$ formed by the totality of photons in a single state as shown in Figure 12. This point is defined as a *hoxel*, and its nature can be illustrated by an example from Lippmann colour photography, in which each light ray is regarded as a monochromatic plane wave with a corresponding wavelength λ_n of a point irradiance $I(r_n)$ of the object, together generating the appearance of colour at the corresponding point by acting as an interference filter for several wavelengths.

Equations 7 and 8 describe the situation in which the quantized energy probability distribution $I(k,\nu)$ in one direction from the object is superposed by many hoxels $H(r_n)$ expressed by $f(r,t) \exp[-2\pi i(-k \bullet r + \nu t)]$ near the object emitting photons individually to form a 2-D illuminance pattern $P(\lambda_n)$ of the object in the direction shown in Figure 13. This 2D pattern is defined as a spatial spectrum (Figure 12), giving a high resolution 2D colour image in digital form (Fan *et al*, 2010).

Although each hoxel may be counted in terms of single photons, so that Heisenberg uncertainty principle applies, the wavefronts still satisfy the Bragg condition, as is demonstrated by holography, where the complex object wavefront is recorded and reconstructed.

As shown in Figure 14, we can treat our visual information of this nature in spatial spectrum language, visualising two enclosed surfaces through which we can see from inside the hoxels across on S1 and from outside the hoxels across on S2. Both inside and outside visual information can be sampled and reconstructed by spatial spectrum collection on either the S1 or the S2 surface via a system

Figure 12. Generation of a hoxel: each light ray represents a plane wave of mean wavelength λ_n radiated from a point object O The final result is a colour image in digital form

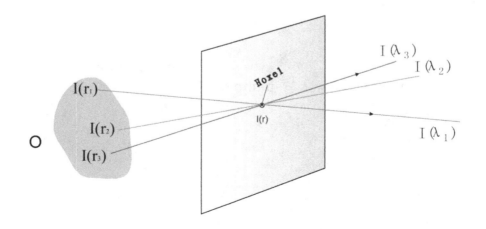

Figure 13. Spatial spectrum: each light ray represents the parallel perspective of each hoxel near the object O

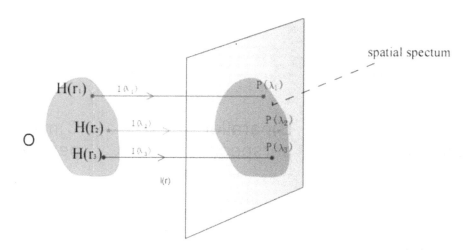

in which M×N individual camera-projectors are arranged in an array centred on the s reference point R.

Illuminance is a measurable quantity in the visible light range. In lay terms it defines the brightness of the light reflected by, or emitted from, an object. We have already defined the temporal and spatial spectrum of light as the illuminance in terms of frequency and spatial direction; we now use this concept

to give a detailed description of the way we recover 3-D imagery from the recorded wavefronts. To begin with we take an arbitrary object. Each point on this object has a corresponding point related to the object perspective in at least one direction, and the point on each perspective is on the surface of the object. The basic way to show the object O (Figure 15) in correct perspective is by use of a pinhole or lens

Figure 14. Visual information in spatial spectrum language

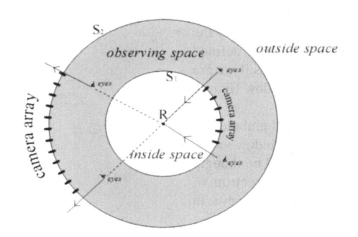

Figure 15. Three basic perspective relationships

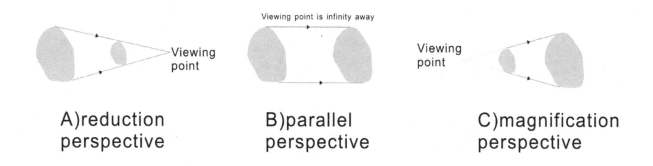

A)reduction
perspective

B)parallel
perspective

C)magnification
perspective

Let us suppose another closed surface S, inside which is the object O, and define S as the spatial spectrum surface of this object, on which its whole spatial spectrum is included as individual perspectives of O through each point P on S. As in a conventional lens imaging process, we define the space inside S as 'object space' and the space outside S as 'observation space' as shown in Figure 16. When considering some point P on S, the observation space becomes the 'image space'. The effective aperture is important in terms of resolution, as this is governed by the size of the aperture relative to its distance from the object. The ultimate limit of resolution is defined by diffraction.

We now come to the method of recovering the 3-D image of the object. To simplify the analysis the spatial spectrum surface defined above is assumed to be a plane, as if we are seeing the object through a window (Figure 17).

Suppose we put a very small pinhole (effectively a point) in front of the window shown in Figure 18. This pinhole could be thought as a sampling unit of the spatial spectrum of the object that can be resolved by the eye. In the 'observation' space, we can only identify a white light point source radiated from this pinhole, even though the object information has been coded as one perspective (i.e. a 2-D image) through the pinhole. By examining the light from the pinhole carefully, as we move around in the observing space we find the hue and luminance of this point source change: the spatial spectrum of the pinhole is continually changing. If we move this pinhole in a raster pattern to scan the whole window, we shall perceive the object appearing from every viewpoint, as if we are open-

Figure 16. Object space, spectrum surface and observation surface

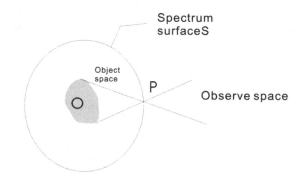

Figure 17. Plane spatial spectrum surface

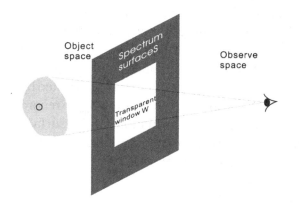

ing the window wide with a large number of touching pinholes. Each pinhole is a hoxel, (equivalent to the pixels of 2-D images). The window may be any shape, of course. Each hoxel can be thought of as a sampling of the quantized energy distribution carried by the photons from the object.

The information focused on the hoxel is expressed in terms of energy distribution in two ways. First, it is a direct perspective projection of the object formed by the illuminance of each corresponding point and can be recorded by photosensitive materials. Secondly, the object is illuminated by a monochromatic source such as a laser, if a reference beam is introduced into the hoxel space, the energy distribution within the hoxel amounts to a volume Fourier transform hologram of the perspective of the object at that point.

To summarise: visual information is provided by light in four dimensions, and is composed of a temporal and spatial spectrum which sampled by photons. The quantized energy distribution of photons is expressed holographically by its wave function or probability amplitude through its combined spatial and temporal spectra; Fourier transforms of this wave function are perceived in a hologram as hoxels. The photographic image can capture images covering the whole visible spectrum, but only in 2-D. Holography, on the other hand, has made possible the recording of the whole spatial spectrum, i.e. with full 3-D representation. Digital holography, which combines modern electronic technology with photonic know-how, will one day open the door to the

Figure 18. Principle of the hoxel

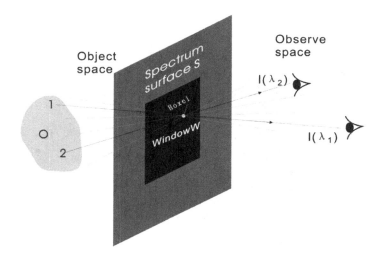

provision of full recording of images in the 4-D world of space and time.

RECORDING AND RECOVERY OF SPATIAL SPECTRA

In order to illustrate the principle more clearly, we choose a window W on a plane S as an example and divide it into equal squares of M×N (M≥1, N≥1) units, corresponding to a spatial spectrum sampling of the object O inside the window W, as shown in Figure 19. The shape of the spectrum surface S is not in fact restricted to a plane; its shape depends on the display methods. The central point S_{mn} of each individual unit is effectively a spatial spectrum sampling of the object O on the surface S; it corresponds to a perspective view I_{mn} of the object O that could be obtained by an M×N digital camera offering the same number of pixels as J×K at their corresponding position S_{mn} and focusing on the same point R of the object O, R being a spatial reference point. The number of M×N units determines the fidelity of the recovered 3D information.

For traditional holography, M×N is so large as to be almost a record of single photons. The spatial coding pattern of a hogel (the term coined for the holographic equivalent of a voxel) can be obtained by recomposing pixels P_{mnjk} of each individual I_{mn} generated by the digital camera at the position S_{mn} to record the original 3-D information on the surface S, i.e. the corresponding digital holographic display could be achieved by printing each individual hogel as the Fourier transform hologram of each spatial coding pattern. The relation M = N = 1 corresponds to 2-D photography; M = 1, N = 2 corresponds to stereoscopic photography (binocular imaging technology); M = 1, N = n corresponds to a stereogram with full horizontal parallax but no vertical parallax.

We may use matching M×N digital projectors to project the corresponding M×N perspectives I_{mn} onto a plane within the original object O and choose a reference point R on each individual I_{mn} in this plane so that each individual pixel P_{mnjk} is superposed at position (j,k) of this plane to form a hoxel H_{jk} as shown

Figure 19. Spatial spectrum recording

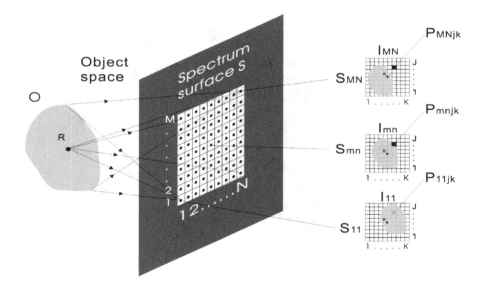

in Figure 20. This plane P_R is called the *reference surface,* and its shape is determined by the recovery and display methods for the 3D information. The images of each I_{mn} projected on P_R may be larger or smaller than the original object. In Figure 20 the optical information on the reference surface P_R is equivalent to the original object O recovered from its aggregate spatial spectrum sampled by M×N numbers at each corresponding situation S_{mn}. The whole process of this sampling and recovering is expressed by the discrete Fourier transform deduced from Equations 1– 4, where the process is in two parts: the digital treatment of each individual perspective I_{mn} which is typically demonstrated as its digital image with pixel numbers J×K or 1024×768; and the number of spatial spectra required to recover the 3-D information of the original object M×N. Colour (derived from the three primaries R,G&B) can be part of the total spatial spectrum. If we divide the reference surface P_R into J×K hoxels in which H_{jk} is the same size as the image pixels, and fit all pixels P_{mnjk} into their corresponding hoxel H_{jk}, the optical information distribution on references surface P_R can be analysed as follows:

If we were to position a window at the surface P_R, we should observe just the M×N point sources which carry the original object information, as shown in Figure 21, whereas the spatial spectrum of each hoxel on P_R is distributed in M×N directions, as shown in Figure 22. Therefore, if the number of sampling densities determined by M×N is the same as the number of projected image pixels J×K, the 3-D information with J×K numbers of hoxels can be directly recovered. But in the present state of technology M×N are much less than J×K. So although the information carried by the M×N spatial spectra with J×K pixels is fully formed on P_R, what we actually observe is only a small portion of M×N out of the abundant information present in the spatial spectrum carrier.

If we place a conventional diffuser screen at P_R, one of the M×N projectors can project

Figure 20. Recovery of 3-dimensional information

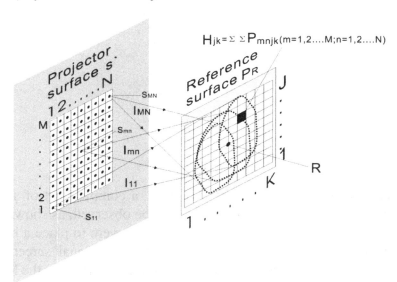

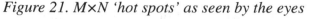

Figure 21. M×N 'hot spots' as seen by the eyes

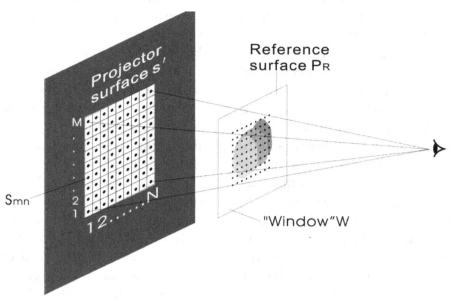

Figure 22. Discrete spatial spectrum distribution of hoxels H_{jk}

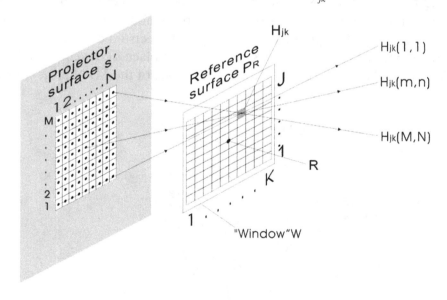

a perspective image of the object from its corresponding spatial spectrum direction displaying the whole image on the screen in 2-D. This is shown in Figure 23. If all possible M×N projectors project M×N clear images from their respective spatial spectrum direc-

tions, a complete set of superposed images will be observed, as shown in Figure 24.

If we were to place a hypothetical 'holographic functional screen' on P_R with the function of spreading the M×N point sources carrying the object information against a

Figure 23. Output with angular spreading by a single M×N spatial spectrum

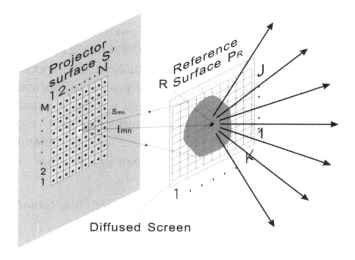

Diffused Screen

uniformly bright background as shown in Figure 25, the output spatial spectrum of each individual voxel H_{jk} would be identical with the reconstructed hogel of a digital hologram. Thus full parallax, full colour 3-D information could be perfectly recovered. The fidelity of the recovery is governed by the density of sampling of M×N. When M = 1and N = n, and we place on P_R the same holographic functional screen with a horizontally spread-ing width D_n and vertical expansion as shown in Figure 26, a full colour 3-D horizontal-parallax-only image is reconstructed (see Figure 25).

The 'holographic functional screen' can be thought of as a kind of optical demodula-tor able to recover a 3D space faithfully through M×N samplings of the spatial spectrum ex-pressed as M×N units of individually pro-jected digital image.

Figure 24. Output with angular spreading by multiple M×N spatial spectra

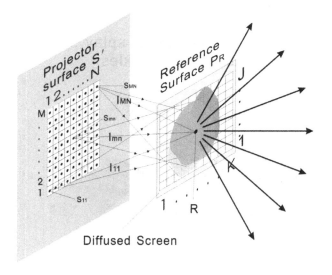

Diffused Screen

As for a method of making such a screen, the key point is that it must expand the spectrum information uniformly with a spatial angle ω_{mn} as shown in Figure 27 to correctly display the recovered 3-D information in the space domain, so that we can perceive the image in a real 3-D space. Some ideas as to how this might be accomplished are given below:

Laser Speckle Technology

According to Chinese Patent No ZL200410022193.7, 'Equipment and method to make digital speckle hologram' there are many ways to fabricate a holographic functional screen. By making use of the speckle patterns on sensitive material such as photoresist, we can generate a master functional screen in low relief, as shown in Figure 28;

Figure 25. Output of spatial spectrum through a holographic functional screen

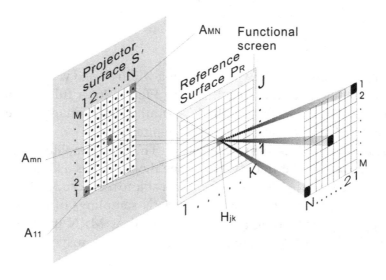

Figure 26. Output for horizontal parallax only

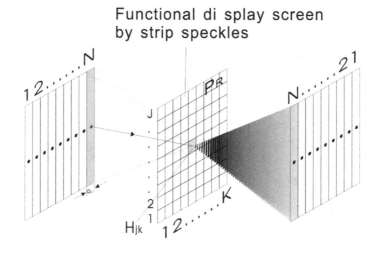

Figure 27. Relation between spreading of single spectrum and integral spectra

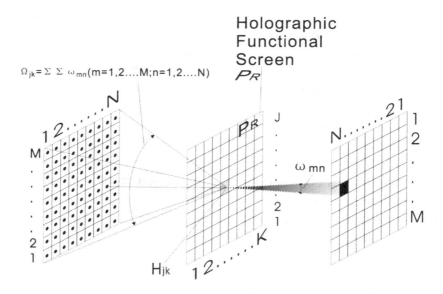

we can then produce this screen in volume economically by existing embossing techniques. The screen's diffusing angle can be held close to ω_{mn} by controlling the shape and size of speckles if we make the mask aperture to subtend an angle $\omega_{mn}/2$ at the screen. If the screen is to display an image with horizontal parallax only, the mask aperture should be a slit with width subtending an angle $\omega_{mn}/2$ at the screen, to generate a 'stripe' speckle screen. The shortcoming of this technology is that we cannot precisely shape the light within the angle ω_{mn}. However, it appears to be satisfactory for general use.

Figure 28. Screen using laser speckle technology

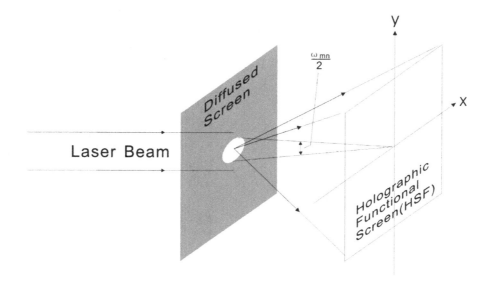

Holographic Lens Array

According to Chinese Patent ZL200510022425.3 'A kind of holographic projection display screen: making method, system and application', there are many ways to make a holographic functional screen. As shown in Figure 29, a micro-lens array can be recorded holographically on photoresist as a master for embossing, provided the angle of field of each micro-lens is controlled to be equal to ω_{mn}. If the screen is to display horizontal parallax only, the mask aperture should be a slit shape with an angle ω_{mn} to the screen (Figure 29).

The advantage of this screen is the precision given by holography to the control of the angle of light spread. Although chromatic dispersion cannot be eliminated it can be made use of for laser projection in three primary colours or for achromatic recording by one of the standard holographic techniques. In addition, the display is, in effect, semi-transparent and appears as if floating in the air; and the reconstruction beam is off-axis.

Computer Generated Hologram (CGH)

The functional screen is also equivalent to an on-axis Fourier transform hologram, focusing an input light beam to a uniform spot at the required angle ω_{mn} with high diffraction efficiency. This hologram master can be made directly by means of e-beam or micro-etching techniques, and copies mass produced using embossing technology.

Microlens Array

The functional screen can also be made by a moulded plastic microlens array in which each microlens produces the required angle ω_{mn}. Each individual microlens can be fabricated with a diameter to match the characteristics of human perception.

Figure 29. Screen using holographic lens array

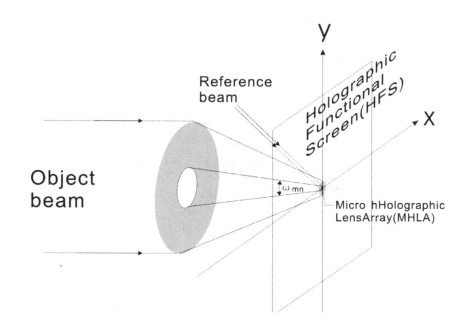

RECORDING AND RECOVERY OF 3-D INFORMATION

The Shannon Sampling Theorem, together with Fourier transform theory, provides the basis for digital technology. We now consider the sampling spatial angle ω_{mn} and its influence on the fidelity of 3-D image reconstruction.

The best angular resolution ω_E of a human eye is roughly 1.5×10^{-4} radians. Each of our eyes can be considered as a spatial spectrum collector that gathers 3-D information by focusing at any point inside the area of hoxels. If $\omega_{mn} = \omega_E$, the complete 3-D information can be recovered with the fidelity inherent in the resolution of the human eye. If the sampling spatial angle $\omega_{mn} >> \omega_E$, this is the criterion for best resolution, expressed as Δh of the hoxel, which is the same size as each pixel projected on the holographic functional screen located at the reference surface P_R. As shown earlier in Figure 17, it can be assumed that each individual point on the functional screen is emitted with its single spectrum information within the sampling spatial angle ω_{mn}. Further information outside the plane of the functional screen can be thought of as the composition of a light spot of size $\Delta Z \times \omega_{mn}$, where ΔZ is the distance behind or in front of the functional screen. When $\Delta h = \Delta Z \times \omega_{mn}$, the space recovering capability with the fidelity of the corresponding hoxel size is (see Figure 30):

$$2\Delta Z = 2\Delta h / \omega_{mn} \qquad (36)$$

Table 2 is compiled for a range of hoxel sizes. If the resulting resolution is acceptable these figures can be increased.

3-D REAL-TIME HOLOGRAPHIC DISPLAY

(In view of the undeveloped state of the relevant technology, the following experiment is at present necessarily conjectural.)

At the simplest level of sophistication, let us consider a straight line array of 30 cameras and projectors installed as if to construct a horizontal parallax only display as shown in Figure 31. Figure 32 is the device for spatial

Figure 30. Recovery of image fidelity

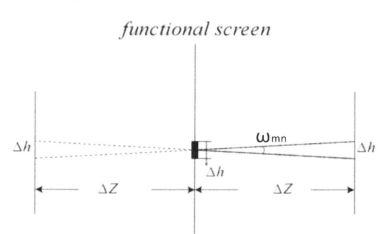

spectrum collection. Each digital camera has a resolution of 480×640 pixels. They are all focused to the same point R of a space framed by each camera. The signals adopted by each individual camera are directly connected to their corresponding projectors as shown in Figure 33. The projected images from each individual projector, also with 480×640 pixels, are precisely superposed on a plane reference surface P_R containing a 480×640mm display screen.

Figure34 shows two sets of pictures of the screen from different viewing angle when the viewer is situated inside the space to be recovered. In our experiment, we obtain a 3-D image straddling the reference surface, with a depth of some 500 mm. The image across the reference surface P_R resembles the original object seen through a large aperture imaging system, in this case about 2 m in diameter; it is a horizontal sampling by 30 spatial spectra through this large 4f system. The quality of the final display is much better than the single projected image of 480×640 pixels because there is 30 times as much information in each individual hoxel, and the luminance is higher; but the important thing is that it is in 3-D, just as if the observer were looking into a real space and observing a real time performance.

Our future task will be to develop practical ways of collecting and recovering the spatial spectra that we have been describing according to the principle shown in Figure 14. The following are some suggestions.

Table 2. Recovering fidelity (2ΔZ) for different values of ω_{mn} and Δh

Δh (mm) ω_{mn} (rad)	0.1	0.2	0.5	1	1.5	2
π /10	0.6	1.2	3.0	6	9	12
π /20	1.2	2.4	6	12	18	24
π /40	2.4	4.8	12	24	36	48
π /90	5.7	11.4	28.5	57	85.5	114
π /180	11.4	22.8	57	114	171	228
π /360	22.8	45.6	114	228	342	556
π /720	45.6	92.2	228	556	684	1112

Figure 31. Schematic diagram of experimental holographic display

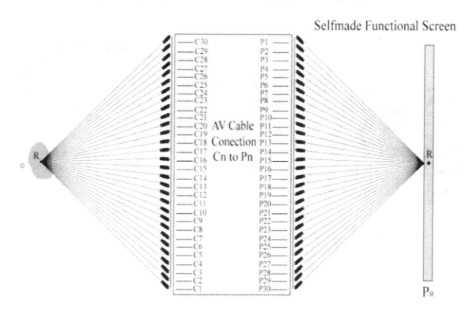

Figure 32. Spatial spectrum collecting system

Curved Surface

Figures 35 and 36 are schematic diagrams for spatial spectrum recording and recovery on a curved surface. The advantage of a curved surface is that it offers a larger scale of spatial information than a plane surface within the same sampling area and density. This method of spatial information collection and recovery would be suitable for holographic movies, TV

Figure 33. Spatial spectrum recovery system

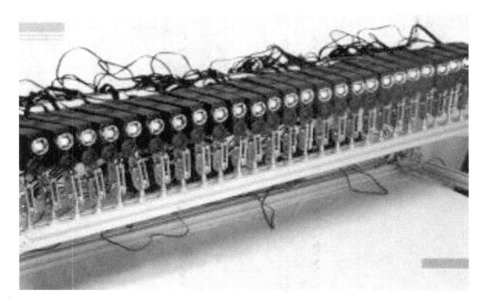

Figure 34. Pictures taken from functional screen from different horizontal viewing angles

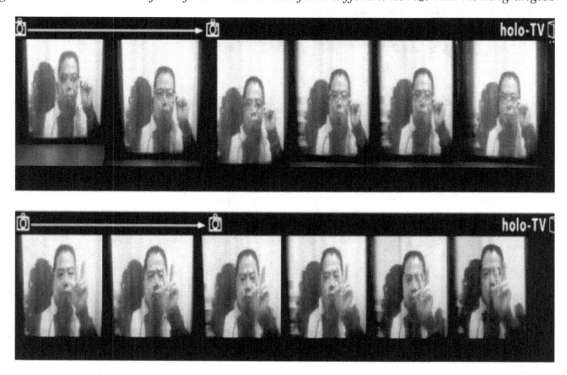

Figure 35. Information recording on a curved surface

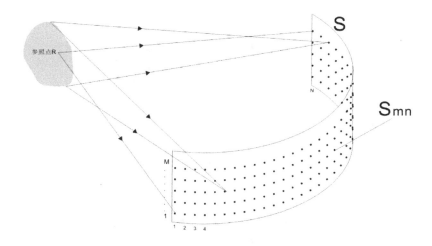

Figure 36. Information recovery from curved surface

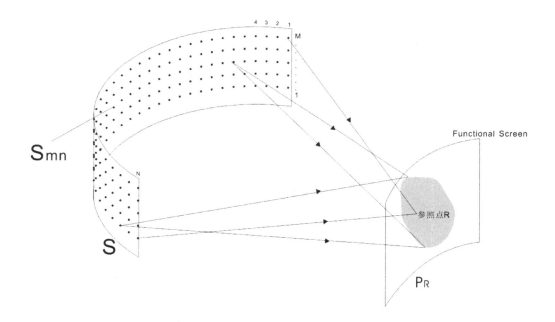

and for upgrading of 2-D displays to 3-D; it could also be used for real-time 3-D medical imaging or for remote sensing and metrology, as well as for 3-D animations.

Recording and recovery of a cylindrical hologram image: This setup offers a display facility giving an all-round (360°) view of an object (see Figures 37 and 38).

Figure 37. Information recording on cylindrical surface (for viewing from inside)

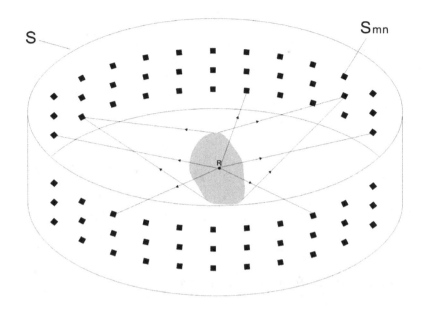

Figure 38. Information recovery from cylinder with 360⁰ panoramic view

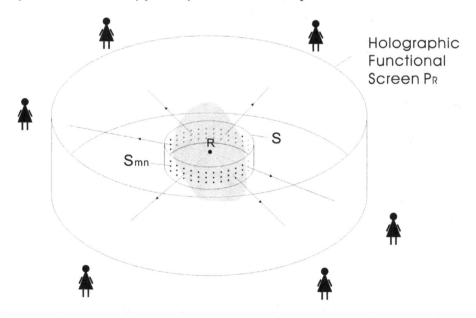

Figure 39. Information recording on a cylinder from outside with 360⁰ view

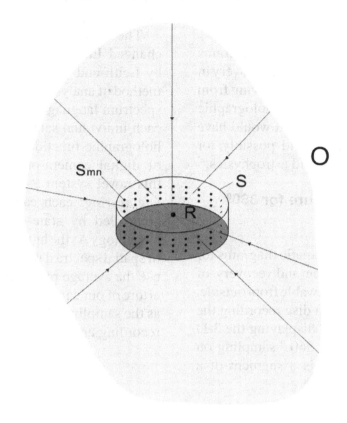

Figure 40. Information recovery from cylinder with 360⁰ image viewable from outside

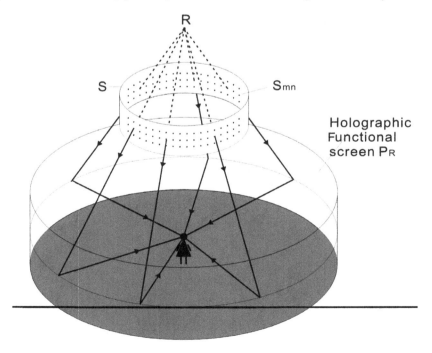

360⁰ Cylinder Form Viewed from Outside

Figures 39 and 40 give are schematic diagrams for spatial spectrum recording and recovery in a 360° cylindrical hologram for viewing from outside. This would be a novel holographic walk-round display facility, and would have applications in museums, and possibly for lectures in earth sciences and astrophysics.

Spherical Shell Structure for 360⁰ Viewable Desk

Figures 41 and 42 are schematic diagrams for spatial spectrum recording and recovery of shell hologram images viewable from outside. The reference surface is a disc recording the complete spatial spectra, displaying the 3-D information collected by 360 ⁰ sampling on a surface (or shell) that is a segment of a sphere. A possible application is the display of athletics events.

The basic theory of holography has changed little since its original exposition by Leith and others. The difference in our method of analysis is that we are using spatial spectrum language, recording and replaying each individual sampled 2-D spectrum on a holographic functional screen with an array of digital camera-projector pairs. We call this novel system *3-D holographic display engineering*: each camera-projector unit is controlled by state-of-the-art information technology As the human eye is also a recorder of spatial spectra it would seem appropriate to use the average pupil diameter D_E as the aperture of our camera-projector system as well as the sampling distance for spatial spectrum recording and projection.

Figure 41. 1nformation recording on surface of segment of sphere for 360⁰ viewing

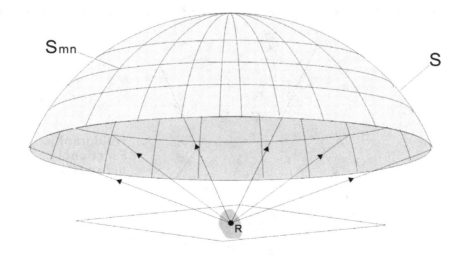

Figure 42. Information recovery from surface of segment of sphere for 360⁰ viewing

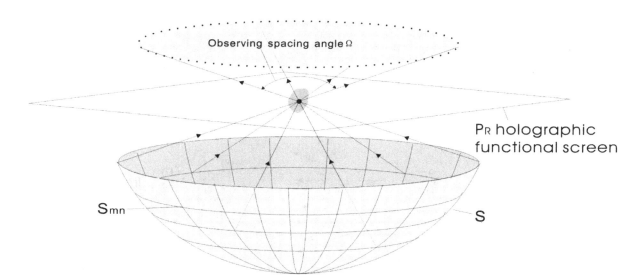

MOUNTING OF THE CAMERA–PROJECTOR ARRAY

Symbols Used

In the subsequent sections the symbols in Table 3 are used throughout.

Choosing the reference point R(R′) As shown in Figure 43, reference point R(R′) is the point determining the 3-D space O(O′) to be reconstructed. It determines the spatial

Table 3. Symbols used

Object 3-D Space	Image 3-D Space
O= object	O′ = image
R= recording mount	R′= imaging mount
C_{mn} = recording camera	P_{mn} = imaging projector
I_{mn} = recorded image	I'_{mn} = recovered image
θ_{mn} = recording angle	θ'_{mn} = projection angle
H_{mn} = sampled hoxel	H_{jk} = recovered hoxel
A_{mn} = axis of recording unit	A'_{mn} = axis of imaging unit
S_{mn} = recording point	S'_{mn} = projection point
ω_{mn} = sampled spatial angle	ω'_{mn} = recovered spatial angle

relationship between the original 3-D spatial information O(O′) to be recovered and the reference point R(R′) when confirmed, i.e. each perspective image $I(I')_{mn}$ formed by the discrete spatial spectrum of the original 3-D information O(O′) to be recovered corresponding to each sampling direction emitted from the reference point R(R′). Here (m,n) is the location of the recording coordinate S_{mn} on an arbitrary spatial spectrum surface corresponding to Figure 19, and I_{mn} is the perspective image of the spatial spectrum of the original 3-D space O to be recovered in Figures 12 and 13 at the position of its hoxel H_{mn} located at S_{mn}, the spatial spectrum direction of this perspective image is along the straight line that combines the reference point R and the recording point S_{mn} to make R the fixed point of each perspective image, and the angle between two adjacent discrete spatial spectra is just the sampling spatial angle ω_{mn} shown in Figure 27.

As shown in Figure 44, if each camera–projector unit $C(P)_{mn}$ has its inherent viewing

Figure 43. Spatial relationship between recovered 3-D space O (O') and reference point R (R')

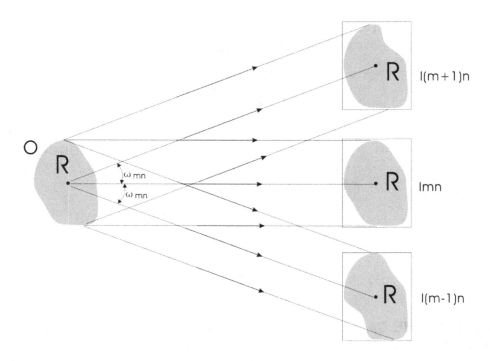

Figure 44. Setup for inside fixing

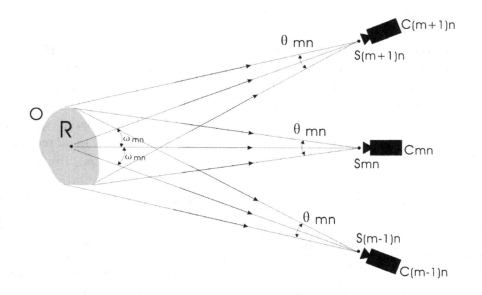

angle θ_{mn} angle, and can collect/project the original 3-D information O(O') to be recovered as a complete viewable scene, the reference point R(R') must be within the original 3-D information O(O'), i.e. all camera-projector units are fixed within the reference point R(R') as shown in Figure 14.

As shown in Figure 45, if each camera-projector unit C(P)$_{mn}$ has its inherent viewing angle θ_{mn} insufficiently large to recover the original 3-D information O(O'), the reference point R(R') can be chosen outside O(O'), i.e. all camera-projector units are mounted outside the reference point R(R') as shown in Figure 14.

Figure 46 is a schematic showing the spatial magnification coefficient M$_s$ of the size relationship between the recovered 3-D space O' and the original 3-D space O. It is equal to the ratio of the spatial spectrum recovery distance d'$_{mn}$ and the spatial spectrum collecting distance d$_{mn}$, i.e. M$_s$ = d'$_{mn}$/d$_{mn}$. R and R' are corresponding reference points with respect to the original 3-D space O and the recovery 3-D space O'.

Recording Spatial Spectrum Information

When the reference point R has been confirmed, each perspective image can be referred to this point. Each camera in Figure 14 records the appropriate discrete spatial spectrum with its own perspective image I$_{mn}$. As shown in Figures 44 and 45, each camera unit C$_{mn}$ at each recording point S$_{mn}$ has its own parameters of scene depth and viewing angle. These are all referred to the point R so that their individual images I$_{mn}$ are generated in the correct directions. For the layout shown in Figure 19, the image of the object O is to be reconstructed inside the original 3D space, except for the reference point R which has the same central point position for each perspective image I$_{mn}$, all other points being rendered as pixels P$_{mnjk}$ at their own perspective position (j,k) along its corresponding direction (m,n) of the spatial spectrum.

For the outside layout of Figure 45, each perspective image I$_{mn}$ is only a portion of the complete perspective image within its own

Figure 45. Setup for outside mounting

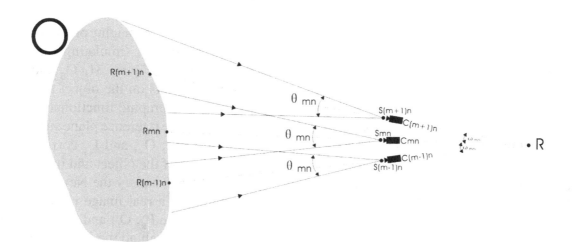

Figure 46. Relationship between original and reconstruction space

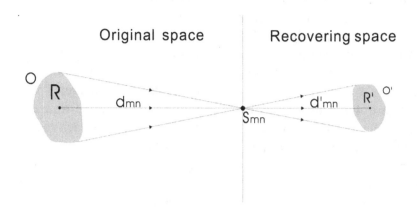

viewing angle θ_{mn} along its corresponding spatial spectrum direction (m,n); the central points of two adjacent perspectives carry the point information R_{mn} within the original 3D space O corresponding to reference point R but positioned with the sampling spatial angle ω_{mn}.

Choosing a Reconstruction Surface

As the image I_{mn} possesses the spatial spectrum information of the original 3D space O, it can be projected by its respective projector unit P_{mn} to the reconstruction surface P_R of the original 3D space O to recover its image information I'_{mn} in the direction (m.n) of its respective spatial spectrum, as shown in Figure 20. The reference surface P_R could be any kind of curved surface. It needs to include all spatial spectrum information from the original 3D space O as light intensity patterns in every direction projected on this surface. It is also necessary to rectify any image distortions (such as keystoning) to match the image of the original 3D space O on P_R along each spatial direction.

The reference surface P_R is a kind of hologram that can recover the entire spatial spectra of the original 3D space. By making use of a new reference point R′ of the recovered 3D space O′, this hologram faithfully records all the perspective image information respective to the original spatial spectrum information in every direction contained in the original 3D space, by projecting each individual light intensity pattern and displaying all these spatial spectra in the form of a digital hologram.

Imaging in Terms of Geometric Optics

Figure 47 depicts the imaging process in terms of geometric optics, simulating the hybrid system shown in Figure 31. O_0 and I_0 are respectively points on the object and image planes; the holographic functional screen is positioned on the reference plane surface, i.e. the image plane; O_{+1} and I_{+1} correspond to points in front of the object and image plane respectively, and satisfy the Newtonian lens laws, and I_{+1} is a real image relative to the reference surface P_R; O_{-1} and I_{-1} correspond to certain points behind the object and image

Figure 47. Simulation of imaging process employing geometric optics

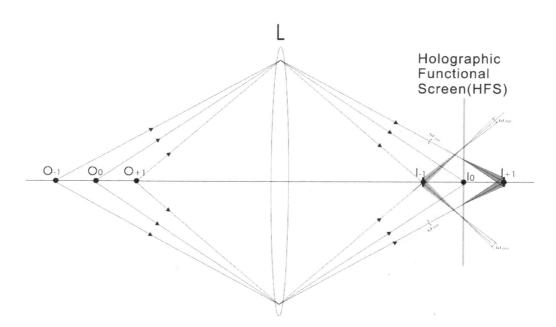

plane, and I_{-1} is a virtual image with respect to the reference plane P_R. Thus all imaging light rays are transformed to a converging light beam distribution with a vertex at the holographic functional screen through which they pass with a solid sampling angle ω_{mn}, to form the holographic display within the reconstruction space.

If the 3-D space O is a dummy space generated by computer aided design (CAD) software, the position of reference point and camera parameters may be chosen using 3-D software to satisfy the corresponding spatial spectrum perspective.

Positioning of Cameras for Spatial Spectrum Recording

The recording of 3-D information differs from the simple treatment of computer simulated light rays by having to obey complicated rules for each camera–projector unit. The following approaches can be adopted:

As shown in Figure 48, fix the reference point R of the original 3D space O, choose M×N points S_{mn} for each spatial spectrum recording position according to the sampling spatial angle ω_{mn}, mount each camera unit C_{mn} so that the rear nodal point of its lens is at the position S_{mn}, set the imaging axis A_{mn} of each camera unit C_{mn} to coincide with the straight line connecting reference point R and spatial spectrum collecting point S_{mn}, i.e. the reference point R corresponds to the centre of each scene of the image I_{mn} recorded by camera units C_{mn}.

The system, when constructed with M×N camera units C_{mn} according to the above, can be used to record 3-D space information from the inside, if each camera unit is rotated 180°. This setup can then be used for recording 3-D information by an outside viewed image. When such a setup is used for spatial spectrum

Figure 48. Alignment of projectors

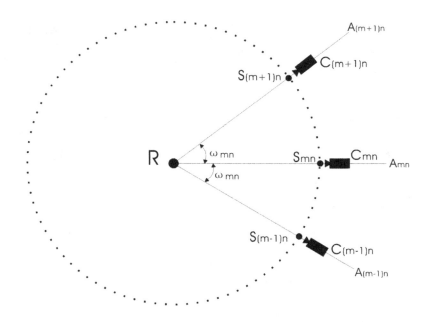

recording, it is necessary to focus each camera unit so that it gives the best resolution of each spatial spectrum.

Positioning of Projectors for Reconstruction of Spatial Spectrum

Having recorded the 3-D information, we now need to reconstruct the image. The M×N individual projector units P_{mn} are positioned as follows: As shown in Figure 48, establish the reference point R' of the reconstructed 3D space O', choose M×N points S'_{mn} of each spatial spectrum projection position, site each projector unit at position S'_{mn} with the lens nodal points in the correct position, and the axis A'_{mn} of each camera unit P_{mn} aligned as described above. The whole system can now be used to reconstruct the 3-D image, if each projector unit is rotated 180° along its imaging axis. The system can also be used

for reconstructing a 3-D space for outside viewing. It will again be necessary to correct for any image distortions.

Choosing the Reference Surface

We need to select the correct reference surface P_R for the reconstruction process and position the holographic functional screen on this surface in order to display the reconstructed spatial information in holographic form (see Figure 50).

Plane Reference Surface

If the reference surface P_R is to be a plane chosen as the projection image plane of the spatial spectrum perspective I'_{mn} as shown in Figure 49, with the exception of the central projector unit P_{mn} all the projector units will produce an image on P_R with keystone distor-

Figure 49. Keystone distortion on a plane reference surface

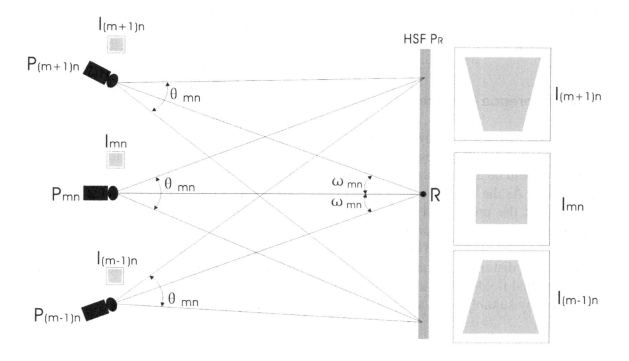

Figure 50. Curved reference surface

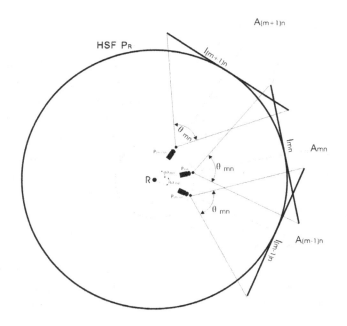

tion. It is necessary to correct this deformation by appropriate software for each projector unit, in advance, to obtain a matched image array; this is important for the setups of Figures 41 and 42.

Curved Reference Surface

For the curved surface P_R shown in Figures 35–40, the requirement is that the curved surface P_R should be a focal surface for each projector unit. As the surface is not flat, it is important that the projector lens apertures are such as to give the appropriate depth of field for the curved surface. There will still be some image distortion towards the edges of the individual fields, and this will have to be corrected by suitable software.

CONCLUSION

Photography has been around for more than 150 years. In earlier times a two-dimensional image was thought of as the last word in realism. Two-image stereoscopy has always been (and remains) on the borderline of gimmickry, at least as far as the general public is concerned. However, the ultimate aim, a communication system using 3-dimensional images was not even contemplated until comparatively recently. Modern optics, including the concept of optical transfer functions and the Fourier approach to imaging, which evolved initially from communications theory, is now in the process of developing into a practical method of display. The concept of a spatial spectrum may herald a modern renaissance in holographic information technology. Some scientific philosophers have even conceived of the whole cosmos in terms of a hologram composed of a spatial and temporal distribution of energy. Light can be defined in terms of mathematical descriptions, the main models being geometrical or ray optics, electromagnetic or wave optics, and the quantum model (photonics). It is the last of these in which the quantized energy probability distribution is defined by Equations 5–8; 5 relates to hoxels and 7 relates to a single spatial spectrum of this nature. We have tamed 2-D image displays by digital treatment of electronic signals and modern optics for digital image processing. It is the time for image information to be treated as a spatial spectrum, and to recover its 3-D information by means of digital vector treatment through hoxels.

Modern display technologies for digital imagery at present ignore the fact that their input is a single spatial spectrum and their output viewing angle is expanded by the individual pixels of the image, i.e. there is a narrow band input and a wide band output of the same spatial spectrum information in terms of colour and brightness. Holography was the first practical technology to recover three-dimensional information of this nature by the interference of coherent light waves. Modern digital holography prints each individual hogel. In this chapter we have made use of the software and hardware resources of modern information technology in the domain of temporal and spatial spectra to reduce the redundancy inherent in conventional holography. The key is to project 3-D displays with sampled spatial spectra in discrete hoxels onto a functional display screen.

As we can now tame the light field into spatial spectra and turn them into hoxels by M×N sets of camera-projector systems, we can realize real-time holographic displays by means of a holographic functional screen with a simple collection of M×N channels. Holography has thus reached a new stage on the way to the realisation of the ultimate dream of full communication in 3-D. We now anticipate a new generation in technology in which the world communicates in three dimensions.

ACKNOWLEDGMENT

We should like to thank the staff of AFC Technology Co. Ltd. for their practical support over some ten years. We also thank Tung H. Jeong, Martin J. Richardson, Dahsung Hsu, and Tianji Wang for their help in useful discussions.

REFERENCES

Benton, S. A. (1969). Hologram reconstruction with extended light sources. *Journal of the Optical Society of America, 59*, 1545.

Bjelkhagen, H. I., & Kostuk, R. K. (2008). Article. *Proceedings of the SPIE, 6912*.

Burckhardt, C. D., & Doherty, E. T. (1969). Beaded plate recording of integral photographs. *Applied Optics, 8*(11), 2329–2331. doi:10.1364/AO.8.002329 PMID:20076020

Caulfield, H. J. (2004). *The art and science of holography: A tribute to Emmett Leith and Yuri Denisyuk*. Bellingham, WA: SPIE Press.

Cross, L. L., & Cross, C. (1992). HoloStories: Reminiscences and a prognostication on holography. *Leonardo, 25*, 421–424. doi:10.2307/1575748

Denisyuk, Y. N. (1962). Photographic reconstruction of the optical properties of an object in its own scattered radiation field. *Soviet Physics, Doklady, 7*, 543–545.

Denisyuk, Y. N. (1963). On the reproduction of the optical properties of an object by the wave field of its scattered radiation. *Opt. Spectrosc. (USSR), 15*, 279–284.

Denisyuk, Y. N. (1965). On the reproduction of the optical properties of an object by the wave field of its scattered radiationII. *Opt. Spectrosc. (USSR), 18*, 152–157.

Fan, F. C., Choi, S., & Jiang, C. C. (2010). Use of spatial spectrum of light to recover three dimensional holographic nature. *Applied Optics, 49*(14), 2676–2685. doi:10.1364/AO.49.002676

Gabor, D. (1948). A new microscopic principle. *Nature, 161*, 77–79. doi:10.1038/161777a0 PMID:18860291

Gabor, D. (1949). Microscopy by reconstructed wavefronts and communication theory. *Proceedings of the Physical Society. Section A, 194*, 454–487.

Goodman, J. W. (1968). *Introduction to Fourier optics*. San Francisco: McGraw-Hill.

Hesselink, L., Orlov, S., & Bashaw, M. C. (2004). Holographic data storage systems. *I to Three Dimensional Displays, 92*, 1231–80.

Huebschman, M. L., Munjuluri, B., & Garner, H. R. (2003). Dynamic holographic 3-D image projection. *Opt. Exp., 11*, 437–445. doi:10.1364/OE.11.000437 PMID:19461750

Huff, L., & Fusek, R. L. (1980). Color holographic stereograms. *Optical Engineering (Redondo Beach, Calif.), 19*, 691–695. doi:10.1117/12.7972589

Ives, F. E. (1903). *U.S. patent no 725567*. Washington, DC: US Patent Office.

Jeong, T. H. (1967). Cylindrical holography and some proposed applications. *Journal of the Optical Society of America, 57*, 1396–1398. doi:10.1364/JOSA.57.001396

Klug, M. A., Halle, M. W., & Hubel, P. M. (1992). Full color ultragrams. *Proceedings of the Society for Photo-Instrumentation Engineers, 1667*, 110–119. doi:10.1117/12.59625

Leith, E. N., & Upatnieks, J. (1962). Reconstructed wavefronts and communication theory. *Journal of the Optical Society of America, 52*(10), 1123–1130. doi:10.1364/JOSA.52.001123

Leith, E. N., & Upatnieks, J. (1963). Wavefront reconstruction with continuous-tone objects. *Journal of the Optical Society of America, 53*(12), 1377–1381. doi:10.1364/JOSA.53.001377

Leith, E. N., & Upatnieks, J. (1964). Wavefront reconstruction with diffused illumination and three-dimensional objects. *Journal of the Optical Society of America*, *54*(11), 1295–1301. doi:10.1364/JOSA.54.001295

Lippmann, M. G. (1891). La photographie des couleurs. *Comptes Rendus Hebdomadaires des Séances de l'Académie des Sciences*, *112*, 274–275.

Lippmann, M. G. (1908). Épreuves reversibles donnant la sensation du relief. *Journal of Physics*, *7*(4), 82–85.

Lohmann, A. W., & Paris, D. (1967). Binary Fraunhofer holograms generated by computer. *Applied Optics*, *6*(10), 1739–1748. doi:10.1364/AO.6.001739 PMID:20062296

Norling, J. A. J. (1953). The stereoscopic art. *SMPTE*, *60*(3), 286–308.

Okoshi, T. (1976). *Three-dimensional imaging techniques*. Oxford, UK: Academic Press.

Richardson, M. (2006). *The prime illusion: Modern holography in the new age of digital media*. London: Holographic Studio.

Shang, X., Fan, F. C., Jiang, C. C., Choi, S., Dou, W., Yu, C., & Xu, D. (2009). Demonstration of a large size real time full color three dimensional display. *Optics Letters*, *34*(24), 3803–3805. doi:10.1364/OL.34.003803 PMID:20016619

Slinger, C. W. (2004). Recent development in computer generated holography: Towards a practical electroholography system for interactive 3D visualization. *Proceedings of the Society for Photo-Instrumentation Engineers*, *5290*, 27–41. doi:10.1117/12.526690

St Hilaire, M., Lucente, P., & Benton, S. A. (1992). Synthetic aperture holography: A novel approach. *Journal of the Optical Society of America. A, Optics and Image Science*, *9*, 1969–1978. doi:10.1364/JOSAA.9.001969

Stetson, K. A., & Powell, R. L. (1966). Hologram interferometry. *Journal of the Optical Society of America*, *56*, 1161–1166. doi:10.1364/JOSA.56.001161

Tay, S. (2008). An updatable holographic three-dimensional display. *Nature*, *451*, 694–698. doi:10.1038/nature06596 PMID:18256667

Zebra Imaging. (2008). Retrieved 24 January 2012 from http:// www.zebraimaging.com

Chapter 9
The Cultural Landscape of Three-Dimensional Imaging

Sean F. Johnston
University of Glasgow, UK

ABSTRACT

This chapter explores the cultural contexts in which three-dimensional imaging has been developed, disseminated, and employed. It surveys the diverse technologies and intellectual domains that have contributed to spatial imaging and argues that it is an important example of an interdisciplinary field. Over the past century-and-a-half, specialists from distinct fields have devised explanations and systems for the experience of 3-D imagery. Successive audiences have found these visual experiences compelling, adapting quickly to new technical possibilities and seeking new ones. These complementary interests, and their distinct perspectives, have co-evolved in lock-step. A driver for this evolution is visual culture, which has grown to value and demand the spectacular. As a result, professional and popular engagements with 3-D have had periods of both popularity and indifference, and cultural consensus has proven to be ephemeral.

DOI: 10.4018/978-1-4666-4932-3.ch009

INTRODUCTION

As the preceding chapters illustrate, the field of 3-D imaging is one that has encouraged a diversity of approaches. Physicists and optical engineers have focused on the properties of light to understand how three-dimensionality can be modelled and reproduced optically. Physiologists have studied binocular vision to reveal how the eyes and brain achieve the miracle of stereopsis. Artists have been intrigued by this curious dimension of imaging that falls between sculpture and painting. And, historical accounts provide a further axis – time – that solidifies these discrete perspectives, placing them in a temporal order that hints at causes and effects.

Each of these perspectives frames the field uniquely, but linking them are the threads of culture. It is generally appreciated that artists and their works flourish or fade in particular cultural environments, but we often overlook the cultural dimensions of science and technology. While scientific attention, for example, may occasionally be prompted by a 'breakthrough' – a sudden revelation triggered by a newly discovered fact –more often than not is inspired by a pre-existing influences. Scientific networks, current topics of discussion and intellectual fashions all play a role in determining where attention is focused and how problems are pursued. Similarly, engineering development may be motivated by identification of a market need, available resources or a particular constellation of skill-sets. The successful marshalling of these factors often requires a particular cultural context. The attitudes and understandings of peers and competitors, purchasers and critics, shape the trajectory of technologies and their social uses.

As this hints, any one of these perspectives – scientific, technological, artistic, economic – may restrict vision, too, in just the way that the Victorian stereoscope provides a sense of reality that is nevertheless unable to view the parts of the scene masked by others.

There are other disciplines, too, that have engaged periodically with the attractions of light and vision. Some of them have had disputed intellectual borders. The link between the eye and brain has both brought together and divided physiologists and psychologists since the turn of the twentieth century (Johnston 2001). The mid-century origins of holography forced together optical specialists and radar engineers, merging the fields of physical optics and communication theory in the process (Johnston, 2006). And the technical subject of imaging has been partitioned through the past century first by the technologies of photographic film and electronic sensors, and later by equally deep analogue-versus-digital divides.

Such collisions of intellectual worlds are at the heart of three-dimensional imaging. The field straddles disciplines, and this uncomfortable position helps explain why it has periodically inhabited a hinterland. Which experts can provide an authoritative voice, guide its path and predict its future? In practice, 3-D imaging has required a melding of expertise, drawing together knowledge, innovation and forecasting. The field is interdisciplinary: more than the sum of its parts, its successful branches have merged disciplinary insights into shared practice. These ongoing processes provide a rich seam for social scientists and scholars of the humanities to explore.

But beyond professional practices, the culture of 3-D imaging has been shaped by its adopters and non-adopters, too. Popular culture, especially since the invention of photography, has been expressed increasingly as a visual culture. The succession of imaging technologies has revealed the role of visual spectacle, and public appetite for new forms of visual surprise. Photography, graphic arts, cinema and television buffeted and contributed nuances to popular visual culture, and popular uptake (and sometimes rejection) was driven by those currents.

This chapter, then, traces the cultural dimensions – professional, popular, and visual – of three-dimensional imaging technologies. In the process, it hints at the integrative vision of cultural studies of imaging, an approach explored elsewhere in greater depth (Johnston, 2015). 3-D imaging and popular culture developed together and shaped each other. Understanding the wider implications of this co-evolution is important for riding the succeeding waves of our technological culture.

BACKGROUND: THE ROLE OF CULTURE

Culture is often overlooked in scientific and technical fields, but much less frequently in overtly artistic and creative fields. 3-D imaging involves all of these, and so a comprehensive appreciation of the field demands a universal glue to bind its components together.

Such a glue is cultural studies. The vague term 'culture' incorporates dimensions that seldom are made explicit in some disciplines. These include notions such as collective belief, community values, popular fashion, customs, philosophical or religious convictions, shared aesthetic sensibilities and forms of creative expression such as music and art.

The relevance of such concepts may not always be obvious, but they can enrich understandings of human activities. This may apply as much to disciplinary experts as to members of the wider public. As noted in the Foreword, physicist Stephen Benton suggested that new technologies evolve inexorably, but historians of technology can find many counter-examples. Indeed, three-dimensional imaging technologies, as a class, have had mixed fortunes. Some achieved significant 'market penetration', to use a phrase popular with economists, but others have been dismissed by consumers with scarcely a second glance. Stereoscopes are an example of a highly popular technology for Victorian audiences, but increasingly dismissed by their children and grandchildren. 3-D cinema has periodically mushroomed in popularity, only to decay again within a decade. And, with the growing union between electronics, computing and optics since the 1980s, a variety of sophisticated imaging technologies have enthused their engineering proponents but sometimes made relatively little impact beyond their specialist communities. Virtual reality headsets, for example, were a flash in the pan during the 1990s. Recognising the historical inevitability of 3-D imaging, then, may be more a matter of philosophical confidence or collective belief.

Explaining popular failures can also be susceptible to myopia. Some of them, of course, have clear technical limitations, as noted in the Introduction. Imaging systems that do not provide all the attributes of stereopsis – or, even worse, force them to compete within our visual system – are likely to be found wanting in some circumstances. But the rise and fall of such technologies reveals the shifting interplay between collective judgement and personal experience. For a time, those technologies were appealing, seductive, exciting – only to be rejected later when they no longer mapped onto other criteria drawn from prevailing fashions and cultural desires. Such technologies did not, or did not for long, resonate with wider community values. Identifying those values may be difficult to discover or demonstrate, though. For example, we may muse as to whether the decline of virtual reality (VR) systems linked in part to physical isolation from others – making the synthetic experience, even when combined with virtual communication with other real players, too different from the conventions of 1990s daily life. Or was it in part because VR became too closely associated with particular gaming cultures, celebrity endorsers or the palette of products from a particular company? Such potential explanations would require considerable work to investigate, but could inform the evaluations made by the next generation of technology companies seeking to revive or improve upon the concept.

The cultural studies approach, admittedly, has some difficulties from the standpoint of science and technology. Its claims may be difficult to quantify: how do we measure the impact of a new idea, for example? One approach is to measure what can be measured – e.g. the number of stereoscopes sold, or the attendance figures for hologram exhibitions – and to make cautious hypotheses beyond them. An additional method is to exemplify, providing cogent anecdotes that characterize community views or popular attitudes. Often focusing on popular culture – mass media, fashion, new technologies and evolving customs – cultural studies can probe deeper societal structures, or alternatively give a perceptive but unfamiliar characterization of changing ways of life. This discipline-crossing approach can be both fertile and disorienting, encouraging the re-examination of our shared perceptions and beliefs. But some have argued that with its rising popularity has come a shift away from analytical precision to mere entertainment for uncritical audiences (Pihlainen, 2012).

Given these limitations and critical approach, the wide attentions of cultural studies may provide a discordant note in some discussions about technology. A few technologies have become polarized by competing views, e.g. biologists pursuing genetic engineering, and physicists dedicated to nuclear power, on the one hand, versus wary publics on the other. These confrontations illustrate more than merely 'competent' and 'incompetent' positions: they represent jarring *cultures* with competing philosophies, norms and behaviours. By contrast, the field of 3-D imaging, a collection of disparate interests having a long history of cooperation, can be better understood and further reconciled by attention to the cultural dimension.

PROFESSIONAL CULTURES

Making sense of the dynamics of three-dimensional imaging technologies – their invention, improvement, promotion and popularisation – is aided by appreciating the distinct working cultures in which they developed.

As discussed in Chapter 1, 3-D imaging exploded in the mid-nineteenth century as a novel variant of photography. Stereoscopy as a science became dominated by scholars just then becoming differentiated as 'physicists'. They adopted new techniques of visual experimentation to better explore, demonstrate and debate stereoscopic vision (Crary 1990). In some cases they adapted their apparatus into practical viewing devices, exhibited them to wider publics, and promoted applications.

Within a decade, this rapidly emerging technology was shared between scientists and commercial photographers. Photographers themselves included a spectrum of expertise from chemists (dabbling empirically in a proliferating variety of exposure and development processes) to mechanical inventors (devising ganged shutters, clever camera bodies and shifting tripods) to optical scientists (calculating improved lens designs and stereoscopic viewers). These cultures rapidly diverged and stabilized. They stratified expertise by defining the criteria that united the different clans of practitioners.

Professional standards clarified their art. Through organisations such as the Royal Photographic Society and periodicals such as *The Photographic Journal* (both founded under different names in 1853), photography became more scientific. Recuperated from amateurs and entrepreneurs, these bodies sought to base photographic techniques more explicitly on chemistry and physics. By the end of the century, company research labs were beginning to emerge, professionalising these new working criteria. Zeiss Optical Works in Jena, Germany, employed physicists not merely to design lenses but to probe the theoretical limits of resolution; at chemical companies such as Gaskell-Deacon works in Lancashire, laws of photographic exposure were teased out; new, dedicated corporate labs sprouted, including those founded by British Thomson-Houston (1894), Du Pont de Nemours (USA, 1903), the National Electric Lamp Association ('NELA', USA, 1908) and Eastman Kodak (1912). National laboratories followed the firms, with the Physikalisch-Technische Reichsanstalt (Germany, 1887), the National Physical Laboratory (UK, 1889) and Bureau of Standards (USA, 1901) each setting up optics divisions.

This new clustering of societies, companies and institutions collectively defined the nature of optical science and industry at the turn of the twentieth century. Within each was a distinct mixture of technical expertise. Kodak combined divisions of chemistry, optics and mechanical engineering; NELA was top-heavy with physicists and electrical engineers and a smattering of opthalmologists and architects. And scientists at the national laboratories, striving to set standards for 'normal' human vision between the world wars, found themselves uncomfortably often face to face with psychologists (Johnston 2001).

Acknowledging the flavours of expertise, stewing together to cook unique working cultures, can help to explain why some imaging technologies prospered or faltered. Gabriel Lippmann's integral photography, discussed in Chapter 8, is a case in point. Lippmann was the doyen of scientific photography. His scheme for three-dimensional imaging, like his concept of interference colour photography (a precursor to what eventually became holography), employed sophisticated physics and careful experimental verifications. Yet his two contributions to the field of photography made little sustained impact on his contemporaries. His peers were impressed by his scientific insights, but did relatively little to extend them either theoretically or experimentally. Contemporary French *physicists* criticized Lippmann's arbitrariness in adopting *chemical* processes; for these technical experts, his interferential photography (Lippmann 1891) was neither, as supporters claimed, 'direct photography', 'natural colour' nor 'objective colour reproduction'. Instead of being an absolute scientific technique, it was just as capricious and subjective as previous methods of generating colour photographs were (Mitchell 2012). The seeming scientific rigour, they suggested, was in fact a fudge.

POPULAR CULTURE AND SPECTACULAR IMAGERY

Popular culture is at least as important in assessing new technologies as professional cultures are. In the case of Gabriel Lippmann's photography, expert criticisms about theoretical elegance were relatively unimportant; what mattered more was popular judgements about the practicality of Lippmann's colour technique. 'Practicality' had a shifting definition, though. Lippmann's system, while eventually meriting the 1908 Nobel Prize, failed to satisfy the reasonable expectations of contemporary culture. His emulsions were much slower than panchromatic emulsions of the time, meaning that still-lifes or posed portraits of Daguerrean rigour were the only option for Lippmann photographic enthusiasts. These were not technical objections alone; after all, photographic processes have always constrained the nature of the images they can record. Fifty years earlier, Louis Daguerre's results had unlocked doors, exciting audiences having few expectations and liberating them to explore the novel applications of photographs. What made colour processes, and especially Lippmann's technique, unacceptable was that they mapped so poorly onto contemporary cultural requirements. In an age being transformed by mechanization and speed, time-exposures were no longer a meaningful representation of reality.

The rising scientific culture was paralleled by a growing popular engagement with imaging technologies. The rise of photography provided a cultural framework for stereoscopes, and these in turn underlay the popular understandings that assessed 3-D motion pictures and holograms decades later. At least four routes contributed to the diffusion: the incorporation of photographs into social life and the home; the uptake of amateur photography; the availability of new technical processes for image reproduction and the consequent expansion of imaging in print media; and, a proliferating variety of images for mass consumption. Together, these contributions promoted a rising visual literacy and appetite for novelty among wider publics.

Over the first fifty years after its invention, the consumption of photographic imagery took place principally in the home. Audiences quickly grew accustomed to the new portraiture and incorporated it into their home lives. The late Victorian parlour increasingly displayed framed family photographs. The rise of the stereoscope, however, transformed the consumption of photographic images. The scientific experiments of the late 1830s blossomed with the Great Exhibition of 1851 and commercial availability soon afterward. The stereoscope successfully combined the appeal of viewing family photographs at home with other forms of home entertainment and education. It broadened the content and transformed the parlour into an outward-looking social environment. A stereo slide collection was akin to a piano or small library; it could be shared to enliven small gatherings, but could also provide private entertainment and even independent study of a newly visible wide world. This novel visual experience shaped viewers' engagement with photography during the second half of the nineteenth century.

The flip side of these expanding cultural uses was that the viewing public became increasingly sensitized to new visual surprises. This was an age of visual tricks, when 'scientific optics' could reveal baffling visual effects for entertainment; the creators and promoters of stereoscopes also introduced optical toys such as the kaleidoscope, zoetrope and magic lantern. The appeal of visual surprise was also behind the international commercial market for stereograms. The unfamiliarity of far-away places proved captivating, and photographers discovered an exploding international market for their easily-transported views. Stereogram scenes also relied increasingly upon episodic novelty and wonder. Current events were a fertile source of views; ceremonies such as the Japanese emperor's birthday, for instance, could reveal unfamiliar cultural practices to curious viewers. Technological accidents such as train wrecks became an important sub-genre (Figure 1a). The aftermath of conflagrations, earthquakes and typhoons proved popular, and military scenes had a similar vicarious fascination (Figure 1b). Such visual interests long preceded the public experiences of the Vietnam War through television, and the rise of internet sites devoted to the *schadenfreude* of unpublishable graphic images.

Thus spectacle, surprise and satiation combined in stereoscope views. But a crucial and unique feature was the visual depth of the images themselves: their stereoscopic nature was central to their appeal. A comparable popular uptake had not occurred with regular photographs (and would not happen until cheap reproduction became possible).

Viewers quickly discovered that stereoscopy was immersive. It required the viewer to look into a device that closed off the field of view and replaced their visual world with a new, detailed and deep one. The most popular versions, constructed from steam-shaped wood veneers, form-fitted the face like a scuba mask and focused attention just as the blinkers of a carriage horse did. Users could examine and savour fascinating scenes with a directed attention and a sense of safe and private – almost voyeuristic – participation.

Figure 1. (a) 'The Great Eastern in the Stereoscope', London Stereoscope Co., c1859-60; (b) 'The Wounded Sentinel, Manchuria', W. B. Kilburn, 1905

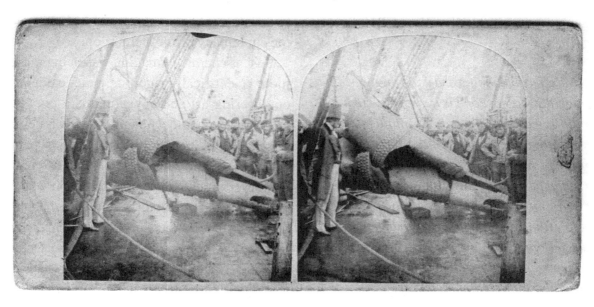

(a)

(b)

And even monochrome or hand-tinted stereo views provided a sense of realism superior to other visual media. As Oliver Wendell Holmes described,

The first effect of looking at a good photograph through the stereoscope is a surprise such as no painting ever produced. The mind feels its way into the very depths of the picture. The scraggy branches of a tree in the foreground run out at us as if they would scratch our eyes out. The elbow of a figure stands forth so as to make us almost uncomfortable. Then there is such a frightful amount of detail, that we have the same sense of infinite complexity which Nature gives us. (Holmes, 1859)

The experience could be sublime, enthralling and visceral. Views of mountain-scapes, tall buildings and monuments employed that perspective to impress upon viewers a sense of height or grand vistas. Given the vividness and depth of imagery, a market for erotic images developed steadily. More subtly, three-dimensionality encouraged depth of interpretation as the eye roamed from foreground details to their wider contexts.

Stereoscopes, then, offered rich opportunities for visual entertainment and education. Their popularity was strong from the 1850s up to the turn of the century as new content drew in new audiences. The last technological innovation to sustain markets was the introduction of printing processes that could reproduce photographic images, offering lower costs and the introduction of reliable colour. The declining appeal of stereoscopes during the Edwardian era, however, illustrates the fac-

tors behind declining popular appeal. Visual interest proved difficult to sustain into the twentieth century as new media challenged the supremacy of the stereoscope.

Just as chemical imaging was being made more scientifically rigorous, practical as an engineering art and appealing as a popular medium, *mechanical* imaging, too, was being improved systematically. Since the European adoption of movable type in the fifteenth century, printed images had relied on wood-cut, engraved metal or, by the eighteenth century, etched stone surfaces (lithographs). The new *halftone* screen process, perfected by the early 1890s, allowed publishers to print photographic images having a full range of grey tones on the same sheet as set type. Just as importantly, the technique mechanized image reproduction. Expert fine engravers were replaced by artisans adopting more readily-mastered skills. The cost and time needed to reproduce images plummeted, and their content moved from artistic to photographic. The invention of the halftone process consequently produced a step-change in public access to photographs. Instead of being limited to a small collection of personal tintype portraits or handmade stereoscope views, fin de siècle audiences could view photographic images in a growing range of newspapers, magazines and cheap stereo views. Print media began a cultural transition towards photographic imagery (Phillips 1995).

Picture periodicals consequently exploded from the turn of the century, offering the consumption of imagery more cheaply and quickly. The falling price of stereo views themselves also arguably made them less appealing.

Just as the introduction of encyclopaedias on CD-ROM and later free internet sites devalued the appeal of printed encyclopaedias, cheaply printed stereo views may have made them too ubiquitous, eroding the sense of exclusiveness felt by middle-class audiences two generations earlier. As the cycles of appeal of twenty-first century products reveal, mass markets often begin with elite purchasers and end when everyone can, and does, have one. Competition, too, played a role. Other technologies provided audiences with new pastimes to replace the stereoscope's appeal. By the First World War, stereoscopy was no longer exhilarating. Visual thrills required constant stimulation, and the stereoscope had used up its stock of visual surprise and technical innovation.

EXTENDING VISUAL CULTURE

Combined with inexpensive reproduction in periodicals, photography transformed visual experience for twentieth-century publics. Ideas introduced by a handful of artists spread internationally with unprecedented speed. Among the most influential was Alexander Rodchenko (1891-1956), the seminal creator of artworks, photographs and graphic art in the early Soviet Union. His contributions to photography were arguably his most influential work. Rodchenko and his peers pioneered ideas that were rapidly taken up in graphic arts around the world.

The first seminal idea was Rodchenko's exploration of unusual perspectives. Indeed, his work challenged the turn-of-the-century pictorialist movement, which sought to turn the craft of photography into an art form modelled on the aesthetic principles of drawing and painting. Artistic notions of perspective, composition and portrayal of reflective mood had become central to its practitioners.

In some respects, camera design until the first decade of the twentieth century had also constrained visual ideas. This, too, was a cultural product, not entirely a deterministic effect of the technology itself. Camera designers sought faithful reproduction of conventional scenes, and their criteria were shaped by contemporary art and urban culture. As a result, serious photographers had been materially encumbered and perceptually channelled by large-format, tripod-mounted plate cameras. The unwieldy hardware was designed to reproduce the horizontal view of human eyes and the traditional perspective of paintings. The designs of equipment consequently favoured these conventions. Pointing upwards or downwards at steep angles was discouraged by problems of camera stability for the long exposures typically used and by the difficulty of loading photographic plates and viewing the ground-glass to focus and compose the picture. As a result, good photographic images aimed to be horizontal, well-focused and carefully posed. Stereo slides had surprised viewers with their three-dimensional content but less frequently by their vertiginous perspectives.

Rodchenko, by contrast, adopted compact, hand-held cameras and pointed them in unfamiliar directions. This was both a technological and cultural shift. To gaze sharply upward or downward was (and still is, in some settings) an unusual social act. Familiar to children, it is culturally discouraged for adults, for whom such views may be interpreted as threatening, provocative or naïve. To stare upwards at city high-rise towers is often the mark of a tourist; to peer down from a high vantage point may be labelled as furtive or voyeuristic. And gazing closely at people in public – especially when recorded candidly via a camera for subsequent examination – can be seen as impolite and intimidating. The unconventional pointing of a camera, then, could recover childlike dimensions and fresh perspectives, both visual and social.

Rodchenko's photographs defamiliarized imagery. The viewer was challenged to recognize and accommodate the odd perspectives. The result could be liberating, revealing unnoticed details and viewpoints, or even frightening. His graphic technique was a new form of spectacle that forced viewers to appreciate the social and even political dimensions of images. It contrasted the introspective tranquillity of pictorialism with an aggressive celebration of the modern social world. Realistic imagery could now be unnerving and even shocking (Tupitsyn, 1996) (see Figure 2).

Rodchenko's avant-garde style of spectacular imagery – especially when applied to industrial subjects – illustrates historian David Nye's notion of the 'technological sublime' (Nye, 1994). The sublime is an experience that transcends the ordinary, often associated with religious revelation, and characteristically evoking exaltation, wonder and fear. The technological sublime is such an experience instantiated in a modern technology or, indeed, imagery that highlights it. For Victorians, stereoscopes had provided that thrill. Rodchenko's graphic arts stretched visual culture further.

Figure 2. Alexander Rodchenko, "Pioneer with a Horn," 1930

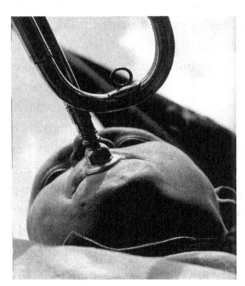

The second unsettling technique promoted by Rodchenko and his peers – photomontage – used photographs as graphic elements in ways that again challenged three-dimensional perception. Borrowing from Cubist art, the combination of visual elements combined jarring perspectives to confuse viewers' sense of depth and point of view. Their creators argued that, in photomontages, the photograph could shock the viewer much more effectively than artistic creations could, because its documentary quality emphasized the reality and currency of a visual event while emphasizing their unfamiliarity – a quality sensed in earlier stereo views. And like later holograms, the photomontage combined disturbing realism with the jarring perception of multiple perspectives.

The new field of photojournalism embraced these graphic approaches and technological tools. Magazines between the world wars became flooded with images, and cross-talk between them generated increasingly dramatic and breath-taking views that sought to surprise and unsettle their viewers. Soviet magazines such as *Kino-Fot* (USSR, 1922) and *LEF* (USSR, 1923) were among the first, quickly followed by *Broom* (Italy, 1924) and *Arbeiter-Illustrierte-Zeitung* (AIZ, Germany, 1924). A growing number of topical magazines such as *Vu* (France, 1928), *Regards* (France, 1932), *Life* (USA, 1936), *Look* (USA, 1937) and *Picture Post* (UK, 1938) further popularized these visual themes. And what had begun as an explicit movement in art and a medium for political education continued in the west to inform photojournalism, advertising and popular culture.

The market for such magazines and visual spectacle expanded rapidly through popular culture. Examples filled wide-circulation magazines aimed consciously at 'everyman' and 'everywoman'. During the 1930s and 1940s, magazine readers were presented with disorienting or surprising photographs, some of which became regular features. The surprise could be elicited not just by viewing the world from unconventional angles, but also by viewing the ultra-small or invisible (Figure 3a). Amateurs were similarly encouraged to add excitement to their photographs by tricks made possible with optical and mechanical tinkering (Fig 3b).

In a sense, these three levels of spectacular imagery – ranging from early Soviet photographic art as propaganda, to photojournal-

Figure 3. (a) Disorienting close-up views, a popular genre during the 1930s and 40s (Korth, 1944), (b) kaleidoscopic pictures at home (Uncredited Photographer, 1940)

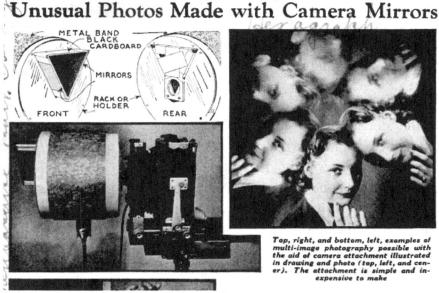

(a)   (b)

istic representations of sublime technologies, and finally to amateur close-up photography – were equally radical. They employed bold new perspectives to deconstruct conventions and to instil new political, economic and perceptual orientations, respectively. Audiences that initially were discomfited were ultimately seduced.

Professional cultures also found themselves adopting the spectacular visuals. Postwar technical magazines employed visual surprise to communicate the power of modernity itself (Figure 4). Illustrations of high-technology visual sciences also frequently featured, with stories on microscopy, stroboscopes and stereoscopy in science. Indeed, striking images offered an apt analogy for the editorialists' aims:

The Technology Review is designed to be an editorial stereoscope for presenting Science's new world picture in the startling clarity of relief. It goes beyond the mere reporting of science and engineering news in plano; it adds the third dimension of interpretation. (Rowlands & Killian, 1937)

Education had been an important theme for the stereoscope from the 1850s, and it sustained it through the early twentieth century: stereo views could tempt students to learn. Visual education was promoted by educationalists during the 1920s, and maps, illustrations, diagrams, graphs, projected images and stereo views were marshalled as their tools; stereo slide publishers aggressively promoted the use of stereoscopes in schools (Bak, 2012). As one teacher argued:

Figure 4. Technical magazines adopting visual surprise: (a) an MIT analog computer (Uncredited Photographer, 1937), (b) a Bell Telephone acoustic laboratory (Uncredited Photographer, 1947)

(a)

(b)

The finest service yet rendered in the school-room has been done by the stereograph. The photograph presents but two dimensions. But the stereocamera and the stereoscope work a miracle. They supply the actuality of binocular vision, and the third dimension is presented to the eye in vivid reality. The person who looks through the stereoscope looks upon the real mountain, looks into the depths of the real canyon, looks upon the actual statue, the actual cathedral… It becomes a game to see who can stand and report in good English what he saw, looking through the window of the stereoscope into the reality beyond. (Good, 1922)

The tactical value of stereo views for wartime education and photo-interpretation were also promoted by stereoscope manufacturers (Figure 5), but children of the 1940s became the principal users of stereoscopes through a revitalized version of the technology. The View-Master, a plastic binocular viewer for stereo view disks that incorporated seven pairs of colour transparencies, was introduced at the New York World's Fair in 1939 and, like the Great Exhibition eighty-eight years earlier, it generated a sales bubble. The vibrant colours, bright images and simple construction made it an appealing children's toy, and parents were reintroduced to stereoscopy as entertainment after a two-generation decline. The today-forgotten visual thrill of seeing bright colour slides – only possible after the marketing of Kodak and Agfa colour films from the late 1930s – was spectacular in just the way that Victorian optical toys, early cinema and Rodchenko's perspectives had

been. Combined with stereoscopy, they captured new audiences. A postwar fashion for amateur stereo photography expanded through the early 1950s particularly in the USA. The Stereo Realist camera, introduced in 1947, triggered competition to produce modern stereo cameras and viewers expanded again as it had first done in Europe at the beginning of the century (Figure 6).

In their own way, motion picture technologies sought to attract audiences by new forms of visual spectacle. Motion itself had proved to be shocking in the earliest public showings. The Lumière brothers' 1896 film of an approaching steam locomotive famously startled audiences unaccustomed to entertainments that threatened to leave the stage. Cinematic innovations focused on providing a more captivating approximation of reality. Sound reproduction, especially when synchronized with the moving image to allow dialogue, transformed the nature of cinema. Colour processes, exhibited experimentally from the turn of the century, sought to transform the monochromatic image into something more lifelike (although the earliest hand-tinted frames could accentuate the visual surprise by showing red fire or unexpected blue below-the-sea views). During the 1930s practical (but relatively expensive) full-colour processes were brought to large audiences, again with the intention of boosting audiences by provoking visual excitement.

Further immersion of viewers within the image could also generate surprise. Wide-screen formats were trialled from the 1920s, touted as a new and natural cinematic experience because it more closely mimicked the

Figure 5. From students to soldiers: stereoscopes as educational technology (Keystone Company, 1937, 1942)

September, 1942   Page 275

The Stereoscope Goes To War

In the August 17 issue of "Life" there is a full-page picture of a second lieutenant looking through a stereoscopic device, and underneath is the following legend: "G-2 STUDENT GETS INFORMATION ABOUT THE ENEMY BY READING AERIAL PHOTOGRAPHS THROUGH AN ARMY STEREOSCOPIC DEVICE."

Keystone stereoscopic vision testing and training equipment, likewise, plays a large part in the selection of men for specialized work in both the Army and in the Navy.

If the use of the stereoscope and stereoscopic pictures is so important to the efficiency of the Army and the Navy and to the successful conduct of the war, it seems hardly necessary to emphasize its importance in the education of our children for their future—whether that future involves war or peace. With its impressive elements of reality, the stereograph brings both to the soldier and to the student factual information that can be obtained in no other way.

With these facts in mind, many schools are buying the Keystone units of stereographs in the social studies and in elementary science.

Each unit covers a limited and specific field of subject matter, and has been prepared, and provided with a teacher's manual, by an outstanding leader in that particular branch of education.

Why not take advantage of this opportunity to equip your schools and your teachers with this valuable teaching aid?

Keystone View Company
Meadville, Penna.

Figure 6. (a) Stereoscopy for merchandising (Sawyer's Inc, 1949), (b) Fred Astaire promotes amateur stereo photography (David White Co, 1950)

(a)

(b)

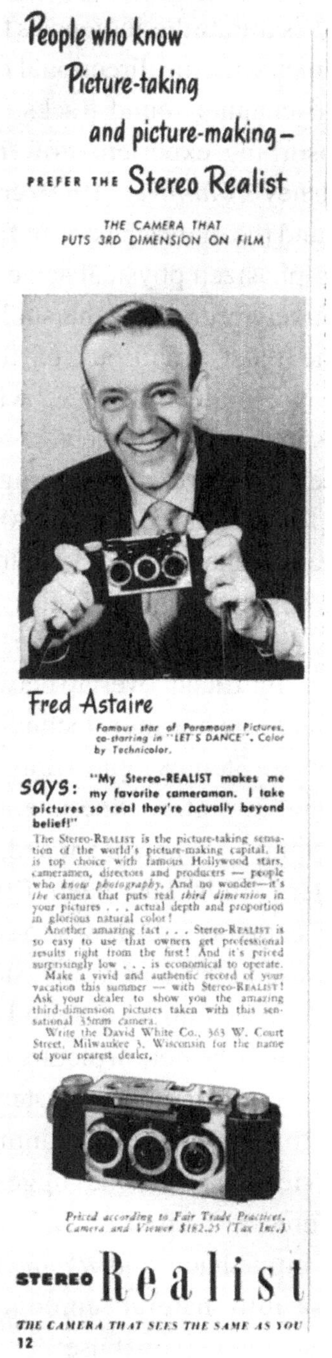

field of view of two eyes. As film producers sought to compete against the looming threat of television during the 1940s, a variety of wide-screen processes and screen geometries were developed to saturate the audience's field of view, sometimes with the directional cues of two- or multi-channel sound tracks. The increasingly absorbing experience of films such as Walt Disney's *Fantasia* (1940) could be spectacular, and the more successful films of the period emphasized physical scale and grandeur, effectively recreating the sublime experience. The most ambitious of these technologies was Cinerama in 1952, which relied on three synchronized projectors to illuminate a screen that covered a viewing arc of 146 degrees for audiences of up to 800.

Motion pictures rehearsed some cultural desires of earlier imaging technologies, and 3-D processes accentuated them further. There had been a significant overlap between early cinema and the parlour technology of the stereoscope. Stereoscopic films had been created and exhibited experimentally since the beginning of cinema; indeed, the Lumière Brothers' train had been filmed with two 35mm cameras and shown to some audiences via two-colour (anaglyphic) stereo glasses. In 1912, an executive of Selig Polyscope Company, an early motion picture firm, mused about the future for cinema. Instead of being relegated to boardwalk entertainments and make-shift viewing rooms, he suggested, the motion pictures of the future would have colour, sound, immediacy and *depth*; they would provide a fully natural reproduction of reality amenable to personal use:

The relief attachment… makes the picture stand out from the screen in a bold, natural relief, instead of being flat. It is the same effect, in short, that our brothers of yesterday secured when they looked at the double picture through what was termed a stereoscope. (Twist, 1912)

For postwar Hollywood, however, 3D movies promised further spectacle and competitive distancing from increasingly familiar and mundane home entertainments. In the same year that Cinerama was debuted, an unexpectedly popular film, *Bwana Devil*, launched a craze in American movie theatres for stereoscopic movies. In 1953 over thirty 3D movies, employing more than a dozen commercial processes, were shown on American screens. And, as with earlier innovations in photography and cinema, amateurs rapidly began to experiment with the new techniques.

Just as printing processes for photographs had contributed to advertising and supporting educational uses of stereo views, motion pictures again provided an impetus to reconsider the older technology. 3-D movies consolidated popular interest in three-dimensional entertainment, and made stereoscopy for adults briefly acceptable again. Zealous promoters could claim that stereoscopic cinema was an inevitable technology that satisfied not just cultural, but universal human, needs. One paper published at the height of the stereo boom outlined therapeutic benefits from viewing stereoscopic motion pictures by its potent stimulation of good binocular vision (Sherman, 1953) (see Figure 7).

Figure 7. Hollywood campaign for '3-D' cinema, 1953

THREE-DIMENSIONAL *lobby display piece designed to promote "House of Wax."*

The cinematic processes, alternately dubbed '3-D', 'Tri-Dee', 'Future Dimension', 'Tru-Stereo' or 'Natural Vision' by their various adopters, shaped the content to the technology and its potential for shock. Where widescreen films sought to awe viewers with sublime airborne travelogues and 'pictorial spectacles—without any attempt at story-telling', as a *New York Times* reviewer put it, 3-D movies tried to surprise and jolt their audiences. The shock could come from the plots, but even more powerfully from the moving objects that seemed to travel up to, or beyond, the viewers' seats. Many mirrored *Bwana Devil*, which featured man-eating (and leaping) lions and the near-tactile experience

of romantic interludes: *House of Wax* contrasted the thrill of a three-dimensional fight with a comically deep Ping-Pong game; and *Dial M for Murder* had audiences avoiding a lunging hand. All of them sought to disorient and involve their viewers in ways that widescreen motion pictures and stereoscopes could not.

Yet the fashion for 3-D movies faltered as audiences first experienced, and then rapidly acclimatized to, the novel visual spectacles. The peak in 1953 was followed by only two releases in 1956. Subsequent booms, about a generation apart, were more modest. The sophisticated widescreen and three-dimensional systems were invariably simplified to improve operation, reliability and profits. In each case, what could be called 'spectacle fatigue' quickly overcame audiences.

CONCLUSION

The environment for 3-D imaging from the 1960s, then, was conditioned by the events of the previous century. A wide variety of technologies had trained experts and conditioned audiences to refine their visual tastes and to enjoy, and expect, new forms of visual surprise. Awe-inspiring scenes, unexpected motion and the surprise of colour vied successively for attention, but the imaging of visual depth repeatedly attracted waves of viewers. Viewed from this perspective, the popularization of holograms looks familiar. Its promotion between 1964 and the early 1970s focused on the bewildering aspects that unsettled audiences: reliance on a mysteriously speckly laser beam; the immaterial image hanging in space; the hologram's ability to generate the appearance of solidity from a featureless flat plate; and, its ability to reveal mysterious scenes in impossible spaces. These compelling magicians' tricks, much like the stereoscopes that so fascinated Victorians, updated the experience for new, more discerning, audiences.

Subsequent holograms mixed some of those earlier technologies back in. The few seconds of motion created by integral holograms made the experience of viewing them much like the earlier zoetrope, but augmented by the sense of image depth. Holographic portraits were startlingly –and even uncomfortably – lifelike, producing a response from 1980s audiences probably similar to those of the 1850s seeing their first framed family photograph. The intricacy, shimmering colour and changing imagery of the first hologram identity cards entranced their first viewers as Victorian optical toys had done, but with far more sophisticated properties to discover.

Each of these experiences separated by time and space was, at least for awhile, exciting, novel and precisely suited to extend the visual grammar of the times.

This examination of imaging technologies suggests that 3-D imaging is a perennial human fascination, but that its requirements vary with cultural context. An awareness of this wider environment seems crucial for twenty-first century imaging research, technological innovation, business launches and corporate investment. Keeping the visual experience fresh and perpetually exciting appears a necessary consideration. Since stereoscopy became culturally important in the 1850s, successive 3-D imaging technologies have impressed their publics either by being technologically advanced, or alternatively by being unexpected, intriguing or surprising. Three-dimensional imaging, like other forms of magic and art, repeatedly adapts itself to the cultural environments that it inhabits.

REFERENCES

Bak, M. A. (2012). Democracy and discipline: Object lessons and the stereoscope in American education, 1870–1920. *Early Popular Visual Culture*, *10*, 147–167. doi:10.1080/17460654.2012.664746

Crary, J. (1990). *Techniques of the observer: On vision and modernity in the nineteenth century*. Cambridge, MA: MIT Press.

David White Co. (1950). People who know picture-taking and picture-making prefer the stereo realist. *Popular Photography*, *26*(1), 12.

Earl, E. W. (Ed.). (1979). *Points of view: The stereograph in America – A cultural history*. New York: Visual Studies Workshop Press.

Good, J. P. (1920). The scope and outlook of visual education. *School Science and Mathematics*, *20*, 82–87.

Holmes, O. W. (1859, June 1). The stereoscope and the stereograph. *Atlantic Monthly*.

Johnston, S. F. (2001). *A history of light and colour measurement: Science in the shadows*. New York: Taylor and Francis. doi:10.1887/0750307544

Johnston, S. F. (2006). *Holographic visions: A history of new science*. Oxford, UK: Oxford University Press. doi:10.1093/acprof:oso/9780198571223.001.0001

Johnston, S. F. (2015). *Holograms: A cultural history*. Oxford, UK: Oxford University Press.

Keystone View Company. (1942). The stereoscope goes to war. *The Educational Screen: The Magazine Devoted Exclusively to the Visual Idea in Education*, *16*, 275.

Korth, F. G. (1944). Take it close up. *Popular Mechanics*, *81*(1), 118–121.

Lippmann, G. (1891). La photographie des couleurs. *Comptes Rendus de L'Academie des Sciences*, *112*, 274–3.

Mitchell, D. J. (2010). Reflecting nature: Chemistry and comprehensibility in Gabriel Lippmann's 'physical' method of photographing colours. *Notes and Records of the Royal Society*, *64*, 319–337. doi:10.1098/rsnr.2010.0072

Nye, D. E. (1994). *American technological sublime*. Cambridge, MA: MIT Press.

Phillips, D. (1996). *Art for industry's sake: Halftone technology, mass photography, and the social transformation of American print culture 1880-1920*. (PhD thesis). Yale University, New Haven, CT.

Photographer, Uknown. (1937). Cover: Simultaneous calculator. *Technology Review*, *39*(3), 1.

Photographer, Unknown. (1940). Unusual photos made with camera mirrors. *Popular Mechanics*, *74*(3), 406.

Photographer, Unknown. (1947). Cover. *Discovery: The Magazine of Scientific Progress*, *8*(11), 1.

Pihlainen, K. (2012). Cultural history and the entertainment age. *Cultural History*, *1*, 168–179. doi:10.3366/cult.2012.0019

Rowlands, J. J., & Killian, J.R., Jr. (1937). Seeing solid: The third dimension at work and play. *Technology Review, 39* (5), 191-5, 182.

Sawyer's Inc. (1949). Sell your products with view-master. *Business Screen*, *10*, 8.

Sherman, R. A. (1953). Benefits to vision through stereoscopic films. *Journal of the Society of Motion Picture and Television Engineers*, *61*, 295–308.

Tupitsyn, M. (1996). *The Soviet photograph, 1924-1937*. New Haven, CT: Yale University Press.

Twist, S. H. (1912, August 17). The picture of the future: A reverie. *New York Clipper*, p. 7.

Chapter 10
The Prime Illusion

Martin Richardson
De Montfort University, UK

ABSTRACT

This final chapter focuses on the potential of three-dimensional imaging. In particular the medium's ability to record three-dimensional objects, as with the holograms made of John Harrison's famous fourth timekeeper "H4" for the Royal Observatory, National Maritime Museum in Greenwich, London, and the strange case of Professor Günter von Hagen and his "BODY WORLDS: The Anatomical Exhibition of Real Human Bodies," who seriously explored its potential but relinquished its further exploration due to negative public opinion of his exhibitions at that time. Holographic stereograms are also discussed, in particular their ability to capture animation, as detailed in Case Study Three: Holograms of David Bowie. The text also explores some future applications of wavefront reconstruction.

CASE STUDY ONE: STOPPING TIME

On the evening of 13th March 2008, between the hours of 6:00pm and 2:00am, five reflection holograms were recorded of John Harrison's fourth timekeeper 'H4', at the Royal Observatory, National Maritime Museum in Greenwich, London. Arguably the most important timekeeper ever made, this watch finally solved one of the greatest scientific problems of its time, that of finding Longitude, and marked the beginning of accurate global positioning. In recent years public awareness of the watch has reached an unprecedented level, together with a string of authoritative writings including the release of Dava Sobels book, 'Longitude', with introduction by NASA astronaut Neil Armstrong, a filmed drama adaptation and even a television sitcom 'Only Fools and Horses' where viewing figures

DOI: 10.4018/978-1-4666-4932-3.ch010

reached a record twenty-four million (Sobel 1996). The watch, its history and its place in history, remain a subject of fascination and curiosity. Now its journey to hologram is traced in this paper through the events of that March evening.

The Time of Day

I'm walking to Greenwich through Black-heath, South London, mass burial graveyard of the unfortunate victims of The 1665 Great Plague which ravaged England from June until November of that year, reaching its peak in September when in one week 12,000 people in London died, from a population of around 500,000 and it was decided that Parliamentary administration should be conducted from Greenwich in order to escape the effects of disease in central London. All this some thirty years before John Harrison's birth, but setting the scene for what was to be center stage for, quite literally, the time of day. I'm on my way to the Greenwich Royal Observatory home of the Harrison Timekeepers to join my friend and fellow holographer, Jeff Blyth, who suggested one winter evening during a meeting of The Royal Photographic Society Holography Group, that a hologram of Harrison's fourth timekeeper 'H4' would be a notable achievement. When it was created, 'H4' represented state-of-the-art, 'leading edge' technology in a highly important scientific and technological field. Our hologram is also 'state of the art' and echoes the work we are doing in a parallel field. Particularly with vulnerable objects, showing holograms instead of the originals could prove to be an important innovation in the conservation world. The timekeepers are rarely out of sight from its secure display case, making this crazy impossible idea all the more reasonable.

A few days before my visit I'd arranged to meet Mr. Jonathan Betts, Senior Specialist in Horology at the National Maritime Museum to put forward our proposal. It was an idea that he greeted with enthusiasm, for he is constantly looking for new and exciting ways to interpret and reveal collections for visitors, and these holograms appear to be the 'next big thing', with very interesting possibilities for exceptionally valuable or vulnerable objects such as 'H4'. Betts is a global authority on the timekeeper, having penned a number of works on the subject. He was the authorative historical advisor to the British Broadcasting Corporation (BBC) for an episode of the television comedy 'Only Fools and Horses'. Screened on the 9th December 1996, it attracted a massive television viewing audience of 24.3 million, a record for a British sitcom. The plot follows these lines: After finding the missing Harrison timekeeper, Del Boy played by actor David Jason, offers it for examination by experts at Sotheby's and all accept it to be the Harrison "lesser watch", a semi-mythical piece whose designs exist although it is unclear whether the watch was ever made. Spivey Del Boy is rewarded for his find to the tune of £6,500,000 at auction, elevating him from his council flat in Peckham, into grand oblivion far beyond our TV screens. It was the last 'Only Fools & Horses' show ever made, leaving the public with a memorable and plausible happy ending.

A Problem Solved

Mr. Betts explained the importance of the timekeeper and described how "Every 15° that one travels eastward, the local time moves one hour ahead. Similarly, traveling West the local time moves back one hour for every 15° of longitude, therefore, if we know the local times at two points on Earth, we can use the difference between them to calculate how far apart those places are in longitude, east or west." This idea was very important to sailors and navigators in the 17th century. They could measure the local time wherever they were by observing the Sun, but navigation required that they also know the time at some reference point, e.g. Greenwich, in order to calculate their longitude. Although accurate pendulum clocks existed in the 17th century, the motions of a ship and changes in humidity and temperature would prevent such a clock from keeping accurate time at sea. Therefore a new type of timekeeper was required to help avoid shipwreck and the loss of life. The loss of life was indeed quadrupled for some cargos lost at sea, for we must remember that this was the time of slavery, that most abhorrent form of trade. In fact Harrison's son William set sail for the West Indies, with 'H4', aboard the ship Deptford on 18 November 1761. They arrived in Jamaica on 19 January 1762, where the watch was found to be only 5.1 seconds slow! It was a remarkable achievement but it would be some time before the Board of Longitude set up to administer and judge and award the longitude prize was sufficiently satisfied to award John Harrison the prize of £20,000. During this period they received more than a few odd but wonderful suggestions. Like squaring the circle or inventing a perpetual motion machine, the phrase 'finding the longitude' became a sort of catchphrase for the pursuits of fools and lunatics. Many people believed that the problem simply could not be solved.

Our meeting concluded with a calendar date in March being set and so the project to record holograms of 'H4' suddenly became real… It was sunset as Jeff and I walked down the hill through Greenwich Park sharing doubts as to the project's feasibility. We saw the observatory perched on the hillside dramatically contrasted with a breathtaking view of the City of London. A golden sunset reflected from the skyscrapers of Canary Wharf and the O2 arena next to the River Thames contrasted the Observatory distinctly curved façade. A vision of this City's immortality. We both stopped to take photographs on our mobile phones and, in that moment of innocence, we addressed our doubts. As TS Eliot once said, "between the thought and the action falls the shadow."

The Observatory's beautiful architecture reflects a moment in history when science purposed that truth was harmony, and beauty science. When it was thought the universe worked like one great mechanism, with all parts in perfectly synchronized rhythm. Time was fixed and time began at Greenwich, the Centre of the Empire. Newtonian principles of the mechanical universe were locked into place with such mathematical certainty that the observatory, along with its stale air, ridged order, and dusty papers, judged all. As sun set I thought of what Harrison did to the way people thought of Time, and how so many of the rules held by these fine astronomers were dissipated when Harrison was vindicated and awarded his long awaited prize. They believed in Newton's description of motion. But for his model to make sense Newton required what he described as an "Absolute, true and mathematical point from which time flows equably, without relation to anything external, an absolute frame of reference." That said, some aspects of Harrison's work are still not fully understood and we flatter ourselves that we are continuing the 'ever-questing' nature of this kind of research, with on-going study of 'H4's secrets. For example, recent research on the diamond pallets in the escapement of 'H4', a project Betts' is carrying out in conjunction with the Cavendish laboratories in Cambridge, to try and discover how on Earth Harrison managed to get smooth curves on the acting surfaces of the pallets (as you know, diamond is of an octahedral crystalline structure and will not normally polish smoothly in 3D). Betts' is continuously re-reading his writings on the development of H4's mechanism in an effort to understand why Harrison arranged some of the devices in the movement the way he did.

The astronomers' fixed stars were once thought of as this unique reference point. Today we know no such reference point is thought to exist. Physicist Ernst Mach argued that the concept of inertia, of stillness, would be meaningless in such a frame, and Einstein was stimulated by this to develop his general theory of relativity. Einstein's' theory is simple: time and relativity depends on your point of view. For example, if you're traveling at the speed of light (a constant) then time moves differently for you than it does for, lets say, a sailing ship at sea. But can time go faster or slower, or even backward, and is the fabric of time itself made from "atoms", quantum flecks modern physicists call chronons? Some scientists believe time is an illusion. If you want to understand time, you have to understand physics. And indeed, if you're really interested in physics, you need to ask difficult questions. Modern Science and Quantum Mechanics make the mathematics of time look strange. Logically there can be no precise way of measuring absolute time, only relative time, and given that the sailing ships Harrison was attempting to save did not travel at the speed of light, (at least not to my knowledge), he did a fine job with his mechanical designs to the extent that 'H4' wandered only thirty seconds during a year-long voyage. How ironic then, but entirely apt, that the Harrison clocks are displayed in the very rooms the astronomers once worked so hard to compete against him. Harrison's was an early technological triumph. And again, how strange that his device to measure time accurately did so much to serve the understanding that modern physics holds today, that time cannot be measured.

13th March 2008

The weather was blustery, with heavy rain. Jeff had selected this particular evening as it was the penultimate to the Equinox when British Summertime began, meaning the clocks would go forward one hour, and we needed to take advantage of the early evening darkness for the recordings due to the holographic light sensitive plates supplied by a remarkable holographer, Mike Medora. That evening Medora was journeying from Trinity Buoy Wharf, home to his company 'Colour Holographic Ltd', a stone's throw from Greenwich, but a good hour's drive through the Blackwall Tunnel from the Isle of Dogs. The holographic camera, which Blyth had designed and built, was unloaded from my car and, knowing it would take some hours to set up position and test, we prepared ourselves for a long night. Waiting at the entrance of the observatory on this stormy night, we were delighted to see a fantastic laser beam illuminating the Meridian emitted from the top of the observatory – fantastic indeed as it lit up individual rain droplets which passed into its path, making each one sparkle like highly polished emeralds, treasure falling from the sky. It is known as 'The Prime Meridian Laser' and had been installed eight years earlier to mark the turning of the Millennium. The laser beam represents a line of '0' degrees longitude from the top of the Royal Observatory and hails a stunning night-time view of London from Greenwich

Park. The thought crossed my mind that I wanted to use this laser to record the hologram. On further investigation however, the meridian argon laser beam turned out not to be of suitable quality.

Our Work Begins

Once inside the Royal Observatory we are led to the Horology Conservation Workshop where we set up the holographic recording system and put into place the tiny German made diode-laser on loan from Klastech GmbH, which pushes out the mighty power of 150 mw. Directing the laser beam toward the Blyth frame, I begin to expand the light and spatially filter the optic while Medora sets up the processing area using trusted chemistry, pre-mixed solutions contained in recycled plastic pint milk cartons. Blyth takes great care in reassuring Betts, and his colleague David Rooney, as to the exact non-hazardous process involved. At eleven p.m., after some hours of preparation, we are ready to make the first recording. We are led from the workshop across the cobbled courtyard that Newton once trod, to 'The Longitude Gallery' where 'H4' is on display. We are marched single file, through a security check and counted one by one; paraded crocodile fashion to tape Closed Circuit Television. 'H4' is positioned inside an extremely large purpose built toughened glass case. Betts, complete with mauve colored protective gloves, slides open the case as goose bumps appear on the back of my neck. 'H4' is released, and placed into its unique titanium transport carrying case. In reverse order we are marched back to the Octagon where Betts' carefully places 'H4' onto the holographic camera and I check laser power, light coverage and, that I'm not dreaming - we are about to make history! Mike Medora

removed an unexposed glass photosensitive plate from its safe box and with the assistance of Betts, gently positioned it above 'H4' to rest just above its surface and Blyth makes final adjustments to the stability of the system using safety razor blades to dampen any vibration, like a waiter would a rocking table, we estimate a twenty second exposure. The room itself, which is currently the horology conservation workshop, was constructed in 1859 and is the very room where the world's 'state of the art' timekeeping trials were carried out; for nearly a century the whole Royal Navy fleet navigated safely because of timekeepers tested in this state-of-the-art facility so, historically there were leading edge scientific resonances with the room itself! Selecting one of the louder 'ticks' echoing the workshop, we start our whispered count down…three…two…one. I raise the makeshift cardboard shutter blocking the laser light and mutter the words "To The Sprit of Harrison". In silence, we observe the beautiful 514 nm wavelength illuminate this marvel of timekeeping mechanics. The hologram is recorded as we watch. The timekeepers' glass reflecting an eerie light to brighten the room as the early morning rain tapped gently on the window. How strange this scene must appear to the outside world, windows flashing with otherworldly light, a bone-chilling storm, perhaps some uncanny experiment was taking place inside conducted by mad scientists? If only they knew…

CASE STUDY TWO: THE STRANGE CASE OF PROFESSOR GÜNTER VON HAGEN'S HOLOGRAMS.

I'm visiting The Chinese Movie Theatre in Hollywood Boulevard. Time-out during holographic recording with Craig Newswanger, one of the foremost creative contributors to the medium. Along with the other tourists I'm queuing outside the main entrance to view the handprints cast in concrete on the 'Boulevard of Stars'. I can see those of Jimmy Stewart, my personal favorite, and Gene Roddenberry, creator and sometime producer of 'Star Trek', and look – there's Boris Karloff, Frankenstein's monster. These concrete negatives of human body parts, imprints of the world's most famous, tell us something about the frailty of their existence fleetingly set into liquid rock, just as Sid Grauman, the movie theatre's founder, intended way back in the silent movie days of 1927. If he were alive today I wonder if he would use holograms of the stars to capture their fleeting existence?

Boris Karloff came to mind again during one wet godforsaken Tuesday afternoon in London's Whitechapel, Jack the Ripper territory: I'm walking to Brick Lane where a gallery is exhibiting the work of Professor Günter von Hagen's 'BODY WORLDS; The Anatomical Exhibition of Real Human Bodies'. I'm late for my appointment. His exhibition, which promises you will discover the 'mysteries under your skin' is an anatomical exhibition of real human bodies, dissected and displayed in a forms many think are macabre mimics of life, composed playing chess, running, riding horseback and – the most memorable for me – a young pregnant woman giving birth and the birth/death is startling. All dead. Today I am meeting the professor and his charming wife to discuss the possibility of replacing these gruesome but fascinating exhibits with holograms.

The gallery is expecting me and as I enter the exhibition hall I notice a hush in the air, and the smell of plastic resins. Plastics and the very faint residue of formaldehyde embalming fluid. Several nights ago von Hagen performed the nation's first live broadcast autopsy. The program, commissioned by Channel 4, was promoted in a suitably ghoulish but fashionable way, using images taken from early woodcut illustrations from 17th century medical surgery and trendy typeface. Many thought it un-godly and since both the public and members of the British Government opinion will prove to make it difficult, if not impossible, to stage another exhibition such as 'Body Worlds', all this was brought to a head by a second autopsy coinciding with a late night opening of the gallery on Halloween. Posters were distributed showing the grim reaper on horse back complete with the death sickle and other horror graphics to raise anything but the professors sincere attempt of demystifying his science. On meeting the Professor I am overshadowed by the height of the man. Topped by a large round black hat that has become his trademark and places him somewhere between a spaghetti western and Dr Jeckle's Hyde all dressed up and ready for business. I'm told that the hat hides a hideous scar from a car crash. He stares at the full colour holograms… a cold still gaze of the scientist in abstract thought. You can see he's talking it all extremely seriously. If he didn't exist, science fiction would have invented him. After a long pause he nods to me and explains in a deep awkward accent that he needs more time to think so another meeting is scheduled two weeks from that date. I have the impression that he, not his wife, is feeling a little overstretched by all the bad press surrounding his curious venture.

I decide that our second meeting will need a more detailed scientific explanation so decide to bring my close friend Dr Hans Bjelkhagen, a Swedish physicist and dedicated academic, to the psychics of full colour holography, important for this project if we are to record veins and flesh. Hans stands at a little over five feet tall with a robust waistline making him barrel shaped. His heavy Swedish accent falls hard on ear but its sentiment sweet. The idea of bringing Von Hagan and Bjelkhagen together to discuss holographie de la Mort appeals to my somewhat warped sense of humor and come around to this idea. I am also accompanied by my 13-year-old daughter Lizzie to see the exhibition and broaden her concept of art, since this is how I view the Professor's handywork. During our meeting I notice von Hagen's staring at Lizzie in a very strange slightly unhinging way and decide if there were to be any further meeting she'd probably be happier either at home or more suitably entertained by the "Mummy; the inside story" exhibition at the British Museum where leading-edge computer technology offers a three-dimensional fly through of a 3,000-year-old mummy of the ancient Egyptian priest Nesperennub aided by stereoscopic eyewear.

The two professors accents engage conversation - a beautiful blend of Germanic/ Swedish techno speak and I fall into a daydream remembering the unhinging feeling many experience when looking into the eyes of a holographic portrait. The recording captures everything, everything in reach of the beautiful red-pulsed laser light. Every hair, pore, glistening tear duct, floating dust particle, time on the watch dial, shine from tooth enamel but especially the iris, lens and pupil, the mechanics of the eye and what we see inside.

I ponder the memory of the portrait made of the movie director/producer Martin Scorsese early one Sunday morning just after he had attended catholic mass in London. He was in town for the premier of the UK release of his new film Casino and to raise money for his next. When he arrived at my studio door I was taken aback by his slight stature, as I am six foot four and towered over him. He was impeccably dressed in a black Armani best, a great sense of order. He fired quick questions throughout the holographic recording session and afterwards he joined me in the darkroom as I processed the glass master plates. At one point he even rolled up his sleeves and helped finish drying the images so we could view them in the red light being emitted from the laser which illuminated his eyes like a possessed demon, giving him a hellish, piercing gaze which was heart stopping. Amazingly a copy from one of these holograms appears on eBay from time to time with a reserve price of £1,200 but no one seems interested?

In the latter half of the ninetieth century it was a common held belief that the retinas of a corpse fix the last image seen, like a photograph, by that poor soul. Murder victims' eyes were often used as a means of trying to identify the culprit. William H. Warner, a prominent British photographer of the 1860's, documented this and this is outlined in Bill Jay's book Cyanide & Sprits:

In April 1863 a young woman, Emma Jackson, was murdered in St Giles, London. Warner immediately sent a letter to detective-officer James F. Thomson at the Metropolitan Police office, Scotland Yard, informing him that 'if the eyes of the murdered person be photographed within a certain time of death, upon the retina will be found depicted the last thing that appeared before them, and that in the present case the features of the murderer would probably be found thereon' he based his assertion on the fact that he had, four years previously, taken a negative of the eye of a calf only a few hours after death and upon microscopic examination of the image found depicted the lines of the pavement on the slaughter-house floor. (Jay 1991)

We have to remember that weird and strange ideas were emerging during this portion of history. In common with those areas emerging alongside primitive technology just as photography was then, and indeed even today it can be myth or science-fiction that so often propels modern technology and none more so than in holographics. Could it be that Heir Professor had come up with the idea of replacing his exhibits with holograms after watching 'Star Wars'?

As it turned out Professor Von Hargan chose not to use holograms after all, but for a moment I came close to creating the world's first holographic morgue. And to some extent, to accepting my own limited fragility. Particularly later when I learnt that Bjelkhagen had already recorded a cadaver for medical purposes in Chicago. On seeing it, it didn't make for easy viewing because as compelling as the holographic image was I would have rather looked the other way – to look in a way not as spectacle but in amazement.

CASE STUDY THREE: THE DAVID BOWIE HOLOGRAMS

Its 5.30pm when I leave the Paramount, a small hotel just off Times Square in New York City. Gliding smoothly across the twisted tarmac in a city limo I arrive at what looks to be an unused storage building in down town Varick Street, noticeable, in this pre-911 City, for its shabby emptiness. No signs, no doorman, no features to memorize and no evidence of what await upstairs. I enter the lobby elevator. The walls are a highly buffed copper metal art Deco design, giving a passenger a pre digital illusion of infinity through reflection upon reflection. I arrive at the penthouse suite and the door slides opens to reveal a hive of activity, a vast open plan office with many people on telephones or tapping away at computer consoles. The reception point resembles an old style ticket booth found in antiquated movie houses. Above it hangs a sign 'CHUNG KING STUDIOS'. I feel outlandish as I announce myself through the small petitioning window to the young woman on the other side. My voice emulates an adolescent boy; my nerves are getting the better of me. Beside the booth are large walls covered in a thick dark brown plate glass, which impressively overshadows a hallway beyond.

One of the glass walls gently slides open and I see now it's an entrance to a long corridor and I make out a figure standing at the other end. The figure is that of David Jones (aka Bowie). Gone is the scarlet red hair, replaced by long blond. The goatee beard however, remains. We start by Bowie introducing me to his co-producer and musician Reeves who is laying down the final tracks to an I.ROM interactive called "The Nomad Soul". This is Bowie's pet project for EIDOs of 'Laura Croft' fame. The groans emitted from a surround sound in the recording studio are that of a female in sexual climax, "a voice for one of the computer animations", he tells me. My tour continues and we move on to an area known as the green room. A place to reflect and talk. I'm not doing much of the latter because I'm too busy absorbing, too much in awe. A huge digital flat screen T.V dominates the green room, standing central – volume down. What looks like his work desk occupies one corner of this room? A small laptop computer can be seen open turned on. Presumably the same tool that transmits and receives his personal E-mail and monitors one of his numerous web sites. Amazingly the green room has no curtains, no blinds. It appears open to the world, his porthole to reality.

After project discussions, we leave the white light of the green room and enter again the recording area where Reeves seems immersed in banks of digital monitors; Apple computers listening too, and recording, sound. I open the first of my two boxes and show the world's first full colour holograms, test shots made with a complex system of lasers but only needing the simple lights available in the studio to see. Like a magician, from the second box I unveil another world first and project a computer-generated hologram into the space before him. (Bowie is speechless, as many people can be when viewing these images.) He calls over Reeves who shares his interest. I look to Reeves for normality and he nods a wink unseen toward Bowie and tells me how he would often visit the Museum of Holography before it closed. For the first time I look around the recording studio and notice that I'm standing in a very large room that appears to have vast quantities of black velvet draped everywhere. The walls, ceilings, floors, presumably to dampen the acoustics. Like a black hole it absorbs so much light I find it hard to see underfoot and on hearing a 'crack' find I'm trampling on Bowie's personal headphones. He leads me out of the room as far as the point where we first shook hands as if he couldn't go any further.

Four weeks later I'm invited to bring my holographic recording system to the photo shoot for the album cover of Bowie's 'hours'. In total my cameraman, John Wiltshire, and I filmed twenty-three minutes of Bowie in various actions directed by me. The resulting dramatic footage is captured as a series of three-dimensional lenticular photographs and a large animated digital hologram of Bowie dancing to the beat of a digitally re-mastered tune. The images show Bowie cradling himself, as if nurturing his nascent future holographic self inside a clinically blue and white birthing room. The recording session was a great success resulting in fifteen minutes of three-dimensional movie footage later used to generate 500,000 CD lenticular inserts.

THE HOLOGRAM AS POPULAR ICON

Dennis Gabor first demonstrated his invention by immortalizing the names of his heroes Huygens, Fresnel and Young. Since then images of Michael Jackson, Ronald Reagan and Andy Warhol, to name but a few, have been holographed in a country that has a taste a cultural saccharine. Home of 'Pop' and 'Hyper Realist' art movements, America thrives on things artificial. For pulsed holography, this vision began in 1967 when the holographers at McDonnell Douglas in St Louis turned up their pulse laser and began creating large format portrait holograms of subjects acting out parades of social events, re-enactments of seedy gambling dens and other fantasy acts constructed as if scenes from a classic 'Film Noire' movie from the 1950's, full of style and dramatic inference together with painstaking attention to detail.

These classic laser transmission holograms were forerunners to today's large format portraits and no doubt inspired others to follow. This directive from the post-hippy period continued momentum across the Atlantic when in 1979 laser stage effects led the rock group the 'Who' to an estranged marriage with holographic science, for they were the first musicians to be catapulted into the public eye as holograms. Again, these were laser transmission works and show drummer Keith Moon drum bashing in a holographic flurry of excitement possibly more interesting than the actual event and in retrospect historically important. Sadly, the whereabouts of these remain unknown.

Twenty-five years on and the world's best-selling holograms are analyzed to assess their respective financial return on images ranging from Anna Maria Nicholson's intimate pulsed hologram of David Byrne from Talking Heads or Richmond Holographic Studios' flamboyant portrait of Boy George from Culture Club to the broader mass distribution of inexpensive full colour embossed stereograms of Prince, Stevie Nicks and Michael Jackson whose face, now holographically preserved, will go on smiling to future generations of pubescent Jacko heart throbs. This novel aspect of music packaging surely appeals to the serious look-a-like fan whose miniature holographic replica is as close to their hero as they're ever likely to get. As the writer Umberto Eco put it, we find "reassurance through imitation" (Eco, 1986).

Further, there seems to be some parallel in the technology used by these musicians and their recording engineer. Both binaural and ring modulator systems evoke a convincing three-dimensional sound quality. There's also the strange case of Holophonics from the Zuccarelli lab in Italy which claims to employ the principle of holography to sound waves, having some of the people believe for some of the time, that a sensitive stereo microphone system could actually record a musical hologram. My own belief is that Zuccarelli's claims are pretentious hogwash. The aphorism "inside every camera is a hologram waiting to escape", adopts new meaning when applied to stereogram holography. It's tempting to be puritanical about stereogram, to think of them as a lesser form of holography, as the poor relation. But the fact is a stereogram, like other time-based media, allows us the luxurious dimension of time and motion. A stereogram may be seen as a mechanism of optical illusion, which, like a flicker book, has a beginning and an end. Unlike other forms of stereo photography, however, our cognitive interaction with a stereogram produces the illusion of movement without the need of a viewing device and makes whatever knowledge we may have about the stereogram uninfluential over the experience it provides.

Technically speaking, the wave front formation encoded in a pulsed hologram can offer us the odd experience of seeing our surroundings in an unusually distorted and monochromatic way. The act of taking a pulsed hologram might be described as a more accurate way of collecting data about our world rather than a single viewpoint as with a photograph. What you see is what there is, or is it? In a paper written by Yu N Denisyuk, we are offered the suggestion that "it is natural to imagine that the property of a three-dimensional hologram is brought about by the fact that its structure somehow aspires to copy a given object's form, but it is very difficult to prove this since no method for predetermining an arbitrary three-dimensional object is known to optics" (Sukhanov & Denisyuk, 1970). To accept a pulsed hologram as illusion is to experience the disappointment of its lifelessness. Most of us have been momentary fooled by a holographic image at some time in our lives; indeed this trickery contributes to its attraction, but we are far more critical when assessing a pulsed portrait for reasons more to do with our own physiological constitution than any fault in holographic technology. As we refine the techniques of holography we are, as it were, also creating an iconoclastic arena, where holographic images such as the holographic replica of Marilyn Monroe look-alike Carilyn Paton, reflects a society's obsession with identity through imitation.

Not since the Renaissance have the visual arts and engineering sciences worked so closely together in their pursuit of the ultimate facsimile of mankind. So what happens when our holographic portraits merge with other parallel technologies, such as artificial intelligence and robotics? Will the resultant avatars be illusions or reality? And how will it affect our own reality if these mimics of us can have an existence that is independent of our own, i.e. when your reflection walks away into its own life? My interest with three-dimensional holographic portraiture, and in particular portraiture of celebrity, concludes that the 'hyper-real' holographic analogue portrait delineates taste in the pursuit of surface reality. However, this is not so with the digital form of hologram as post-production manipulation offers the computer aided artist opportunity of intervening the verisimilitude. Other parallel technologies, such as artificial intelligence and robotics, seem equally determined in their pursuit of the facsimile of mankind, all of which rests comfortably in the entertainment arena. Not since the renaissance period have two such diverse disciplines such as photonics and engineering, aimed in parallel toward the same goal. On one hand we have the counter-culture personalities attracted to the practice of holographic illusion, and on the other we have the engineering pursuit of the ultimate Robot. Both using technology initially developed for warfare, both moving closer to the ultimate illusion of life. Each understanding little regarding the effect these breakthroughs have on popular culture and each feeding results back into the very chaos from which they attempt to escape.

Representation of the figure has held a special place in the world of art. Artists often create images of the ideal, particularly in commercial art and advertising, to which a vast majority of us attempt to aspire and copy. Many contemporary visions ignore imperfec-

tion – the supreme female beauty becomes Venus, Goddess of Love. The Human body can also be seen as loathsome, to be covered and hidden. When this happens the representation of the figure goes into abeyance, and here needs to be a new medium that again can explore and change these perceptions before the body once again becomes acceptable. The measuring of human figures for their ideal proportion is a myth. The superficial barriers between disabled and able-bodied people are real because of this myth. This project aims to go some way toward dismantling this myth. It will create holograms using subject such as Aimee Mullins. Aimee Mullin's legs were amputated at the age of one, but with the help of the most advanced artificial legs, she has set world records in running the 100-meter, 200-meter dash and the long jump at the 2000 Para-Olympics Games in Sydney, Australia. Of the field, Mullins is working on a successful career. *People* magazine named her one of the "50 Most Beautiful People."

COMPUTER-SYNTHETIC HOLOGRAMS

The development of new techniques for generating large full color, full parallax computer generated holograms at Zebra Imaging in Texas, where large three-dimensional images are made in tiled 60cm x 60cm increments. So far, the largest image created at Zebra was of Ford's P2000 Prodigy concept car in which ten such hologram tiles made up one very large computer-generated color reflection hologram. This hologram was unveiled by the then United States President Bill Clinton at the winter 1998 Detroit Motor Show to much acclaim. Digital technology is changing holography. It's liberating those who make a living from the medium from the restrictions of their nineteenth-century darkrooms. Upon first seeing a digital hologram, I was amazed because I had never seen anything like in art or holographic science. It made me feel nauseous but intuitively I knew this was an image a new generation of artists could relate to, and one that I felt deeply impressed by. History will determine which colour technique will survive and which falls to the sidewalk on this long journey into holographic hyper reality. That said, the holographic system which employs actual white light as its primary source, and ultra high resolution LCD screens as receiver via Internet broadcast, will outlive all the above-mentioned regimes. Whatever the outcome, we cannot remove ourselves from one fact that is entirely astonishing. That it is part of human creative nature to evolve a timeless form of illusion that will no doubt go on to fascinate and influence millennia after millennia and true colour holography takes us one step closer to this.

Holograms are going to be very exciting, I think. You can really, finally, with holograms, pick your own atmosphere. They will be televising a party and you want to be there, and with holograms you will be there. You will be able to have this 3-D party in your house. You will be able to pretend you are there, and walk in with the people. You can even rent a party. You can have anybody famous that you want sitting right next to you.... (Warhol 1977)

Children adore illusion and the more convincing the illusion the more determine they are in attempting to revile the magician's secret. As a child I remember being transfixed by magic performed both on stage and early evening television; especially the illusions kept secret by men of mystery - magicians

from the Orient as the idea of these strange-looking people travelling from such distant places appealed to my sprit of adventure. Dramatic in appearance, these entertainers appeared dressed in fantastically outrageous robes with baggy sleeves, long pencil like beards, crowned by exotic hats and I fell for it all. Even if they were lowbrow versions of the 'real' thing, turning out to be Joe Soap from Slough, and not the real Ali Bongo from deepest China to which their stage persona eluded. To the pure all things are pure. Of cause many of the illusions I so eagerly observed were in fact based in age-old Chinese science; combustible gunpowder's, refractive optics, precise measurements and alike. The deception however was a positive one that sparked in my mind a curiosity of optical illusions, of how mirrors work and the transitory delight of tricks performed with light. As the use of holography has grown so too has the artistic interaction with the medium adding to its inventory of objects and people. A conceptual framework for the critical analysis of holographic imagery, has bearing on the traditional aesthetic qualities attributed to other media, such as photography and sculpture; photography for its 'realistic' documentation and sculpture for its 'spatial' or 'three-dimensional' qualities, both of which are inherent in the aesthetic of holography, and are examined in the following essays through the context of mixed media art.

Analysis of three-dimensional spatial imaging compels us to make decisions about 'real-time' and 'real-space', for our primeval instincts are as exact in their recognition of the unfamiliar as the familiar, but far less exact when a three-dimensional image gives no clues as to its origins being holographic. We primarily recognize the object rather than the hologram, particularly if we are not familiar with the medium. We generally continue to employ the criteria of recognition, i.e. the comparison between it, ourselves and our spatial surroundings, our orientation to it and its orientation to us, even though it is an illusion that we are assessing and not tactile information. In this sense holography may be used as a cognitive tool which offers a new insight into the way we perceive and evaluate our surroundings through a manipulation and distortion of its inherent qualities of wave front formation that produce a realistic image. The success of an artistic hologram depends less on this realisation (consciously or unconsciously) or the craftsmanship put into it, than on the level of mystery that the image sustains through its visual charge. Holography's perfect 'realism' make it the ideal medium for the reproduction of sculptural artwork already existing in other media, offering a new slant to the concept of the multiple of mass production. The juxtaposition between reality and holographic illusion is a technique used by artists, including myself, who assemble actual objects alongside their holographic recordings. One of the earliest examples of this type of work was made by American artist Rick Silberman, most famous for his holographic artwork titled 'The Meeting'. c.1979. 'The Meeting' consists of a hologram of a mass-produced wine glass projected over a real broken wine glass which occupies the same space as the holographic image. Silberman's illusion, possibly through its own rarity, raised the value of the mundane glass to an item of interest.

My own investigation with this type of work began with the mixed-media piece 'The Value of a Disposable Item' (circa 1980), better know as 'Behind Glass'. A crumpled tin can provided the image for exploring the triplicate nature of three representations of the same object, in different form; two crumpled cans were presented alongside the holographic and photographic recordings. When planning this work I took into consideration its presentation as a piece, in particular the contradiction of the entire three-dimensional assemblage being framed behind glass. The flattened cans became almost two-dimensional like the photograph, leaving the liberated holographic images unconstrained to project themselves into space beyond the glass. The three Denisyuk holograms in 'The Value of a Disposable Item' provide evidence of a pictorial space receding beyond the actual space of the work, which is immediately countered by the flatness of the photographic image, in a simultaneous perception. Holograms seem to become externalized through their juxtaposition with both real objects and photographs, particularly photographs which seem to carry more weight than the floating holographic replica. Having accepted the content as 'possible', the viewer is led by the realism of the details and of the space they occupy to accept the illusion as reality. Needless to say, many viewers find this acceptance troublesome. During holographic exhibitions one often sees people reaching out to touch the holograms, as if testing the reality of their vision. They are forced by comparison not only to become familiar with something previously alien to them, but to experience its visual dynamics as they move around it. For many, the only barrier between the holographic reality and out own is the glass plate through which we must view the hologram for it is the dividing line between this world and another.

In the early 1960s, new ways for artists using photography began to evolve and sine then a body of work began to grow up which was perhaps misleadingly labeled 'Conceptual Art'. What began to matter was not so much the actual work but the idea it embodied. Conceptual Art was seen by some of its practitioners as political, as embodying a stance against exploitative art galleries, art as commodity, as objects that were bought and sold. Needless to say, conceptual artists did not make things that were shown in galleries and bought and sold. However, this attitude of mind did produce two other ways of thinking about and making art that had consequences for artists using photography. One was the increased analytical and critical attitude towards society, and the other was the analysis of the means of communication in visual terms. The latter is paralleled in academic circles and universities by the studies of linguistics, structuralism philosophy and other disciplines which attempt to analyze dispassionately the very way the emotive parts of human culture, such as language, as put together and work. Photographic language itself is analyzed, but often by people who are labeled artists, and photographs taken by people labeled artists are often used to form visual essays that are then displayed in art galleries and sometime published in books.

At the same time as this conceptual art seemed to imply some criticism of the idea of art as commodity, meanwhile managing to fit very nicely into a system it was criticizing, another body of artists were making great sculptures and manipulating landscape in the wildness, and the kind of art they were engaged in rapidly became labelled 'land art' or 'earth art'. The materials used were real. That is stones, dirt, wood, earth, as found in real places. Sometime real materials are brought to the art gallery. Sometimes in remote places sculptures are made using elaborate earth-moving equipment, sometimes simply by moving materials out manually and sometimes using a kind of high technology. Examples here range from Robert Smithson's

'Spiral Jetty', a spiral of earth winding its way into the great Salt Lake, Utah, to Walter de Maria's 'Lightning Field' 1971-1977, a field of lightning rod in the South West, New Mexico. Naturally much of this work is documented and therefore shown in the cities that have the audience for advanced art by and through photographs. Parallel with this activity, which can only be assessed by means of some kind of documentation of an artist's walk in a remote place, the memory of the activity being communicated by means of the artist's labeled photographs of the event. In a sense this is a highly sophisticated version of the tourist photograph. Instead of the snapshot of New York City, the artist provides a captioned photograph of a remote place in the wilderness to which he has journeyed and where he may have made from the materials he found, some kind of arrangement of wood, stone or earth.

LANDSCAPE HOLOGRAPHY

Landscape tradition has always been diversified and open to new interpretations readily adaptable to new media such as holography. Holography's ability to recreate depth and volume makes it well suited to landscape matter. It can do beyond the mere recording of nature to produce a unique integration of light and space. (Barilleaux, 1984)

Holography's history is a continuous evolution involving all kinds of new interests and rarified problems and it will not be long before holograms of landscapes will be made in formats the size of a cinema screen. But what of the aesthetic problems of giant depth and scale, when the holographic image is larger than the room it is being projected into?

Let's consider the visible reality of a definite scene, for example the gigantism of the New York skyline and select one building from it, the Empire State building. If this building is standing on a flat plateau in from of me, its position determines whether I can realize its shape properly. If, for example, I see it from the side as an oblong, I have no means of knowing that the Empire State building is in front of me; I see only the oblong surface. The eye and equally the photographic lens, acts from a particular position and from there takes in only as much portion of the field as the lens allows. If we then move our positions, we may see from a few degrees to one side the cube or the oblong as a building, whereas before five of its faces were screened by the sixth, therefore only the sixth was visible, but since this fact might equally well conceal something quite different since it may be the base of a pyramid, or the side of a canyon wall, our view of this building had not been selected characteristically. We have, therefore, already established one important principle in arranging hologenic subject matter. If I wish to record the Empire State building it is not merely to bring the building within the coherence of my laser; it is rather the question of my position relative to the building, or where I place by holographic plate to it and how it appears to me, bearing in mind the how and where it is to be played back considered its 'site' for the final display hologram.

This aspect gives very little information as to the overall form of the Empire State building. However, if we were to position the holographic plate on one of the top axes of the building to reveal three surfaces of the corner of the building and their relationship to one another, it will show enough to make it fairly unmistakable what the subject actually is, since our fiend of vision is full of a solid object. Our eyes see this field from only two stationary points, since the eyes perceive the rays of light that are reflected from the object only by projecting them onto a place surface, the retina.

The second aspect gives a much truer picture of the Empire State building than the first and represents a more interesting 'hologenic' scene. The reason for this is the second shows more of the first; three facades instead of only one. There is no formula to help choose the most interesting characteristic aspects of landscape holography; it is a question of feeling. In display holography the aspects that best show the characteristics of a particular subject are not by any means always chosen. In the earlier part of these essays I discussed the conditions which arise from the fact that when a holographic representation of a three-dimensional image in space is exhibited, the very format of the medium, a two-dimensional surface, is unharmonious when framed. It was first demonstrated that an object could be recorded characteristically or otherwise according to what view is chosen and it should be noted again that an artistically prescribed hologram will involved, or should involve the use of fantasy, whereas if a scientific hologram were to be shown, the scientist making the hologram would place the plate in front of the object at such a distance that nothing would be left out of the picture. The edges of the hologram itself circumscribe its own reality, and so a 'life-size' landscape would be indistinguishable from the real world. It may be convenient to summarize briefly here what has been said in the above paragraphs. Firstly, that by reproducing a landscape hologram from an unusual and striking angle the artistic forces the spectator to take a keener interest, which goes beyond a scientific recording. The scene thus holographer shows vitality in its impression. Secondly however, the artist not only directs the attention to the scene itself but also to its formal qualities. Stimulated by the provocative unfamiliarity of the aspect, the spectator looks more closely and observes. It is a way of learning how to see our world afresh.

CASTING IN LIGHT

How does the three-dimensional image being projected from the two-dimensional plane fill space as analogous to a piece of light sculpture? As most 'large-scale' sculpture is designed for landscape, does it follow that large format holograms must be designed for specific sites? A landscape hologram can be without any distortion or violation in the reality it is projecting. It is however in the design of landscape sculpture to occupy or dominate its surroundings. Holographic landscapes on the other hand are more like preservations. They may be allowed to appear simply as themselves or attempt to capture, as the great landscape artists have in the past, the true nature of the poetry in the landscape. The process of man in his environment remains here the central theme, a mastering of the intellectual no-man's-land, rather than man's emotional dependency on it. If we imagine a landscape hologram being shown in an art gallery, it is possibly more avant-garde that the realization of a scientific exercise. It could be more of a naturalist's demonstration than a fine art environment, a visual pastime for inanimate armchair tourists.

There are numerous types of holograms. It is important to learn the basic differences between the various types and what terms are used in referring to them so that you will understand immediately what someone means if they say, for example, they have just made a reflection hologram or transmission hologram or in-line hologram. Holograms can differ in the way in which they are produced and they can incorporate and store the information for playback. The latter difference is the simplest to explain so we'll begin with that.

Under normal conditions we will be using a silver halide type film so we will talk about that specific case. The holographic information is coded in the emulsion according to the localized microscopic differences in the absorption of light or by the amount of silver halide converted to silver atoms during exposure and development. This is referred to as an absorption hologram. The absorption pattern on the film corresponds with the amount of light incident on the plate during exposure. If that same hologram is put through a bleaching process it will then be termed a phase hologram. The absorption index by changing the different residues of silver to corresponding thickness of transparent substance. The hologram is then played back by the refraction of the reference beam dictated by changes of refraction in the emulsion. In a phase hologram the reference beam is phase modulated in order to reconstruct the wavefronts of the original objects. In absorption holograms the reference beam is diffracted by the small patterns of exposed emulsion in the form of silver residue.

The first hologram ever made by Dennis Gabor, in 1947, was an in-line, plane, transmission type. Remember at this time the laser was still yet to be invented, so Gabor had to make do with the quasi-coherent light gained by squeezing light from a mercury vapor lamp through a pinhole and then color filtering it (he used the 0.546 micron mercury green line). In-line means that the reference beam and object beam are coming from the same direction or are the same beam. Gabor had to do this in order to maintain the little coherency he had gained. All in-line holograms are also single beam set-ups. The same beam acts both as reference and object beam. This was made possible by using a transparency as the object. The light that went through the transparency before reaching the plate was modulated by the transparency; the light, which went through it and was not affected by the transparency, was the reference beam. The diffracted light and reference light interfered on the emulsion of the hologram and thus fulfilled one basic requirement for the construction of a hologram. When the reference beam was later shown back through the hologram at the same angle relationship it had with the plate in the reconstruction stage an image appeared. A poor image due to the lack of coherent light, but worse still the reference beam shone directly into the viewer's eye, thus greatly compromising the viewing of the reconstructed object. Although it was a poor image it was there in all its dimensionality. A new medium had been born, alas, a little prematurely and in 1948, was placed on the shelf until the advent of the laser.

WAVEFRONT CAPTURE

As I mentioned above in order to playback a hologram the reference beam must be shone back through the hologram at the same angle relationship as it had in construction. This is where the term transmission hologram arises. Transmission merely means that the reference beam must be transmitted through the hologram in order for the image to be reconstructed.

From 1960 Leith and Upatnieks at the University of Michigan removed Gabor's brain child from the shelf and gave holography its rebirth. Like Gabor they did their early experiments with a filtered mercury arc lamp. Leith and Upatnieks invented the off-axis reference beam with all its great advantages which they did not even appreciate at the time. After the development of the continuous wave gas laser in 1960 by Ali Javan et al. Leith and Upatnieks started using the laser and discovered the three dimensionality of the images. They performed these experiments as an adjunct to their work in side-looking microwave radar. They independently invented off-axis holography only to find that Gabor had proposed holography 12-14 years earlier.

The term off-axis means that the reference beam and object beam are not coming from the same direction. Naturally in order to perform this feat we must have two different beams, thus the term twin beam. Because the laser gives a homogenous beam of coherent light we can extract a beam from the original beam as I mentioned earlier. This is done with the aid of a beam splitter, which could be nothing more than a piece of optical glass. A part of the original beam goes through the glass and a part is reflected at the same angle as its incident. This allows one to bring in the reference beam from an infinite number of angles in relation to the object directed beam, thus avoiding the inconvenience in play back of having to look directly in the reference beam as with the in-line, transmission hologram.

This is a good time to point out the differences between a plane hologram and a volume hologram. As the angle difference between the object beam (or the wavefronts bouncing off the object) and the reference beam changes, so does the spacing of the patterns in the emulsion. As long as the angle difference remains less than 90 degrees the hologram is called a plane hologram. Plane meaning that the holographic information is primarily contained in the two-dimensional plane of the emulsion. Although the emulsion does have a thickness, usually around seven microns or 7/millionths of a meter, the spacing between fringes is large enough, when the angle is less than 90 degrees, for us to imagine that the depth of the emulsion isn't really being utilized in the recording of the hologram. At 90 degrees, which is really a convenient but arbitrary point, the angle is great enough and fringe spacing has become small enough for us to say that the recording process is taking place throughout the volume of thickness of the emulsion. A point to remember is that although there are different thicknesses of emulsion put on celluloid or glass plates seven microns is an average. One can use the same emulsion say seven microns thick, and make both plane and volume holograms depending on the angle difference between reference and object beam.

A very important point for differentiation occurs as the reference beam swings around its arc of possible positions. In a plane transmission hologram the reference beam is hitting the film from the same side as the object beam. In a volume reflection hologram the reference beam has made an arc clear around so that it hits the film from the opposite side as the modulated object beam. When 180 degrees difference is reached you are constructing an in-line, volume, reflection hologram. A transmission type hologram means that the reference beam must be transmitted through the hologram, in order to decode the interference patterns and render the reconstructed image. The light which is used for playback must be coherent or semi-coherent or the image will not be sharp. If a non-coherent source, such as the light from a common, unfiltered slide projector is used, then the hologram will diffract all the different wavelengths. The interference pattern or grating etched in the emulsion is not particular as to which wavelengths it bends or focuses; therefore, you end up with an unclear overlapping spectrum of colors which somewhat resemble your object. A hologram will playback just as well with laser light of a different color or wavelength than the light with which it was made. However, the object will appear to be of a different size and/or distance from the plate. For example, a hologram of an object made with neon or red light will playback that object smaller or seemingly further away if a blue color laser is used.

THE STEREOGRAPHIC MULTIPLEX HOLOGRAM AND LANDSCAPE

In addition to the previously mentioned types of holograms commonly made today there is the multiplex hologram and its this type of hologram that could certainly be used in Landscape recording. A multiplex hologram is the holographic storage of photographic information. In the first stage a series of photographs or a certain amount of motion picture footage of the subject is exposed. The number of stills or frames taken depends on how much of an angle of view you want of the subject in your finished hologram. For example if you want a 360 degree view of the subject you might expose 3 frames per degree of movement around the subject (usually the camera remains stationary and subject rotates) this will result in the exposure of 1080 frames. When your film is developed you proceed to the holographic lab and (using a laser) make a series of "slit" holograms using each frame of film as a subject for each slit of holographic film. The slits are usually about one millimeter wide and are packed so closely that there is no "dead space" in between. Also the hologram is bleached so that the strips disappear. Usually a multiplex hologram yields horizontal not vertical parallax although a digital hologram can offer both. This is because the camera usually moves around (or the subject moves around in front of the camera) and doesn't usually pass over the subject. Also, psychologically, horizontal parallax is much more desirable and the lack of horizontal parallax, to humans, is much more noticeable than the lack of vertical parallax. The multiplex hologram is usually, though not always, made on flexible film coated with the same holographic emulsion as the plates. The procedure can be totally mechanized so that a machine can expose a slit hologram per each frame of footage at a very rapid pace. The advantage of this type of hologram is that you can now have a hologram of almost anything you can capture on ordinary film without the need of the expensive, clumsy pulse ruby laser. The disadvantage is that it is not truly a hologram but photographic information holographically stored.

Landscape is certainly the most frequently depicted subject in photography and its attraction is obvious. The grandeur and color of the far-stretching scene, its openness and its unlimited freedom are hard to resist and, at first glance, seem to offer the greatest opportunity for making beautiful pictures. Yet landscape, for all its beauty, is probably the most difficult subject from the point of view of the rectangle of a photograph. The extent to which a photograph can disappoint as a record may only be measured by the visual totality of a holographic recording.

The greatest problem in landscape photography is the wide divergence between the way we look and see and the way a camera records our vision. The eye doesn't rest for an instant. It roams constantly, sweeping the panorama in front, focus shifting endlessly between objects far and near, yet appraising the scene in its entirety. The camera cannot do that. It only sees one part at a time. A part that totally excludes the rest of the panorama. Landscape photography in fact demands not merely a direct enthusiastic response; it needs careful consideration of a scene's photographic potentiality. Perhaps a seriousness and limited characteristic of a potentially fine landscape hologram can be underlined by the anchoring of an image to a point of visual reference. In photography, many successful landscape photographs contain a foreground element: a tree, a group of rocks, a flowerbed, a hinge or whatever, which, through its proximity and relative size, gives a visual clue to the vast space stretching beyond.

An American holographic artist who has made multi-colored holograms of landscape rocks but not yet landscapes is all too aware of the huge potential for landscape holograms. John Kaufman lives virtually on the San Andreas fault line and his work reflects the geological forces surrounding him! 'The rocks in these holograms seem to have a weight and, although stationary, retain a sense of potential energy'. Kaufman's optical devices such as overlapping planes and colour are used to lead the viewer from place to place. Convergence of lines and atmospheric perspective can also add to the plasticity of the rendering by emphasizing the space and its dimensions. Photography can use the recession of tones, whereas this may not be so obvious in holography. How does one make a hologram of the sky? For the sky is obviously another very important part of landscape and it needs very careful consideration. Very few landscape pictures do not contain the sky and for many it is a dominant feature. Of all the elements in landscape holography perhaps light itself is the most crucial. The expressive characteristic of light in all its infinite variations, from the delicate translucence of very early morning to the harsh unrelenting quality of the midday glare. These graduations determine an overall characteristic of the picture. Although a pulsed laser in holography has relatively fixed capabilities when it comes to lighting objects normally, it can be seen from the examples given in the exhibition that a mood may be rendered from different variations of lighting, but practically impossible to render any favorable normal natural lighting appearance. In other words, the lighting can only be used locally. The lighting effects used by Dutch art-holographer Rudie Berkhout are amongst the finest to date. His work relies almost totally upon the way we make visual associations between points of light in space, which we mentally unite into a form. Small, flat, luminous shapes combine to produce larger images which appear to occupy space. The shapes are isolated by careful location and color control; this enables Rudie to build a complex form from simple parts.

All artists have been concerned with three-dimensional reality since the days of Velasquez and in modern times the analytic cubism of Picasso tried again to capture the three-dimensions of Velasquez. Now with the genius of Gabor, the possibility of a new Renaissance in art has been realized with the use of holography." (Dali, 1972)

Let us consider the disadvantages that 'hologenics' contend, beginning with the way a traditional frame is used as a border to a hologram, a boundary to three-dimensional spatial imagery. Frames mark out a different type of space, a different type of reality from the one within. They direct that which is outside to stand apart, and at the same time the projected or receding image marks off the special kind of reality of the hologram, metaphorical to the way force fields are often depicted in science fiction and similar to the psychological space described by Heidegger. Personal distance is the term originally used by Heidegger to designate the distance consistently separating the members of society. It might be thought of as a small protective sphere or bubble that we maintain between others and ourselves. The distance of 1½ to 2½ feet is the normal distance. The sense of closeness or intimacy with a holographic image is predetermined by the size of the material used, but if we stay with the aforesaid distances at which one can hold or grasp the other person, we find in the normal situation that visual distortion of the hologram is no longer apparent, but there is a feeling if intruding into someone else's space, especially if the holographic image is a pulsed hologram portrait of a living person. Keeping somebody at arm's length is one way of expressing the far phase of personal distance. It extends from appoint that is just outside easy touching distance by one person to a point where two people can touch fingers if they extend both arms. This is the limit of physical domination in the very real sense. Beyond it a person cannot easily get his hands on somebody else. Subjects of personal interest or involvement can be discussed at this distance. Social distance, or the boundary line between the far phase of personal distance and the close phase of social distance marks the limit of domination. Intimate visual detail of the fact is not perceived and nobody touches or expects to touch another person unless there is some special effort. In the traditional methods used by fine arts for producing the illusion of space, one finds an essential factor in obtaining the illusion of depth is the way in which the sense of space is conveyed. To quote Miriam Milman on trompe l'oeil art:

A whole body of empirical and mathematical methods (has) been employed to produce in the work the illusion of that third dimension which is essential to the world around us. However, the spectator has to exercise his own will in order to enter this 'fictional' world. Once he has entered he finds himself in a context where content may strike his more or less fruitfully. (Milman, 1983)

The two most important geometric rules that dominate traditional perspective depictions are: (1) horizontal and vertical lines parallel to the plane of the picture are depicted as horizontal and vertical lines (equal distance along these lines are show as equal distances in the pictures, and (2) parallel lines which recede out of the plane of the picture are depicted as lines passing through a single point (the vanishing point). The most ancient known relics – e.g. the painting in the Lascaux (France) and Altamira (Spain) caves and the small sculpture known as the Venus of Willendorf (Austria) – date back to the thirtieth millennium B.C. Also of early Paleolithic origin are carvings on household articles of wood and mammoth tusk. These and other, later, relics of ancient culture testify to the fact that, right from the start, attempts to fix the surrounding world in images developed in two directions. The first, sculpture, was essentially a method of creating a three-dimensional copy of a real object. The other,

painting, transformed the three-dimensional world into its two-dimensional image.

The solid representation produced by sculpture was far better for creating the illusion of the reality of objects, but the laborious process of making a sculpture, even today, can be undertaken by only a small number of talented artists. The development of painting, and of its use to create visual likenesses, was through the utilization of the laws of perspective by painters outlined previously and described for the first time by an Italian architect, Filippo Brunelleschi, in the middle of the fifteenth century. These laws were adhered to in the works of such famous artists as Piero della Francesca, Albrecht Durer and others. In paintings made in accordance with the laws of perspective, it is possible to see whether an object is situated in the foreground of in the background and to locate the point where all the lines describing the depth of the depicted space converge. Nevertheless, even this innovation did not solve the main problem, since the three-dimensional scene remained two-dimensional in the picture. Further perfection of the painting technique and the application of innovatory methods, such as the use of an improved camera obscura were aimed at the inclusion of greater detail.

Any work of art, irrespective of the epoch in which it was created, is entirely subjective in its manner of presenting information about real objects. It is precisely because of the artist's individual perception of the surrounding world, inherent in his way of 'viewing' real objects, that he can express differences in shape, color, tone and shadow – differences that make the finished work a work of art. A marked step in solving the complicated problem of detailed representation of the surrounding world was made by photography. In taking photographs the lens is used to construct the image of an object in the plane of the recording material. The process of image formation is here of an objective nature and, therefore, reflects the reality with greater authenticity. (Strictly speaking, even in photography the subjective factor is still present in the selection of viewing angle, magnification, etc.). To quote Graham Saxby, 'Inside every camera there is a three-dimensional image trying to escape.' (Saxby, 1988).

One of the most convincing two-dimensional depictions of perspective rules was made in 1952 by the Dutch artist M. C. Escher. In his studies for the lithograph 'House of Stairs', Escher's sole aim was to depict an infinite extent of space. He used no other means than the laws of classical perspective. The viewer sees the infinite extent of space through a square window. As the space is divided up into similar cubes of bars running apparently in three dimensions, a suggestion of the whole is achieved. The 'exercise of the will', suggested by Miriam Milman, is an interesting point but applies only in the context of traditional pictorial perspective laws as exemplified by Escher. In holography these pictorial laws can be ignored, as it is in the very nature of the medium to offer a real perspective space within which one can observe genuine spatial relationships. The subject matter forms the basis by which one expands and amplifies one's ideas and thoughts; it is a mantra of concentration and opportunity. Whether the subject matter is pre-visualized or spontaneously evoked is irrelevant. It is the meaning attached to the subject which should be approached critically. When assessing a creative hologram it is not easy to achieve a balance between technical skill and artistic merit, unless one concludes that these values go beyond personal taste. These values have to be balanced if the work

is to communicate a clear presentation that avoid the stigma of 'high tech art', a label so easily attached to holography. Holographic images must not be allowed to seduce the viewer through their technical facility alone; on the other hand, a holographic image of artistic merit can suffer by poor techniques.

3-D ILLUSION AND FANTASY

In Umberto Eco's book, *Faith in Fakes*, the first chapter, 'Travels in Hyperreality', subtitled 'The Fortress of Solitude', he describes a multiplex hologram of two beautiful girls crouched facing each other. They are enclosed in a cylinder of transparent plastic. They touch each other sensually, and kiss each other's breasts lightly with the tips of their tongues. Eco describes his feelings of disappointment that one looking into the cylinder, 'the girls are no longer there' (Eco, 1995). He reasons that holography could only prosper in America, a country obsessed with realism, and if a reconstruction is to be credible it must be absolutely iconic, a perfect likeness, a 'real' copy of the reality being presented, the perfect fantasy. Eco's hypothesis, in this case the link between holographic image and fantasy, is vital to our understanding of how the fantasies of the unconscious come into play as we set out to understand the holographic aesthetic. Fantasy is essential to holographic imagery since it is only in fantasy that a hologram's illusory world can be translated into convincing information on which we may begin to base an aesthetic, remembering that a hologram is something quite different in nature to that which it depicts. If a holographic image does not exist and yet the precept persists, the appropriate judgement is in all likelihood that we are subject to an hallucination. 'Hallucinary perception involves the non-existence of an object and its analysis shows that it involves to be a subject to an alternative world. It also shows that the subject is the subject of an alternative reality and does not quite work in the real word. It can only be intermittent' (Eco, 1995).

Mathematically speaking, space can be broken down into curves. It is possible to break down any space into three types of motion. East/West, North/South and Up/Down. By combining the three mutually perpendicular types of motion one can trace out any possible curve in space. No more than three directions are needed, and no less than three directions will do – hence we call our space three-dimensional. Fundamental percepts such as up/down, left/right cannot be applied to these holographic images and no longer function in the process of identification, if identification is based on the orientation of universal gravity. The figure in the work appears to be floating, similar to a nineteenth-century magical illusion. Alfred Hitchcock was known for his experimentation with new technology. He had a history of using special effects and I believe it would not be at all surprising to find him working with holography had he been alive today. During his film career he handled 3-D in his own masterful style. For example, every shot from the film 'Dial M for Murder' in 3-D was carefully composed, so that with proper camera movement and very precise convergence, virtually error-free 3-D was achieved. Hitchcock's daring use of unusual techniques makes this a landmark 3-D film. Many if the effects have never been equaled. 'Dial M for Murder' contains three outstanding examples of three-dimensional virtuosity. First are the tight macro-closeups on wrist watches and second the suspenseful ultra-close shot of a telephone dial as Ray Milland's right index

finger pokes into the number six hole. Both shots were faked. With the 3-D camera systems available at the time the film was made, it was possible to achieve the type of tight close-up Hitchcock felt he needed to enhance the story. For virtually the same technical reasons, he used a similar device in 'Spellbound' in 1945 for a scene in which Ingrid Bergman was held at gunpoint. Hitchcock's solution to having both the gun in the foreground and Miss Berman in the background in focus, was to have a giant hand hold a giant gun. The same giant prop technique was applied to 'Dial M for Murder', making it possible to manage ultra-closeups without inflicting eyestrain, and demonstrating Hitch's undisputed genius for outwitting the limitations of the mechanical devices he worked with.

The third example of 3-D brilliance in 'Dial M for Murder' was the considerable use of rear screen projection which Hitchcock resorted to for cost-saving reasons. Rear screen was avoided in most three dimensional films because it can be easily detected by viewers, but in 'Dial M for Murder' the visual blending was managed so well that rear screen is totally unobtrusive, well-executed shots cannot be fully appreciated. Hitchcock's most obvious and effective 3-D moment comes during the murder itself. As the murderous intruder attempts to strange Grace Kelly, she is forced back across a desk, and her frantic hand thrusts out at the audience as she tries to reach the scissors to defend herself. A response that comes from one's instinctive responses for survival, rather than premeditated attack. Incorporating suspense into holograms is very important. 'Perpetual Fear' from the series "VORTEX", reveals my head at floor level apparently, decapitated. My features confronted by a hand, severed at its wrist, fingers torn to the knuckles. The hand's substance seems as real as the head which turns in fear from the mutilated hand, fearful of impending doom. The mouth is open stretching wide to scream revealing a brilliant light, beaming from my mouth by means of a mirror – a scream, unable to sound, it looks like a molten furnace. (The technical word used by holographers for the radiance of light reflected from the mirror in the head's mouth is, in fact, noise.). The absurd nature of the "VORTEX" series heightened by visual play of reduction techniques drew attention to the expressions manifested by surrealist imagery. Not that a hologram is surrealist within the meaning attached to it by art historians, but the background of surrealism does serve as a contrast, and a selection of surrealist work could be made for this purpose. I have chosen two works by Man Ray. For me, Man Ray provides a constant source of inspiration, particularly in his unusual perspective forms taken from the body. A fine example of this is 'Femme aux longs cheveux', 1929. It is a young woman laid on her back on a plinth, the long hair sweeping forward, the head tilting down, defying gravity.

A prominent surrealist painter writing in 1933 imagined the following scene: A man is staring dreamily at a luminous point, thinking it a star, only rudely to awaken when he realizes it is merely the tip of a burning cigarette. This man is then told that the cigarette end is in fact the only visible point of an immense psycho-atmospheric anamorphic object, knowledge our writer assures us, that will instantly cause the . . . point of burning ash to recover all its irrational glamour and its most incontestable and dizzying powers of seduction. These objects are complex reconstructions made in the dark of an original object chosen in the dark.' Our writer, Salvador Dali, goes on to imagine the story he will tell his now rapt listener about the history of this particular object whose burning tip alone can be seen. 'Ten years earlier, Man Ray has made the following image: a strange construction rises from the bottom edge of the photograph towards the top of its frame. The top of the pyramid is a cigarette, its ash just the edge of the sheet, its other end clenched in the teeth of a barely-seem mouth at the apex of this construction's human base, for we are able to read as the support of the cigarette a face rotated 180 degrees, its humanness hardly recognizable from this position, the mass of falling hair that fills the bottom half of the frame a swirling formless field.' (Krauss, 1985)

I have discussed the basis of aesthetics but I have offered no solution for the most vexing problem, the problem of objectifying and the problem of value. Any detailed treatment of such issues would require being at least as complicated as those I have already given, but I shall conclude with a few suggestions. First, the question of holographic objectivity. Even if we leave evaluation aside we find that our account of aesthetic description presents an awkward problem, for the meanings of many aesthetic descriptions have been explained without reference to their justification. Nothing has been said that will provide rules for the application of a hologenic aesthetic. It must not be assumed, however, that the use of aesthetic descriptions is entirely arbitrary or subjective. A hologram may be seen in several incompatible ways. It cannot be held, therefore, that the possibility of rival interpretations of a hologram demonstrates subjectivity or criticism. It might be thought that, in arguing that aesthetic evaluation in the art of holography depends on descriptions which need have no possible bearing on everyday reality, even though this art captures reality more convincingly than any other medium. I have in effect made it possible to describe the holograms through the use of my comparisons with surrealism and mixed-media artwork. They do, however, exist in parallel planes within which the condition of their acceptance involves a decision of desire. How, in this new medium, do we secure the objectivity or critical judgment that will make our decisions definitive?

My theory argues that the objectivity of an aesthetic judgment is founded no differently from the objectivity of judgment's ascribing to secondary qualities in the form of technical excellence, scientific knowledge, method and result. The definition of a hologram is inevitably subject to cultural definition and, at its very best; the notion of definition is a notion of its limits. As the artist Rudie Berkhout once said, 'the rules and tolerances of holography are narrow and restrictive, the necessary attention to detail can intimidate holographers into an intense approach that is not always positive. Such intensity can result in the process imposing its limits on the individual's vision.' (Berkhout, 1989). I suggest an overall definition of three-dimensional imaging could be regarded as a limiting function, but it is that limiting function to imaging that determines the very possibility of its meaning.

REFERENCES

Barilleaux, R. P. (Ed.). (1984). Holography (re)defined: Innovation through tradition. New York: Museum of Holography.

Berkhout, R. (1989). Holography: Exploring a new art realm and shaping empty space with light. *Leonardo, 22*, 313–316. doi:10.2307/1575385

Dalí, S., & Knoedler, M. (1972). *Holograms conceived by Dali*. New York: M. Knoedler.

Eco, U. (1986). *Travels in hyper reality – Essays*. San Diego, CA: Harcourt Brace Jovanovich.

Eco, U. (1995). *Faith in fakes: Travels in hyperreality*. London: Vintage.

Jay, B. (1991). *Cyanide and spirits: An inside-out view of early photography*. Portland, OR: Nazraeli Press.

Krauss, R. (1985). Corpus delicti. *October, 33*, 31-72.

Milman, M. (1983). *Trompe-L'oeil painting: The illusions of reality*. Paris: Skira.

Saxby, G. (1988). *Practical holography*. Bristol, UK: Institute of Physics Press. doi:10.1887/0750309121

Sobel, D. (1996). *Longitude: The true story of a lone genius who solved the greatest scientific problem of his time*. London: Fourth Estate.

Sukhanov, V. I., & Denisyuk Yu, N. (1970). On the relationship between spatial frequency spectra of a three-dimensional object and its three-dimensional hologram. *Optics and Spectroscopy, 28*(1), 63–66.

Warhol, A. (1977). *The philosophy of Andy Warhol: From A to B and back again*. Toronto, Canada: Harvest.

ADDITIONAL READING

Barnouw, E. (1981). *The Magician and the Cinema*. Oxford: Oxford University Press.

Benton, S. A. (1969). Hologram reconstructions with external incoherent sources. *Journal of the Optical Society of America, 59*, 1545–1546.

Benton, S. A., Mingace, H. S. Jr, & Walter, W. R. (1980). One-step white-light transmission holography and recent advances in holography. *Proceedings of the Society for Photo-Instrumentation Engineers, 215*, 156–161. doi:10.1117/12.958435

Berger, J. (1972). *Ways of Seeing*. London: Pelican Books.

Berkhout, R. (1996). Investigating the use of HOEs in the holographic image making process. Practical Holography X. *Proceedings of the Society for Photo-Instrumentation Engineers, 2652*, 204–212. doi:10.1117/12.236062

Bjelkhagen, H. I., Jeong, T. H., & Vukicevic, D. (1996). Color reflection holograms recorded in a panchromatic ultrahigh-resolution single-layer silver halide emulsion. *The Journal of Imaging Science and Technology, 40*(2), 134–146.

Bjelkhagen, H. I., & Vukicevic, D. (1994). Lippmann color holography in a single-layer silver-halide emulsion. *Proceedings of the Society for Photo-Instrumentation Engineers, 2333*, 34–48. doi:10.1117/12.201881

Coren, S., Ward, L., & Enns, J. (1979). *Sensation and Perception* (5th ed.). Fort Worth: Harcourt College Publishers.

Crenshaw, M. (1987). Pseudo-colour holography – An artist's perspective. Practical Holography II. *Proceedings of the Society for Photo-Instrumentation Engineers, 747,* 104–107. doi:10.1117/12.939800

Grand, S. (2003). *Growing Up With Lucy: How to Build an Android in Twenty Easy Steps.* London: Weidenfield & Nicholson.

Itten, J. (1970). *The Elements of Color* (E. Van Hagen, Trans.). New York: John Wiley & Sons.

Jones, M. (Ed.). (1990). *FAKE? The Art of Deception.* Berkeley: University of California Press.

Kaufman, J. (1983). Pre-visualization and pseudocolour image plane reflection holograms. *Proceedings of the International Symposium on Display Holography, 1,* pp. 195-207. Lake Forest College, IL.

Klug, M. A., Klein, A., Plesniak, W. J., Kropp, A. B., & Chen, B. (1997). Optics for full-parallax holographic stereograms. *Proceedings of the Society for Photo-Instrumentation Engineers, 3011,* 78–88. doi:10.1117/12.271340

Kubota, T. (1986). Recording Of high quality colour holograms. *Applied Optics, 25,* 4141–4145. doi:10.1364/AO.25.004141 PMID:18235757

Lieberman, L. (1992). Paint with light: artistic manipulation of color in multicolor reflection holograms. *Proceedings of the Society for Photo-Instrumentation Engineers, 1600,* 224–228. doi:10.1117/12.57798

Lieberman, L. (1992). Holo madness and the search For the 'hologenic' image. *Leonardo, 25,* 481–485. doi:10.2307/1575760

Markov, V. B., Mironyuk, G. I., & Yavtushenko, I. G. (1982). *Holography and its Applications in Museum Work.* Studies and Documents on the Cultural Heritage. Report CLT/84/WS.16. Moscow: UNESCO.

Norden, H. (1977). *Illusions.* London: Thames & Hudson.

Oliveira, R. M., Carretero, L., Madrigal, R., Garzon, M. T., Pinto, J. L., & Fimia, A. (1997). Experimental study of color control on reflection holography. *Proceedings of the Society for Photo-Instrumentation Engineers, 3011,* 224–230. doi:10.1117/12.271355

Ornstein, R. E. (1972). *The Psychology of Consciousness.* New York: W. H. Freeman.

Orr, E. D Trayner (1989). Deep image reflection holograms in black and white and additional colours. *Proceedings of the International Symposium on Display Holography, 3,* pp. 379-88. Lake Forest College, IL.

Parker, W. (1992). Exploring holography through science and art. *Leonardo, 25,* 487–492. doi:10.2307/1575761

Pilbeam, P. (2003). *Madame Tussaud and the History of Waxworks.* London: Hambledon and London.

Richardson, M. (1987). Mixed media: holography within art. *Leonardo, 20*(3), 251–255. doi:10.2307/1578168

Rowland, M. (2003). *The Philosopher at the End of the Universe – Philosophy Explained Through Science Fiction Films.* London: Ebury Press.

Rush, M. (1999). *New Media in Late Twentieth Century Art*. London: Thames & Hudson.

Scruton, R. (1974). *Art and Imagination: a Study in the Philosophy of Mind*. London: Routledge & Kegan Paul.

Standage, T. (2002). *The Mechanical Turk*. New York: Walker & Co.

Walker, J. L., & Benton, S. A. (1989). In-situ swelling for holographic colour control. *Practical Holography III*, ed. by S A Benton. *Proceedings of the SPIE, 1051*, 192-199.

Wood, G. (2002). *Living Dolls: A Magical History of the Quest for Mechanical Life*. London: Faber & Faber.

APPENDIX: CONTINUING DEVELOPMENTS IN HOLOGRAPHY

Holographic Data Storage

The capability of coherent light to form holograms provides a convenient way to address a storage medium in three dimensions, while only scanning the beams in two dimensions. Holography records the information from a three-dimensional object in such a way that a three-dimensional image may subsequently be constructed. Holographic memory uses lasers for both reading and writing the blocks of data into the photosensitive material. Recording the interference pattern between a discretely modulated coherent wave front and a reference beam on a photosensitive material forms a digital hologram. The three main companies involved in developing holographic memory are InPhase, Polaroid and Optware in Japan. Holography breaks through the density limits of conventional storage by going beyond recording only on the surface, to recording through the full depth of the medium. Unlike other technologies that record one data bit at a time, holography allows a million bits of data to be written and read in parallel with a single flash of light. This enables transfer rates significantly higher than current optical storage devices. Combining high storage densities, fast transfer rates, with durable, reliable, low cost media, make holography poised to become a compelling choice for next-generation storage and content distribution needs. In addition, the flexibility of the technology allows for the development of a wide variety of holographic storage products that range from handheld devices for consumers to storage products for the enterprise. Imagine 2GB of data on a postage stamp, 20 GB on a credit card, or 200 GB on a disk.

Integrated Holographic Optics

As the demand for lighter, more compact systems increases, the use of precision optical components will grow. The company policy of constantly improving and developing new processes and technologies for component manufacture and ultra-precision machining. Currently Thales Optics at the forefront of this leading edge technology.

Holographic Video

The Spatial Imaging group at the MIT Media Lab developed new technology and interfaces for holographic and other spatial displays. Professor S Benton, who passed away November 9, 2003, directed it. Work on electronic capture and display of 3-D images has continued as a part of the work of the Media Lab's Object Based Media Group. The holographic video project is developing a real-time imaging system that can render and display computer generated holograms at near-video rates. So far, we have demonstrated two prototype displays: the Mark-I and the Mark-II. The Mark-1 display is capable of rendering full color 25x25x25mm images with a 15 degree view zone at rates around 20 frames per second. The Mark-II display provides 150x75x150mm images with a 36 degree view zone at rates of around 2.5 frames per second.

Our ongoing research examines methods to reduce computation time, ways to generate more realistic images, hologram compression and encoding schemes, new specialized hardware for optical modulation and computation, and new display architectures.

Holographic Reality

Currently, virtual reality and holographic technology are evolving separately. However, they are expected to converge before the end of the decade. Thus far, application of holographic technology has been relatively limited and generally focused on the areas of art, entertainment and security. However, the depth and scope of holographic technology applications is expanding. It is on the threshold of rapid growth in several new areas including: data storage and retrieval; information search; artificial intelligence; communication; navigation; medicine; sports; education; and manufacturing.

Holographic Printers for Personal Use

In 2013 a new image processing approach to construct parallax-images from sampled viewpoints was developed by Pioneer, based in Japan. It uses a new technique to reconstruct full parallax three-dimensional data of object scenes to make holographic stereograms on photopolymer material.

Holographic Projections / HOLO Screens

The HOLO Screen is a revolutionary holographic rear projection screen for Point of Sale applications in shops, window displays, airports, banks and other high-traffic areas. The HOLO Screen features an advanced holographic film that displays images rear projected from 30°-35°. All other light is ignored. The result is remarkably bright and sharp images – even in brightly lit environments. The transparent display allows viewers to look at – and see through – the screen. This is especially useful in shop environments, as the screen becomes an integral part of the in store layout. The HOLO Screen is the ideal Point of Sale display: it displays sharp, crisp images in brightly lit environments – and takes up a minimum of valuable retail space. Apparently suspended in mid-air, the transparent HOLO Screen gives an impression of almost-3D depth. The unique projection concept means that the HOLO Screen can be hung or mounted in the best locations, where most people will be exposed to the messages e.g. in shop windows, above counters, in-store hallways, escalators and in café environments.

Holographic Telecoms

Telecoms systems contain an awkward mixture of optics and electronics. A purely optical system would permit the very high data rates needed by the Internet, but at the moment the switching and routing, as well as the "last mile" to the customer, still depend on slower electronic components. Speaking at the Institute of Physics Congress on Monday 24 March, Professor Robert

Denning from Oxford University explained how his novel holographic approach to making 3-dimensional photonics crystals could allow optical components to be built that remove this bottleneck. The team uses holographic lithography to make the 3-dimensional photonic crystals. Professor Denning said: Holograms are usually made by making two beams of light interfere with each other and then storing the resultant intensity pattern via a light induced chemical change in some medium, photographic film for example. Holographic lithography is just a fancy name for defining the pattern of the photonic crystal via the intensity variations caused when four laser beams interfere. The trick is to find the right chemical reactions to make this possible.

Related References

To continue our tradition of advancing media and communications research, we have compiled a list of recommended IGI Global readings. These references will provide additional information and guidance to further enrich your knowledge and assist you with your own research and future publications.

Abadía, M. F., Quintana, M. S., & Calvo, P. Á. (2013). Application of topographical capture techniques for modelling virtual reality: From the static object to the human figure. In I. Management Association (Ed.), Geographic information systems: Concepts, methodologies, tools, and applications (pp. 970-990). Hershey, PA: Information Science Reference. doi: doi:10.4018/978-1-4666-2038-4.ch060

Abdel Alim, O. A., Shoukry, A., Elboughdadly, N. A., & Abouelseoud, G. (2013). A probabilistic neural network-based module for recognition of objects from their 3-d images. [IJSDA]. International Journal of System Dynamics Applications, 2(2), 66–79. doi:10.4018/ijsda.2013040105

Abdelouahad, A. A., El Hassouni, M., Cherifi, H., & Aboutajdine, D. (2013). A new image distortion measure based on natural scene statistics modeling. In I. Management Association (Ed.), Geographic information systems: Concepts, methodologies, tools, and applications (pp. 616-630). Hershey, PA: Information Science Reference. doi: doi:10.4018/978-1-4666-2038-4.ch037

Agina, A. M., Tennyson, R. D., & Kommers, P. (2013). Understanding children's private speech and self-regulation learning in web 2.0: Updates of Vygotsky through Piaget and future recommendations. In P. Ordóñez de Pablos, H. Nigro, R. Tennyson, S. Gonzalez Cisaro, & W. Karwowski (Eds.), Advancing information management through semantic web concepts and ontologies (pp. 1–53). Hershey, PA: Information Science Reference.

Ahmad, A., Demian, P., & Price, A. (2013). Creativity with building information modelling tools. International Journal of 3-D Information Modeling (IJ3DIM), 2(1), 1-10. doi:10.4018/ij3dim.2013010101

Akoumianakis, D., Milolidakis, G., Vlachakis, G., Karadimitriou, N., & Ktistakis, G. (2013). Retaining and exploring digital traces: Towards an excavation of virtual settlements. In S. Dasgupta (Ed.), Studies in virtual communities, blogs, and modern social networking: Measurements, analysis, and investigations (pp. 232–252). Hershey, PA: Information Science Reference.

Al-Suqri, M. N., Al-Hinai, K. A., & Al-Hashmi, K. M. (2013). Towards Arab digital libraries: Opportunities, challenges, and requirements. In T. Ashraf, & P. Gulati (Eds.), Design, development, and management of resources for digital library services (pp. 50–57). Hershey, PA: Information Science Reference.

Alam, S. S., Marcenaro, L., & Regazzoni, C. S. (2013). Opportunistic spectrum sensing and transmissions. In M. Ku, & J. Lin (Eds.), Cognitive radio and interference management: Technology and strategy (pp. 1–28). Hershey, PA: Information Science Reference.

Alberto, N. J., Bilro, L. M., Antunes, P. F., Leitão, C. S., Lima, H. F., & André, P. S. … Pinto, J. D. (2013). Optical fiber technology for ehealthcare. In M. Cruz-Cunha, I. Miranda, & P. Gonçalves (Eds.), Handbook of research on ICTs and management systems for improving efficiency in healthcare and social care (pp. 180-200). Hershey, PA: Medical Information Science Reference. doi: doi:10.4018/978-1-4666-3990-4.ch009

Aldeias, C., David, G., & Ribeiro, C. (2013). Preservation of data warehouses: Extending the SIARD system with DWXML language and tools. In J. Ramalho, A. Simões, & R. Queirós (Eds.), *Innovations in XML applications and metadata management: Advancing technologies* (pp. 136–159). Hershey, PA: Information Science Reference.

Alfirevic, N., Draskovic, N., & Pavicic, J. (2013). Toward a customer-centric strategy implementation model: The case of a European mid-sized glass-packaging producer. In H. Kaufmann, & M. Panni (Eds.), *Customer-centric marketing strategies: Tools for building organizational performance* (pp. 476–497). Hershey, PA: Business Science Reference.

Alfred, C. (2013). Contexts and challenges: Toward the architecture of the problem. In I. Mistrik, A. Tang, R. Bahsoon, & J. Stafford (Eds.), *Aligning enterprise, system, and software architectures* (pp. 250–274). Hershey, PA: Business Science Reference.

Ali, A. (2013). Integrating corporate education in Malaysian higher education: The experience of open university Malaysia. In B. Narasimharao, S. Kanchugarakoppal, & T. Fulzele (Eds.), *Evolving corporate education strategies for developing countries: The role of universities* (pp. 321–338). Hershey, PA: Information Science Reference. doi:10.4018/978-1-4666-2845-8.ch023

Ali, S. O., & Albadri, F. (2013). Multimedia use in UAE education context. In F. Albadri (Ed.), *Information systems applications in the Arab education sector* (pp. 80–92). Hershey, PA: Information Science Reference.

Alshehri, A. M. (2013). Employing emerging technologies in educational settings: Issues and challenges. In J. Keengwe (Ed.), *Research perspectives and best practices in educational technology integration* (pp. 148–162). Hershey, PA: Information Science Reference.

Amine, K., & Djellab, R. (2013). Industrial and urban applications of Eulerian and Chinese walks. In R. Farahani, & E. Miandoabchi (Eds.), *Graph theory for operations research and management: Applications in industrial engineering* (pp. 271–279). Hershey, PA: Business Science Reference.

Andonegui, J., Eguzkiza, A., Auzmendi, M., Serrano, L., Zurutuza, A., & Pérez de Arcelus, M. (2013). E-ophthalmology in the diagnosis and follow-up of chronic glaucoma. In M. Cruz-Cunha, I. Miranda, & P. Gonçalves (Eds.), *Handbook of research on ICTs and management systems for improving efficiency in healthcare and social care* (pp. 88–108). Hershey, PA: Medical Information Science Reference. doi:10.4018/978-1-4666-3990-4.ch005

Antoniadis, D. N. (2013). Teams and complexity: Merging theories towards a finite structure. In S. Banerjee (Ed.), *Chaos and complexity theory for management: Nonlinear dynamics* (pp. 202–239). Hershey, PA: Business Science Reference.

Archondakis, S. (2013). Static telecytological applications for proficiency testing. In V. Gulla, A. Mori, F. Gabbrielli, & P. Lanzafame (Eds.), *Telehealth networks for hospital services: New methodologies* (pp. 228–239). Hershey, PA: Medical Information Science Reference. doi:10.4018/978-1-4666-2979-0.ch015

Atyabi, A., Luerssen, M., Fitzgibbon, S. P., & Powers, D. M. (2013). The use of evolutionary algorithm-based methods in EEG based BCI systems. In G. Fornarelli, & L. Mescia (Eds.), *Swarm intelligence for electric and electronic engineering* (pp. 326–344). Hershey, PA: Engineering Science Reference.

Azar, A. T. (2013). Overview of biomedical engineering. In I. Management Association (Ed.), Bioinformatics: Concepts, methodologies, tools, and applications (pp. 1-28). Hershey, PA: Medical Information Science Reference. doi: doi:10.4018/978-1-4666-3604-0.ch001

Bandera, J., Rodríguez, J., Molina-Tanco, L., & Bandera, A. (2013). Gesture learning by imitation architecture for a social robot. In J. Garcia-Rodriguez, & M. Cazorla Quevedo (Eds.), *Robotic vision: Technologies for machine learning and vision applications* (pp. 211–230). Hershey, PA: Information Science Reference. doi:10.4018/978-1-4666-4607-0.ch014

Baum, J. (2012). Object design in virtual immersive environments. In I. Management Association (Ed.), Virtual learning environments: Concepts, methodologies, tools and applications (pp. 402-418). Hershey, PA: Information Science Reference. doi: doi:10.4018/978-1-4666-0011-9.ch213

Bekebrede, G., Harteveld, C., Warmelink, H., & Meijer, S. (2013). Beauty or the beast: Importance of the attraction of educational games. In C. Gonzalez (Ed.), *Student usability in educational software and games: Improving experiences* (pp. 138–160). Hershey, PA: Information Science Reference.

Ben Jabra, S., & Zagrouba, E. (2013). A robust embedding scheme and an efficient evaluation protocol for 3d meshes watermarking. In M. Sarfraz (Ed.), *Intelligent computer vision and image processing: Innovation, application, and design* (pp. 83–100). Hershey, PA: Information Science Reference. doi:10.4018/978-1-4666-3906-5.ch007

Ben Youssef, B. (2013). Integration of a visualization solution with a 3-d simulation model for tissue growth. In C. Rückemann (Ed.), *Integrated information and computing systems for natural, spatial, and social sciences* (pp. 388–407). Hershey, PA: Information Science Reference.

Benito, R. V., Vega-Colado, C., Coco, M. B., Cuadrado, R., Torres-Zafra, J. C., & Sánchez-Pena, J. M. … López-Miguel, A. (2013). New electro-optic and display technology for visually disabled people. In M. Cruz-Cunha, I. Miranda, & P. Gonçalves (Eds.), Handbook of research on ICTs for human-centered healthcare and social care services (pp. 687-718). Hershey, PA: Medical Information Science Reference. doi: doi:10.4018/978-1-4666-3986-7.ch036

Bhowmik, M. K., Saha, P., Majumder, G., & Bhattacharjee, D. (2013). Decision fusion of multisensor images for human face identification in information security. In S. Bhattacharyya, & P. Dutta (Eds.), *Handbook of research on computational intelligence for engineering, science, and business* (pp. 571–591). Hershey, PA: Information Science Reference.

Birkenbusch, J., & Christ, O. (2013). Concepts behind serious games and computer-based trainings in health care: Immersion, presence, flow. In K. Bredl, & W. Bösche (Eds.), *Serious games and virtual worlds in education, professional development, and healthcare* (pp. 1–14). Hershey, PA: Information Science Reference. doi:10.4018/978-1-4666-3673-6.ch001

Bishop, M. C., & Yocom, J. (2013). Video projects: Integrating project-based learning, universal design for learning, and Bloom's Taxonomy. In E. Smyth, & J. Volker (Eds.), *Enhancing instruction with visual media: Utilizing video and lecture capture* (pp. 204–220). Hershey, PA: Information Science Reference. doi:10.4018/978-1-4666-3962-1.ch015

Blundell, B. G. (2013). On volume based 3d display techniques. In M. Khosrow-Pour (Ed.), *Managing information resources and technology: Emerging applications and theories* (pp. 257–267). Hershey, PA: Information Science Reference. doi:10.4018/978-1-4666-3616-3.ch017

Bonanni, L., Seracini, M., Xiao, X., Hockenberry, M., Costanzo, B. C., & Shum, A. … Ishii, H. (2012). Tangible interfaces for art restoration. In B. Falchuk, & A. Fernandes-Marcos (Eds.), Innovative design and creation of visual interfaces: Advancements and trends (pp. 47-58). Hershey, PA: Information Science Reference. doi: doi:10.4018/978-1-4666-0285-4.ch004

Bonnet, S., Jallon, P., Bourgerette, A., Antonakios, M., Rat, V., Guillemaud, R., & Caritu, Y. (2012). Ethernet motion-sensor based alarm system for epilepsy monitoring. [IJEHMC]. *International Journal of E-Health and Medical Communications*, *3*(3), 45–53. doi:10.4018/jehmc.2012070104

Bösche, W., & Kattner, F. (2013). Fear of (serious) digital games and game-based learning? Causes, consequences and a possible countermeasure. In P. Felicia (Ed.), *Developments in current game-based learning design and deployment* (pp. 203–218). Hershey, PA: Information Science Reference.

Boutellier, R., Heinzen, M., & Raus, M. (2013). Paradigms, science, and technology: The case of e-customs. In H. Rahman (Ed.), *Cases on progressions and challenges in ICT utilization for citizen-centric governance* (pp. 323–351). Hershey, PA: Information Science Reference.

Bruce, N. D., & Tsotsos, J. K. (2013). Attention in stereo vision: implications for computational models of attention. In M. Pomplun, & J. Suzuki (Eds.), *Developing and applying biologically-inspired vision systems: Interdisciplinary concepts* (pp. 65–88). Hershey, PA: Information Science Reference.

Bujokas de Siqueira, A., Rothberg, D., & Prata-Linhares, M. M. (2012). Redrawing the boundaries in online education through media literacy, OER, and Web 2.0: An experience from Brazil. In A. Okada, T. Connolly, & P. Scott (Eds.), *Collaborative learning 2.0: Open educational resources* (pp. 164–182). Hershey, PA: Information Science Reference. doi:10.4018/978-1-4666-0300-4.ch009

Burkle, M. (2013). e-Learning challenges for polytechnic institutions: Bringing e-mobility to hands-on learning. In I. Association (Ed.), Digital literacy: Concepts, methodologies, tools, and applications (pp. 1008-1025). Hershey, PA: Information Science Reference. doi: doi:10.4018/978-1-4666-1852-7.ch052

Cabezas, I., & Trujillo, M. (2013). Methodologies for evaluating disparity estimation algorithms. In J. Garcia-Rodriguez, & M. Cazorla Quevedo (Eds.), *Robotic vision: Technologies for machine learning and vision applications* (pp. 154–172). Hershey, PA: Information Science Reference.

Calderero, F., & Marqués, F. (2013). Image analysis and understanding based on information theoretical region merging approaches for segmentation and cooperative fusion. In S. Bhattacharyya, & P. Dutta (Eds.), *Handbook of research on computational intelligence for engineering, science, and business* (pp. 75–121). Hershey, PA: Information Science Reference.

Carmona, R., & Navarro, H. (2012). An image-space approach for collision detection between multiple volumes and a surface. [IJCICG]. *International Journal of Creative Interfaces and Computer Graphics, 3*(1), 16–27. doi:10.4018/jcicg.2012010102

Carrozzino, M., Evangelista, C., Neri, V., & Bergamasco, M. (2012). Interactive digital storytelling for children's education. In J. Jia (Ed.), *Educational stages and interactive learning: From kindergarten to workplace training* (pp. 231–252). Hershey, PA: Information Science Reference. doi:10.4018/978-1-4666-0137-6.ch014

Castiglioni, I., Gilardi, M. C., & Gallivanone, F. (2013). e-Health decision support systems for the diagnosis of dementia diseases. In A. Moumtzoglou, & A. Kastania (Eds.), e-Health technologies and improving patient safety: Exploring organizational factors (pp. 84-97). Hershey, PA: Medical Information Science Reference. doi: doi:10.4018/978-1-4666-2657-7.ch006

Ch'ng, E. (2013). The mirror between two worlds: 3D surface computing for objects and environments. In D. Harrison (Ed.), *Digital media and technologies for virtual artistic spaces* (pp. 166–185). Hershey, PA: Information Science Reference. doi:10.4018/978-1-4666-2961-5.ch013

Chaka, C. (2013). Digitization and consumerization of identity, culture, and power among gen mobinets in South Africa. In R. Luppicini (Ed.), *Handbook of research on technoself: Identity in a technological society* (pp. 399–418). Hershey, PA: Information Science Reference.

Chambel, T., & Martinho, J. (2012). Interactive visualization and exploration of video spaces through colors in motion. In A. Lugmayr, H. Franssila, P. Näränen, O. Sotamaa, J. Vanhala, & Z. Yu (Eds.), *Media in the ubiquitous era: Ambient, social and gaming media* (pp. 171–187). Hershey, PA: Information Science Reference.

Chauhan, K., & Mahapatra, R. K. (2013). Information seeking behavior in digital environments and libraries in enhancing the use of digital information. In T. Ashraf, & P. Gulati (Eds.), *Design, development, and management of resources for digital library services* (pp. 289–299). Hershey, PA: Information Science Reference.

Chen, C., Huang, P., Gwo, C., Li, Y., & Wei, C. (2013). Mammogram retrieval: Image selection strategy of relevance feedback for locating similar lesions. In C. Wei (Ed.), *Modern library technologies for data storage, retrieval, and use* (pp. 51–59). Hershey, PA: Information Science Reference. doi:10.4018/978-1-4666-2928-8.ch004

Chen, R. I., Wang, X., & Hou, L. (2013). Augmented reality for collaborative assembly design in manufacturing sector. In I. Management Association (Ed.), Industrial engineering: Concepts, methodologies, tools, and applications (pp. 1821-1832). Hershey, PA: Engineering Science Reference. doi: doi:10.4018/978-1-4666-1945-6.ch097

Chen, Y. (2013). Construction of digital statistical atlases of the liver and their applications to computer-aided diagnosis. In J. Wu (Ed.), *Technological advancements in biomedicine for healthcare applications* (pp. 68–79). Hershey, PA: Medical Information Science Reference.

Cheng, B., Stanley, R. J., De, S., Antani, S., & Thoma, G. R. (2013). Automatic detection of arrow annotation overlays in biomedical images. In J. Tan (Ed.), *Healthcare information technology innovation and sustainability: Frontiers and adoption* (pp. 219–236). Hershey, PA: Medical Information Science Reference. doi:10.4018/978-1-4666-2797-0.ch014

Cherry, D., Li, M., & Qi, M. (2012). A distributed storage system for archiving broadcast media content. In I. Management Association (Ed.), Grid and cloud computing: Concepts, methodologies, tools and applications (pp. 669-679). Hershey, PA: Information Science Reference. doi: doi:10.4018/978-1-4666-0879-5.ch307

Chou, C. C., & Hart, R. K. (2013). A case study of designing experiential learning activities in virtual worlds. In E. McKay (Ed.), *ePedagogy in online learning: New developments in web mediated human computer interaction* (pp. 227–242). Hershey, PA: Information Science Reference. doi:10.4018/978-1-4666-3649-1.ch014

Chu, S., & Hsiao, C. (2012). Optimizing techniques for OpenCL programs on heterogeneous platforms. [IJGHPC]. *International Journal of Grid and High Performance Computing*, *4*(3), 48–62. doi:10.4018/jghpc.2012070103

Clarke-Midura, J., & Garduño, E. (2013). Students' perceptions of a 3d virtual environment designed for metacognitive and self-regulated learning in science. In K. Nettleton, & L. Lennex (Eds.), *Cases on 3D technology application and integration in education* (pp. 1–25). Hershey, PA: Information Science Reference. doi:10.4018/978-1-4666-2815-1.ch001

Constant, J. (2012). Digital approaches to visualization of geometric problems in wooden Sangaku tablets. In A. Ursyn (Ed.), *Biologically-inspired computing for the arts: Scientific data through graphics* (pp. 240–253). Hershey, PA: Information Science Reference. doi:10.4018/978-1-4666-0942-6.ch013

Curran, K., McFadden, D., & Devlin, R. (2013). The role of augmented reality within ambient intelligence. In K. Curran (Ed.), *Pervasive and ubiquitous technology innovations for ambient intelligence environments* (pp. 73–88). Hershey, PA: Information Science Reference.

Cuzzocrea, F., Murdaca, A. M., & Oliva, P. (2013). Using precision teaching method to improve foreign language and cognitive skills in university students. In A. Cartelli (Ed.), *Fostering 21st century digital literacy and technical competency* (pp. 201–211). Hershey, PA: Information Science Reference.

D'Alba, A., Najmi, A., Gratch, J., & Bigenho, C. (2013). Virtual learning environments. The oLTECx: A study of participant attitudes and experiences. In R. Ferdig (Ed.), *Design, utilization, and analysis of simulations and game-based educational worlds* (pp. 35–50). Hershey, PA: Information Science Reference. doi:10.4018/978-1-4666-4018-4.ch003

Da Silva, M. P., & Courboulay, V. (2013). Implementation and evaluation of a computational model of attention for computer vision. In M. Pomplun, & J. Suzuki (Eds.), *Developing and applying biologically-inspired vision systems: Interdisciplinary concepts* (pp. 273–306). Hershey, PA: Information Science Reference. doi:10.4018/978-1-4666-3994-2.ch022

Dallalzadeh, E., & Guru, D. S. (2013). Shape features of overlapping boundary for classification of moving vehicles. [IJCVIP]. *International Journal of Computer Vision and Image Processing*, *3*(1), 42–54. doi:10.4018/ijcvip.2013010104

Danahy, J., Wright, R., Mitchell, J., & Feick, R. (2013). Exploring ways to use 3d urban models to visualize multi-scalar climate change data and mitigation change models for e-planning. [IJEPR]. *International Journal of E-Planning Research*, *2*(2), 1–17. doi:10.4018/ijepr.2013040101

Danihelka, J., Hak, R., Kencl, L., & Zara, J. (2013). 3D talking-head interface to voice-interactive services on mobile phones. In J. Lumsden (Ed.), *Developments in technologies for human-centric mobile computing and applications* (pp. 130–144). Hershey, PA: Information Science Reference.

Daskalaki, A., Giokas, K., & Koutsouris, D. (2012). Surgeon assistive augmented reality model with the use of endoscopic camera for line of vision calculation. [IJRQEH]. *International Journal of Reliable and Quality E-Healthcare*, *1*(4), 25–42. doi:10.4018/ijrqeh.2012100103

Davis, C. (2013). Flipped or inverted learning: Strategies for course design. In E. Smyth, & J. Volker (Eds.), *Enhancing instruction with visual media: Utilizing video and lecture capture* (pp. 241–265). Hershey, PA: Information Science Reference. doi:10.4018/978-1-4666-3962-1.ch017

Davy-Jow, S. L., Decker, S. J., & Schofield, D. (2013). Virtual forensic anthropology: Applications of advanced computer graphics technology to the identification of human remains. In I. Association (Ed.), *Image processing: Concepts, methodologies, tools, and applications* (pp. 832–849). Hershey, PA: Information Science Reference. doi:10.4018/978-1-4666-3994-2.ch042

De, S., Bhattacharyya, S., & Chakraborty, S. (2013). Multilevel image segmentation by a multiobjective genetic algorithm based OptiMUSIG activation function. In S. Bhattacharyya, & P. Dutta (Eds.), *Handbook of research on computational intelligence for engineering, science, and business* (pp. 122–162). Hershey, PA: Information Science Reference.

De Pauw, P. (2013). Communication systems in automotive systems. In O. Strobel (Ed.), *Communication in transportation systems* (pp. 33–69). Hershey, PA: Information Science Reference. doi:10.4018/978-1-4666-2976-9.ch002

DeCesare, J. A. (2013). Navigating multimedia: how to find internet video resources for teaching, learning, and research. In E. Smyth, & J. Volker (Eds.), *Enhancing instruction with visual media: Utilizing video and lecture capture* (pp. 11–26). Hershey, PA: Information Science Reference. doi:10.4018/978-1-4666-3962-1.ch002

Dee, M. (2013). Teaching digital natives using technology: Learning requirements, multimedia design elements, and effectiveness. In V. Wang (Ed.), *Handbook of research on technologies for improving the 21st century workforce: Tools for lifelong learning* (pp. 157–177). Hershey, PA: Information Science Publishing.

Dee, M. (2013). Using information and communication technology to maximize workforce readiness. In V. Bryan, & V. Wang (Eds.), *Technology use and research approaches for community education and professional development* (pp. 209–224). Hershey, PA: Information Science Reference.

Dehlinger, H. (2012). Drawings from small beginnings. In A. Ursyn (Ed.), *Biologically-inspired computing for the arts: Scientific data through graphics* (pp. 255–266). Hershey, PA: Information Science Reference. doi:10.4018/978-1-4666-0942-6.ch014

Dehlinger, H. (2012). Line drawings that appear unsharp. In B. Falchuk, & A. Fernandes-Marcos (Eds.), *Innovative design and creation of visual interfaces: Advancements and trends* (pp. 149–162). Hershey, PA: Information Science Reference. doi:10.4018/978-1-4666-0285-4.ch011

Delcourt, J., Mansouri, A., Sliwa, T., & Voisin, Y. (2013). An evaluation framework and a benchmark for multi/hyperspectral image compression. In M. Sarfraz (Ed.), *Intelligent computer vision and image processing: Innovation, application, and design* (pp. 56–70). Hershey, PA: Information Science Reference. doi:10.4018/978-1-4666-3906-5.ch005

Denman, S., Lin, F., Chandran, V., Sridharan, S., & Fookes, C. (2013). Improved subject identification in surveillance video using super-resolution. In R. Farrugia, & C. Debono (Eds.), *Multimedia networking and coding* (pp. 315–358). Hershey, PA: Information Science Reference.

Desai, A. M., & Forrest, E. (2013). Mobile marketing: The imminent predominance of the smartphone. In H. El-Gohary, & R. Eid (Eds.), *e-Marketing in developed and developing countries: Emerging practices* (pp. 97–115). Hershey, PA: Business Science Reference. doi:10.4018/978-1-4666-3954-6.ch007

Devyatkov, V., & Alfimtsev, A. (2013). Human-computer interaction in games using computer vision techniques. In I. Association (Ed.), *Image processing: Concepts, methodologies, tools, and applications* (pp. 1210–1231). Hershey, PA: Information Science Reference. doi:10.4018/978-1-4666-3994-2.ch061

Dickerson, J., Winslow, J., & Lee, C. Y. (2013). Teacher training and technology: Current uses and future trends. In V. Wang (Ed.), *Handbook of research on technologies for improving the 21st century workforce: Tools for lifelong learning* (pp. 243–256). Hershey, PA: Information Science Publishing.

Dima, A. M. (2013). Challenges and opportunities for innovation in teaching and learning in an interdisciplinary environment. In S. Buckley, & M. Jakovljevic (Eds.), *Knowledge management innovations for interdisciplinary education: Organizational applications* (pp. 347–365). Hershey, PA: Information Science Reference.

Ding, Y. (2013). Artificial higher order neural networks for modeling combinatorial optimization problems. In M. Zhang (Ed.), *Artificial higher order neural networks for modeling and simulation* (pp. 44–57). Hershey, PA: Information Science Reference.

Dingli, A., Attard, D., & Mamo, R. (2012). Turning homes into low-cost ambient assisted living environments. [IJACI]. *International Journal of Ambient Computing and Intelligence, 4*(2), 1–23. doi:10.4018/jaci.2012040101

Donelli, M. (2013). Design and optimization of microwave circuits and devices with the particle swarm optimizer. In G. Fornarelli, & L. Mescia (Eds.), *Swarm intelligence for electric and electronic engineering* (pp. 1–17). Hershey, PA: Engineering Science Reference.

Drigas, A., Kouremenos, D., & Vrettaros, J. (2013). Learning applications for disabled people. In I. Association (Ed.), *Digital literacy: Concepts, methodologies, tools, and applications* (pp. 1090–1103). Hershey, PA: Information Science Reference.

Driouchi, A. (2013). ICTs, youth, education, health, and prospects of further coordination. In *ICTs for health, education, and socioeconomic policies: Regional cases* (pp. 126–145). Hershey, PA: Information Science Reference. doi:10.4018/978-1-4666-3643-9.ch006

Dugas, D. P., DeMers, M. N., Greenlee, J. C., Whitford, W. G., & Klimaszewski-Patterson, A. (2013). Rapid evaluation of arid lands (REAL): A methodology. In D. Albert, & G. Dobbs (Eds.), *Emerging methods and multidisciplinary applications in geospatial research* (pp. 143–161). Hershey, PA: Information Science Reference.

El-said, S. A. (2013). Reliable face recognition using artificial neural network. [IJSDA]. *International Journal of System Dynamics Applications, 2*(2), 14–42. doi:10.4018/ijsda.2013040102

Eri, Z. D., Abdullah, R., Jabar, M. A., Murad, M. A., & Talib, A. M. (2013). Ontology-based virtual communities model for the knowledge management system environment: Ontology design. In M. Nazir Ahmad, R. Colomb, & M. Abdullah (Eds.), *Ontology-based applications for enterprise systems and knowledge management* (pp. 343–360). Hershey, PA: Information Science Reference.

Estrela, V. V., & Coelho, A. M. (2013). State-of-the art motion estimation in the context of 3D TV. In R. Farrugia, & C. Debono (Eds.), *Multimedia networking and coding* (pp. 148–173). Hershey, PA: Information Science Reference.

Falcinelli, F., & Laici, C. (2013). ICT in the classroom: New learning environment. In P. Pumilia-Gnarini, E. Favaron, E. Pacetti, J. Bishop, & L. Guerra (Eds.), *Handbook of research on didactic strategies and technologies for education: Incorporating advancements* (pp. 48–56). Hershey, PA: Information Science Reference. doi:10.4018/978-1-4666-4502-8.ch001

Farley, H. (2013). Facilitating immersion in virtual worlds: An examination of the physical, virtual, social, and pedagogical factors leading to engagement and flow. In B. Tynan, J. Willems, & R. James (Eds.), *Outlooks and opportunities in blended and distance learning* (pp. 189–203). Hershey, PA: Information Science Reference.

Farley, H., & Steel, C. (2012). Multiple sensorial media and presence in 3d environments. In G. Ghinea, F. Andres, & S. Gulliver (Eds.), *Multiple sensorial media advances and applications: New developments in MulSeMedia* (pp. 39–58). Hershey, PA: Information Science Reference.

Fernández-Madrigal, J., & Blanco Claraco, J. L. (2013). Advanced SLAM techniques. In *Simultaneous localization and mapping for mobile robots: Introduction and methods* (pp. 336–389). Hershey, PA: Information Science Reference.

Fernández-Madrigal, J., & Blanco Claraco, J. L. (2013). Maps for mobile robots: Types and construction. In *Simultaneous localization and mapping for mobile robots: Introduction and methods* (pp. 254–297). Hershey, PA: Information Science Reference.

Fernández-Madrigal, J., & Blanco Claraco, J. L. (2013). Robotic bases. In *Simultaneous localization and mapping for mobile robots: Introduction and methods* (pp. 28–59). Hershey, PA: Information Science Reference.

Ferrer-Roca, O., & Méndez, D. G. (2012). Health 4.0 in the i2i era. [IJRQEH]. *International Journal of Reliable and Quality E-Healthcare*, *1*(1), 43–57. doi:10.4018/ijrqeh.2012010105

Fialho, A. S., Cismondi, F., Vieira, S. M., Reti, S. R., Sousa, J. M., & Finkelstein, S. N. (2013). Challenges and opportunities of soft computing tools in health care delivery. In M. Cruz-Cunha, I. Miranda, & P. Gonçalves (Eds.), *Handbook of research on ICTs and management systems for improving efficiency in healthcare and social care* (pp. 321–340). Hershey, PA: Medical Information Science Reference. doi:10.4018/978-1-4666-3990-4.ch016

Fleury, M., & Al-Jobouri, L. (2013). Techniques and tools for adaptive video streaming. In D. Kanellopoulos (Ed.), *Intelligent multimedia technologies for networking applications: Techniques and tools* (pp. 65–101). Hershey, PA: Information Science Reference.

Fui, L. H., & Isa, D. (2013). Feature selection based on minimizing the area under the detection error tradeoff curve. In W. Samuelson Hong (Ed.), *Modeling applications and theoretical innovations in interdisciplinary evolutionary computation* (pp. 16–31). Hershey, PA: Information Science Reference. doi:10.4018/978-1-4666-3628-6.ch002

Fukami, T., & Wu, J. (2013). Image fusion method and the efficacy of multimodal cardiac images. In J. Wu (Ed.), *Technological advancements in biomedicine for healthcare applications* (pp. 47–54). Hershey, PA: Medical Information Science Reference.

Gaber, T. (2013). Digital rights management: Open issues to support e-commerce. In H. El-Gohary, & R. Eid (Eds.), *e-Marketing in developed and developing countries: Emerging practices* (pp. 69–87). Hershey, PA: Business Science Reference. doi:10.4018/978-1-4666-3954-6.ch005

García, M., Lloret, J., Bellver, I., & Tomás, J. (2013). Intelligent IPTV distribution for smart phones. In D. Kanellopoulos (Ed.), *Intelligent multimedia technologies for networking applications: Techniques and tools* (pp. 318–347). Hershey, PA: Information Science Reference.

García-Crespo, Á., Paniagua-Martín, F., López-Cuadrado, J. L., Carrasco, I. G., Colomo-Palacios, R., & Gómez-Berbís, J. M. (2012). Public information services for people with disabilities: An accessible multimedia platform for the diffusion of the digital signature. In I. Management Association (Ed.), Digital democracy: Concepts, methodologies, tools, and applications (pp. 594-609). Hershey, PA: Information Science Reference. doi: doi:10.4018/978-1-4666-1740-7.ch030

García-Rodríguez, J., García-Chamizo, J. M., Orts-Escolano, S., Morell-Gimenez, V., Serra-Pérez, J. A., & Angelolopoulou, A. … Viejo, D. (2013). Computer vision applications of self-organizing neural networks. In J. Garcia-Rodriguez, & M. Cazorla Quevedo (Eds.), Robotic vision: Technologies for machine learning and vision applications (pp. 129-138). Hershey, PA: Information Science Reference. doi: doi:10.4018/978-1-4666-2672-0.ch008

Garonce, F. V., & Lacerda dos Santos, G. (2013). Web conferencing as a pedagogical tool: Results from a Brazilian experience. In H. Yang, & S. Wang (Eds.), *Cases on e-learning management: Development and implementation* (pp. 371–382). Hershey, PA: Information Science Reference.

Gavrilova, M. L., & Monwar, M. (2013). Biometric image processing. In *Multimodal biometrics and intelligent image processing for security systems* (pp. 26–46). Hershey, PA: Information Science Reference. doi:10.4018/978-1-4666-3646-0.ch003

Gavrilova, M. L., & Monwar, M. (2013). Fuzzy fusion for multimodal biometric. In *Multimodal biometrics and intelligent image processing for security systems* (pp. 98–111). Hershey, PA: Information Science Reference. doi:10.4018/978-1-4666-3646-0.ch007

Gavrilova, M. L., & Monwar, M. (2013). Multimodal biometric system and information fusion. In *Multimodal biometrics and intelligent image processing for security systems* (pp. 48–68). Hershey, PA: Information Science Reference. doi:10.4018/978-1-4666-3646-0.ch004

Gavrilova, M. L., & Monwar, M. (2013). Overview of biometrics and biometrics systems. In *Multimodal biometrics and intelligent image processing for security systems* (pp. 6–25). Hershey, PA: Information Science Reference. doi:10.4018/978-1-4666-3646-0.ch002

Gesquière, G., & Manin, A. (2013). 3D visualization of urban data based on CityGML with WebGL. In I. Association (Ed.), *Image processing: Concepts, methodologies, tools, and applications* (pp. 1410–1425). Hershey, PA: Information Science Reference. doi:10.4018/978-1-4666-3994-2.ch070

Gibbs, W. J. (2012). New media and digital learners: An examination of teaching and learning. In S. Ferris (Ed.), *Teaching, learning and the net generation: Concepts and tools for reaching digital learners* (pp. 289–309). Hershey, PA: Information Science Reference.

Gil, J., Díaz, L., Granell, C., & Huerta, J. (2013). Open source based deployment of environmental data into geospatial information infrastructures. In I. Management Association (Ed.), Geographic information systems: Concepts, methodologies, tools, and applications (pp. 952-969). Hershey, PA: Information Science Reference. doi: doi:10.4018/978-1-4666-2038-4.ch059

Giuliani, G., Ray, N., Schwarzer, S., De Bono, A., Peduzzi, P., & Dao, H. … Lehmann, A. (2013). Sharing environmental data through GEOSS. In D. Albert, & G. Dobbs (Eds.), Emerging methods and multidisciplinary applications in geospatial research (pp. 266-281). Hershey, PA: Information Science Reference. doi: doi:10.4018/978-1-4666-1951-7.ch018

Gonçalves, A. G., & De Martino, J. M. (2013). Expressive audiovisual message presenter for mobile devices. [IJHCR]. *International Journal of Handheld Computing Research*, *4*(1), 70–83. doi:10.4018/jhcr.2013010105

Gonçalves, P. J., Almeida, R. J., Pinto, J. R., Vieira, S. M., & Sousa, J. M. (2013). Image based classification platform: Application to breast cancer diagnosis. In M. Cruz-Cunha, I. Miranda, & P. Gonçalves (Eds.), *Handbook of research on ICTs and management systems for improving efficiency in healthcare and social care* (pp. 595–613). Hershey, PA: Medical Information Science Reference. doi:10.4018/978-1-4666-3990-4.ch031

Graça, J. I. (2012). "Casa do Conhecimento" (Knowledge House): Open innovation case in an urban context. In H. Rahman, & I. Ramos (Eds.), *Cases on SMEs and open innovation: Applications and investigations* (pp. 198–211). Hershey, PA: Business Science Reference.

Griffiths, M., & Chinnasamy, S. (2013). Malaysia's internet governance dilemma. In Z. Mahmood (Ed.), *Developing e-government projects: Frameworks and methodologies* (pp. 248–266). Hershey, PA: Information Science Reference. doi:10.4018/978-1-4666-4245-4.ch012

Grois, D., & Hadar, O. (2013). Advances in region-of-interest video and image processing. In R. Farrugia, & C. Debono (Eds.), *Multimedia networking and coding* (pp. 76–123). Hershey, PA: Information Science Reference. doi:10.4018/978-1-4666-3994-2.ch063

Grois, D., & Hadar, O. (2013). Region-of-interest processing and coding techniques: Overview of recent trends and directions. In D. Kanellopoulos (Ed.), *Intelligent multimedia technologies for networking applications: Techniques and tools* (pp. 126–155). Hershey, PA: Information Science Reference.

Guan, P. P., & Yan, H. (2013). A hierarchical multi-level image thresholding method based on the maximum fuzzy entropy principle. In S. Bhattacharyya, & P. Dutta (Eds.), *Handbook of research on computational intelligence for engineering, science, and business* (pp. 241–272). Hershey, PA: Information Science Reference. doi:10.4018/978-1-4666-3994-2.ch016

Guelzim, I., Hammouch, A., Mouaddib, E. M., & Aboutajdine, D. (2013). Edge detection by maximum entropy: Application to omnidirectional and perspective images. In M. Sarfraz (Ed.), *Intelligent computer vision and image processing: Innovation, application, and design* (pp. 146–159). Hershey, PA: Information Science Reference. doi:10.4018/978-1-4666-3906-5.ch011

Guo, Z., & Guo, Z. (2013). Learner engagement in computer-mediated Chinese learning. In B. Zou, M. Xing, Y. Wang, M. Sun, & C. Xiang (Eds.), *Computer-assisted foreign language teaching and learning: Technological advances* (pp. 104–117). Hershey, PA: Information Science Reference.

Gupta, J. P., Singh, N., Dixit, P., Semwal, V. B., & Dubey, S. R. (2013). Human activity recognition using gait pattern. [IJCVIP]. *International Journal of Computer Vision and Image Processing, 3*(3), 31–53. doi:10.4018/ijcvip.2013070103

Gupta, M., Jin, S., Sanders, G. L., Sherman, B. A., & Simha, A. (2012). Getting real about virtual worlds: A review. [IJVCSN]. *International Journal of Virtual Communities and Social Networking, 4*(3), 1–46. doi:doi:10.4018/jvcsn.2012070101

Gupta, M. M., Bukovsky, I., Homma, N., Solo, A. M., & Hou, Z. (2013). Fundamentals of higher order neural networks for modeling and simulation. In M. Zhang (Ed.), *Artificial higher order neural networks for modeling and simulation* (pp. 103–133). Hershey, PA: Information Science Reference.

Gwilt, I. (2013). Data-objects: Sharing the attributes and properties of digital and material culture to creatively interpret complex information. In D. Harrison (Ed.), *Digital media and technologies for virtual artistic spaces* (pp. 14–26). Hershey, PA: Information Science Reference. doi:10.4018/978-1-4666-2961-5.ch002

Ha, H., Fleites, F. C., & Chen, S. (2013). Content-based multimedia retrieval using feature correlation clustering and fusion. [IJMDEM]. *International Journal of Multimedia Data Engineering and Management, 4*(2), 46–64. doi:10.4018/jmdem.2013040103

Hackley, D. C., & Leidman, M. B. (2013). Integrating learning management systems in K-12 supplemental religious education. In A. Ritzhaupt, & S. Kumar (Eds.), *Cases on educational technology implementation for facilitating learning* (pp. 1–22). Hershey, PA: Information Science Reference. doi:10.4018/978-1-4666-3676-7.ch001

Hai-Jew, S. (2012). Adding self-discovery learning to live online conferences: Using digital poster sessions in higher education. In U. Demiray, G. Kurubacak, & T. Yuzer (Eds.), *Meta-communication for reflective online conversations: Models for distance education* (pp. 265–281). Hershey, PA: Information Science Reference.

Hai-Jew, S. (2012). High-touch interactivity around digital learning contents and virtual experiences: An initial exploration built on real-world cases. In S. Long (Ed.), *Virtual work and human interaction research* (pp. 127–147). Hershey, PA: Information Science Reference. doi:10.4018/978-1-4666-0963-1.ch008

Harnett, B. (2013). Patient centered medicine and technology adaptation. In I. Management Association (Ed.), *User-driven healthcare: Concepts, methodologies, tools, and applications* (pp. 77-98). Hershey, PA: Medical Information Science Reference. doi: doi:10.4018/978-1-4666-2770-3.ch005

Hartsell, T., & Wang, S. (2013). Introduction to technology integration and leadership. In S. Wang, & T. Hartsell (Eds.), *Technology integration and foundations for effective leadership* (pp. 1–17). Hershey, PA: Information Science Reference.

Harty, D. (2012). Drawing//digital//data: A phenomenological approach to the experience of water. In A. Ursyn (Ed.), *Biologically-inspired computing for the arts: Scientific data through graphics* (pp. 337–355). Hershey, PA: Information Science Reference. doi:10.4018/978-1-4666-0942-6.ch019

Haskell, C. (2013). 3D GameLab: Quest-based pre-service teacher education. In Y. Baek, & N. Whitton (Eds.), *Cases on digital game-based learning: Methods, models, and strategies* (pp. 302–340). Hershey, PA: Information Science Reference.

Hatzistergos, M. S. (2013). Materials characterization techniques for solar cell devices: Imaging, compositional and structural analysis. In S. Anwar, H. Efstathiadis, & S. Qazi (Eds.), *Handbook of research on solar energy systems and technologies* (pp. 294–307). Hershey, PA: Engineering Science Reference.

Hegedüs, P., Orosz, M., Hosszú, G., & Kovács, F. (2013). Multicast over location-based services. In I. Management Association (Ed.), *Geographic information systems: Concepts, methodologies, tools, and applications* (pp. 609-615). Hershey, PA: Information Science Reference. doi: doi:10.4018/978-1-4666-2038-4.ch036

Heinzel, A., Fechete, R., Söllner, J., Perco, P., Heinze, G., & Oberbauer, R. et al. (2012). Data graphs for linking clinical phenotype and molecular feature space. [IJSBBT]. *International Journal of Systems Biology and Biomedical Technologies, 1*(1), 11–25. doi:10.4018/ijsbbt.2012010102

Ho, C. (2013). Stereo-vision-based fire detection and suppression robot for buildings. In I. Management Association (Ed.), Geographic information systems: Concepts, methodologies, tools, and applications (pp. 783-797). Hershey, PA: Information Science Reference. doi: doi:10.4018/978-1-4666-2038-4.ch048

Ho, C. M. (2012). Virtual museums: Platforms, practices, prospect. In H. Yang, & S. Yuen (Eds.), *Handbook of research on practices and outcomes in virtual worlds and environments* (pp. 117–144). Hershey, PA: Information Science Publishing.

Hoonakker, P., Cartmill, R. S., Carayon, P., & Walker, J. M. (2013). Development and psychometric qualities of the SEIPS survey to evaluate CPOE/EHR implementation in ICUs. In J. Tan (Ed.), *Healthcare information technology innovation and sustainability: Frontiers and adoption* (pp. 161–179). Hershey, PA: Medical Information Science Reference. doi:10.4018/978-1-4666-2797-0.ch010

Hsieh, K. (2013). Museum without walls: Digital technology and contextual learning in the museum environment. In H. Yang, & S. Wang (Eds.), *Cases on formal and informal e-learning environments: Opportunities and practices* (pp. 241–260). Hershey, PA: Information Science Reference.

Hsu, W. H. (2013). Creating open source lecture materials: A guide to trends, technologies, and approaches in the information sciences. In S. Hai-Jew (Ed.), *Open-source technologies for maximizing the creation, deployment, and use of digital resources and information* (pp. 253–280). Hershey, PA: Information Science Reference.

Hua, G. B. (2013). Current and emerging IT tools and applications. In *Implementing IT business strategy in the construction industry* (pp. 141–166). Hershey, PA: Business Science Reference. doi:10.4018/978-1-4666-4185-3.ch007

Hua, J., & Kuang, W. (2013). Image denoising via 2-D FIR filtering approach. In J. Tian, & L. Chen (Eds.), *Intelligent image and video interpretation: Algorithms and applications* (pp. 184–198). Hershey, PA: Information Science Reference.

Hyyppä, J., Zhu, L., Liu, Z., Kaartinen, H., & Jaakkola, A. (2012). 3D city modeling and visualization for smart phone applications. In R. Chen (Ed.), *Ubiquitous positioning and mobile location-based services in smart phones* (pp. 254–296). Hershey, PA: Information Science Reference. doi:10.4018/978-1-4666-1827-5.ch010

Intagorn, S., & Lerman, K. (2013). Mining geospatial knowledge on the social web. In M. Jennex (Ed.), *Using social and information technologies for disaster and crisis management* (pp. 98–112). Hershey, PA: Information Science Reference. doi:10.4018/978-1-4666-2788-8.ch007

Isaacs, J. P., Blackwood, D. J., Gilmour, D., & Falconer, R. E. (2013). Real-time visual simulation of urban sustainability. [IJEPR]. *International Journal of E-Planning Research, 2*(1), 20–42. doi:10.4018/ijepr.2013010102

Iwaki, S. (2013). Multimodal neuroimaging to visualize human visual processing. In J. Wu (Ed.), *Biomedical engineering and cognitive neuroscience for healthcare: Interdisciplinary applications* (pp. 274–282). Hershey, PA: Medical Information Science Reference.

Jacquemin, C., Ajaj, R., Le Beux, S., d'Alessandro, C., Noisternig, M., Katz, B. F., & Planes, B. (2012). Organ augmented reality: Audio-graphical augmentation of a classical instrument. In B. Falchuk, & A. Fernandes-Marcos (Eds.), *Innovative design and creation of visual interfaces: Advancements and trends* (pp. 131–147). Hershey, PA: Information Science Reference. doi:10.4018/978-1-4666-0285-4.ch010

Jamil, M. S., Sheikh, M. A., & Li, L. (2013). A finite element study of buckling and upsetting mechanisms in laser forming of plates and tubes. In J. Davim (Ed.), *Dynamic methods and process advancements in mechanical, manufacturing, and materials engineering* (pp. 140–156). Hershey, PA: Engineering Science Reference.

Jiang, Y., Wang, S., Tan, R., Ishida, K., Ando, T., & Fujie, M. G. (2013). Motor cortex activation during mental imagery of walking: An fNIRS study. In J. Wu (Ed.), *Biomedical engineering and cognitive neuroscience for healthcare: Interdisciplinary applications* (pp. 29–37). Hershey, PA: Medical Information Science Reference.

Jonas, P. M., & Bradley, D. J. (2013). Videagogy: Using humor and videos to enhance student learning. In E. Smyth, & J. Volker (Eds.), *Enhancing instruction with visual media: Utilizing video and lecture capture* (pp. 138–147). Hershey, PA: Information Science Reference. doi:10.4018/978-1-4666-3962-1.ch010

Kamba, M. A. (2013). The influence of social networking and library 2.0 as a gateway for information access and knowledge sharing in Africa. In M. AI-Suqri, L. Lillard, & N. AI-Saleem (Eds.), Information access and library user needs in developing countries (pp. 30-42). Hershey, PA: Information Science Reference. doi: doi:10.4018/978-1-4666-4353-6.ch003

Kanovic, Ž. S., Rapaic, M. R., & Jelicic, Z. D. (2013). The generalized particle swarm optimization algorithm: Idea, analysis, and engineering applications. In G. Fornarelli, & L. Mescia (Eds.), *Swarm intelligence for electric and electronic engineering* (pp. 237–258). Hershey, PA: Engineering Science Reference.

Karmakar, G., Dooley, L. S., Karmakar, N. C., & Kamruzzaman, J. (2013). Object analysis with visual sensors and RFID. In I. Association (Ed.), *Image processing: Concepts, methodologies, tools, and applications* (pp. 1492–1507). Hershey, PA: Information Science Reference. doi:10.4018/978-1-4666-3994-2.ch073

Karpinsky, N., & Zhang, S. (2013). 3D shape compression using holoimage. In I. Association (Ed.), *Image processing: Concepts, methodologies, tools, and applications* (pp. 939–956). Hershey, PA: Information Science Reference. doi:10.4018/978-1-4666-3994-2.ch047

Karray, H., Kherallah, M., Ben Halima, M., & Alimi, A. M. (2012). An interactive device for quick arabic news story browsing. [IJMCMC]. *International Journal of Mobile Computing and Multimedia Communications*, *4*(4), 62–82. doi:10.4018/jmcmc.2012100104

Kastania, A., & Moumtzoglou, A. (2012). Quality implications of the medical applications for 4G mobile phones. [IJRQEH]. *International Journal of Reliable and Quality E-Healthcare*, *1*(1), 58–67. doi:10.4018/ijrqeh.2012010106

Katkar, G., & Ghosekar, P. (2012). TexRet: A texture retrieval system using soft-computing. [IJISSC]. *International Journal of Information Systems and Social Change*, *3*(1), 37–46. doi:10.4018/jissc.2012010104

Katsura, S. (2013). Preservation and reproduction of human motion based on a motion-copying system. In J. Wu (Ed.), *Technological advancements in biomedicine for healthcare applications* (pp. 375–384). Hershey, PA: Medical Information Science Reference.

Katzenbeisser, S., Liu, H., & Steinebach, M. (2013). Challenges and solutions in multimedia document authentication. In I. Association (Ed.), *Digital rights management: Concepts, methodologies, tools, and applications* (pp. 1586–1605). Hershey, PA: Information Science Reference.

Ketabdar, H., Haji-Abolhassani, A., & Roshandel, M. (2013). MagiThings: Gestural interaction with mobile devices based on using embedded compass (magnetic field) sensor. [IJMHCI]. *International Journal of Mobile Human Computer Interaction*, *5*(3), 23–41. doi:10.4018/jmhci.2013070102

Khalid, M. S., & Hussain, R. M. (2013). Learn in your avatar: A teacher's story on integrating virtual worlds in teaching and learning. In S. D'Agustino (Ed.), *Immersive environments, augmented realities, and virtual worlds: Assessing future trends in education* (pp. 78–88). Hershey, PA: Information Science Reference.

Khan, I., & Sudirman, S. (2012). Analysis of the performance of Eigenfaces technique in recognizing non-Caucasian faces. [IJCVIP]. *International Journal of Computer Vision and Image Processing*, *2*(4), 37–50. doi:10.4018/ijcvip.2012100104

Kharrat, A., Gasmi, K., Ben Messaoud, M., Benamrane, N., & Abid, M. (2013). Medical image classification using an optimal feature extraction algorithm and a supervised classifier technique. In Y. Wang (Ed.), *Advances in abstract intelligence and soft computing* (pp. 43–56). Hershey, PA: Information Science Reference.

King, T. M., Ganti, A. S., & Froslie, D. (2013). Towards improving the testability of cloud application services. In S. Tilley, & T. Parveen (Eds.), *Software testing in the cloud: Perspectives on an emerging discipline* (pp. 322–339). Hershey, PA: Information Science Reference.

Kinsner, W., Schor, D., Fazel-Darbandi, R., Cade, B., Anderson, K., & Friesen, C. et al. (2013). The T-Sat1 nanosatellite design and implementation through a team of teams. [IJCINI]. *International Journal of Cognitive Informatics and Natural Intelligence*, 7(1), 32–57. doi:10.4018/jcini.2013010102

Kishore, V. V., & Satyanarayana, R. V. (2013). A multi-functional interactive image processing tool for lung CT images. [IJBCE]. *International Journal of Biomedical and Clinical Engineering*, 2(1), 1–11. doi:10.4018/ijbce.2013010101

Klausner, A., Trachtenberg, A., Starobinski, D., & Horenstein, M. (2013). An overview of the capabilities and limitations of smartphone sensors. [IJHCR]. *International Journal of Handheld Computing Research*, 4(2), 69–80. doi:10.4018/jhcr.2013040105

Kliazovich, D., Granelli, F., Fonseca, N., & Bouvry, P. (2013). Architectures and information signaling techniques for cognitive networks. In T. Lagkas, P. Sarigiannidis, M. Louta, & P. Chatzimisios (Eds.), *Evolution of cognitive networks and self-adaptive communication systems* (pp. 286–300). Hershey, PA: Information Science Reference. doi:10.4018/978-1-4666-4189-1.ch012

Koch, A., Dipanda, A., & Bourgeois-République, C. (2013). Direct 3D Information Determination in an Uncalibrated Stereovision System by Using Evolutionary Algorithms. In M. Sarfraz (Ed.), *Intelligent computer vision and image processing: Innovation, application, and design* (pp. 101–112). Hershey, PA: Information Science Reference. doi:10.4018/978-1-4666-3906-5.ch008

Koumpis, A. (2012). Scenes from the future: Reflections from the service factory workshop sessions. In *Management information systems for enterprise applications: Business issues, research and solutions* (pp. 266–289). Hershey, PA: Business Science Reference. doi:10.4018/978-1-4666-0164-2.ch012

Kovács, J., Bokor, L., Kanizsai, Z., & Imre, S. (2013). Review of advanced mobility solutions for multimedia networking in IPv6. In D. Kanellopoulos (Ed.), *Intelligent multimedia technologies for networking applications: Techniques and tools* (pp. 25–47). Hershey, PA: Information Science Reference.

Kranz, M., Murmann, L., & Michahelles, F. (2013). Research in the large: Challenges for large-scale mobile application research- A case study about NFC adoption using gamification via an app store. [IJMHCI]. *International Journal of Mobile Human Computer Interaction*, 5(1), 45–61. doi:10.4018/jmhci.2013010103

Krishnamoorthi, R., & Devi, S. S. (2013). Rotation invariant texture image retrieval with orthogonal polynomials model. In M. Sarfraz (Ed.), *Intelligent computer vision and image processing: Innovation, application, and design* (pp. 239–261). Hershey, PA: Information Science Reference. doi:10.4018/978-1-4666-3906-5.ch017

Kritikou, Y., Paradia, M., & Demestichas, P. (2013). Cognitive techniques for the development of services in broadband networks: The case of vocabulary learning management systems. In T. Lagkas, P. Sarigiannidis, M. Louta, & P. Chatzimisios (Eds.), *Evolution of cognitive networks and self-adaptive communication systems* (pp. 336–360). Hershey, PA: Information Science Reference. doi:10.4018/978-1-4666-4189-1.ch014

Kuehler, M., Schimke, N., & Hale, J. (2013). Privacy considerations for electronic health records. In I. Management Association (Ed.), User-driven healthcare: Concepts, methodologies, tools, and applications (pp. 1387-1402). Hershey, PA: Medical Information Science Reference. doi: doi:10.4018/978-1-4666-2770-3.ch069

Kulikova, M., Jermyn, I., Descombes, X., Zhizhina, E., & Zerubia, J. (2013). A marked point process model including strong prior shape information applied to multiple object extraction from images. In M. Sarfraz (Ed.), *Intelligent computer vision and image processing: Innovation, application, and design* (pp. 71–82). Hershey, PA: Information Science Reference. doi:10.4018/978-1-4666-3906-5.ch006

Kuo, D., Wong, D., Gao, J., & Chang, L. (2013). A 2D barcode validation system for mobile commerce. In W. Hu, & S. Mousavinezhad (Eds.), *Mobile and handheld computing solutions for organizations and end-users* (pp. 1–19). Hershey, PA: Information Science Reference.

Kwon, O., & Park, R. (2013). Watermarking for still images using a computation of the watermark weighting factor and the human visual system in the DCT domain. In K. Kondo (Ed.), *Multimedia information hiding technologies and methodologies for controlling data* (pp. 286–304). Hershey, PA: Information Science Reference.

Lakafosis, V., Gebara, E., Tentzeris, M. M., DeJean, G., & Kirovski, D. (2013). Near field authentication. In N. Karmakar (Ed.), *Advanced RFID systems, security, and applications* (pp. 74–99). Hershey, PA: Information Science Reference.

Langran, E. (2013). Caste, class, and IT in India. In I. Association (Ed.), *Digital literacy: Concepts, methodologies, tools, and applications* (pp. 976–994). Hershey, PA: Information Science Reference.

Laramée, E. A., Thokala, K. C., Webb, D., Kang, E., Kolodziej, M., Niewiarowski, P., & Xiao, Y. (2012). Seeing the unseen. In A. Ursyn (Ed.), *Biologically-inspired computing for the arts: Scientific data through graphics* (pp. 105–124). Hershey, PA: Information Science Reference. doi:10.4018/978-1-4666-0942-6.ch007

Lavin, J., & Bai, X. (2013). Developing 3D case studies for authentic learning experiences. In K. Nettleton, & L. Lennex (Eds.), *Cases on 3D technology application and integration in education* (pp. 150–173). Hershey, PA: Information Science Reference. doi:10.4018/978-1-4666-2815-1.ch007

Law, C. L., Roe, P., & Zhang, J. (2013). BioCondition assessment tool: A mobile application to aid vegetation assessment. In D. Tjondronegoro (Ed.), *Tools for mobile multimedia programming and development* (pp. 228–247). Hershey, PA: Information Science Reference.

Lawrence, L., & Abdel Nabi, D. (2013). The compilation and validation of a collection of emotional expression images communicated by synthetic and human faces. [IJSE]. *International Journal of Synthetic Emotions*, *4*(2), 34–62. doi:10.4018/ijse.2013070104

Lawrence, S. A. (2013). Teacher education in online contexts: Course design and learning experiences to facilitate literacy instruction for teacher candidates. In R. Hartshorne, T. Heafner, & T. Petty (Eds.), *Teacher education programs and online learning tools: Innovations in teacher preparation* (pp. 216–243). Hershey, PA: Information Science Reference.

Lazakidou, A., Petridou, M., & Iliopoulou, D. (2013). Computational modeling and simulations in life sciences. [IJSBBT]. *International Journal of Systems Biology and Biomedical Technologies*, *2*(2), 1–7. doi:10.4018/ijsbbt.2013040101

Lee, C., & Ho, Y. (2013). Technical challenges of 3D video coding. In R. Farrugia, & C. Debono (Eds.), *Multimedia networking and coding* (pp. 124–147). Hershey, PA: Information Science Reference.

Lee, C. K., & Sidhu, M. S. (2013). Computer-aided engineering education: New learning approaches and technologies. In V. Wang (Ed.), *Handbook of research on teaching and learning in K-20 education* (pp. 317–340). Hershey, PA: Information Science Reference. doi:10.4018/978-1-4666-4249-2.ch019

Lee, L. (2012). A new leaf. In A. Ursyn (Ed.), *Biologically-inspired computing for the arts: Scientific data through graphics* (pp. 278–288). Hershey, PA: Information Science Reference. doi:10.4018/978-1-4666-0942-6.ch016

Leni, P., Fougerolle, Y. D., & Truchetet, F. (2013). The Kolmogorov Spline network for authentication data embedding in images. In B. Igelnik, & J. Zurada (Eds.), *Efficiency and scalability methods for computational intellect* (pp. 96–114). Hershey, PA: Information Science Reference. doi:10.4018/978-1-4666-3942-3.ch005

Li, P., Lee, S., & Hsu, H. (2012). Fusion on citrus image data from cold mirror acquisition system. [IJCVIP]. *International Journal of Computer Vision and Image Processing*, *2*(4), 11–24. doi:10.4018/ijcvip.2012100102

Li, X., Lin, Z., & Wu, J. (2013). Language processing in the human brain of literate and illiterate subjects. In J. Wu (Ed.), *Biomedical engineering and cognitive neuroscience for healthcare: Interdisciplinary applications* (pp. 201–209). Hershey, PA: Medical Information Science Reference.

Li, Y. (2013). A survey of digital forensic techniques for digital libraries. In C. Wei (Ed.), *Modern library technologies for data storage, retrieval, and use* (pp. 82–101). Hershey, PA: Information Science Reference. doi:10.4018/978-1-4666-2928-8.ch007

Li, Y., & Wei, C. (2013). Digital image authentication: A review. In C. Wei (Ed.), *Modern library technologies for data storage, retrieval, and use* (pp. 102–128). Hershey, PA: Information Science Reference. doi:10.4018/978-1-4666-2928-8.ch008

Liang, S., Li, R., & Baciu, G. (2012). Cognitive garment panel design based on BSG representation and matching. [IJSSCI]. *International Journal of Software Science and Computational Intelligence*, *4*(1), 84–99. doi:10.4018/jssci.2012010104

Liao, H. (2013). Biomedical information processing and visualization for minimally invasive neurosurgery. In J. Wu (Ed.), *Technological advancements in biomedicine for healthcare applications* (pp. 36–46). Hershey, PA: Medical Information Science Reference.

Lin, H. K., & Hsieh, M. (2012). The establishment and usability evaluation on a markerless AR-based hairstyle simulation system. [IJOPCD]. *International Journal of Online Pedagogy and Course Design*, *2*(2), 100–109. doi:10.4018/ijopcd.2012040107

Lonbom, K. C. (2013). Listening to images: exploring alternate access to a digital collection. In C. Cool, & K. Ng (Eds.), *Recent developments in the design, construction, and evaluation of digital libraries: Case studies* (pp. 93–102). Hershey, PA: Information Science Reference. doi:10.4018/978-1-4666-4422-9.ch082

Lowe, G. (2013). Introduction to olfaction: Physiology. In T. Nakamoto (Ed.), *Human olfactory displays and interfaces: Odor sensing and presentation* (pp. 1–43). Hershey, PA: Information Science Reference.

Ltifi, H., Ben Ayed, M., Trabelsi, G., & Alimi, A. M. (2012). Perspective wall technique for visualizing and interpreting medical data. [IJKDB]. *International Journal of Knowledge Discovery in Bioinformatics*, *3*(2), 45–61. doi: doi:10.4018/jkdb.2012040104

Majumder, S., & Das, T. S. (2013). Watermarking of data using biometrics. In S. Bhattacharyya, & P. Dutta (Eds.), *Handbook of research on computational intelligence for engineering, science, and business* (pp. 623–648). Hershey, PA: Information Science Reference.

Mandal, S., Das, A. K., Bhowmick, P., & Chanda, B. (2013). A unified algorithm for identification of various tabular structures from document images. In C. Wei (Ed.), *Modern library technologies for data storage, retrieval, and use* (pp. 1–28). Hershey, PA: Information Science Reference. doi:10.4018/978-1-4666-2928-8.ch001

Mantoro, T., Mili$\check{s}$ic, A., & Ayu, M. (2013). Online authentication using smart card technology in mobile phone infrastructure. In I. Khalil, & E. Weippl (Eds.), *Contemporary challenges and solutions for mobile and multimedia technologies* (pp. 127–144). Hershey, PA: Information Science Reference.

Marcialis, G. L., Coli, P., & Roli, F. (2013). Fingerprint liveness detection based on fake finger characteristics. In C. Li (Ed.), *Emerging digital forensics applications for crime detection, prevention, and security* (pp. 1–17). Hershey, PA: Information Science Reference. doi:10.4018/978-1-4666-4006-1.ch001

Marlow, C. M. (2012). Making games for environmental design education: Revealing landscape architecture. [IJGCMS]. *International Journal of Gaming and Computer-Mediated Simulations*, *4*(2), 60–83. doi:10.4018/jgcms.2012040104

Martin, F., Lehmann, M., & Anderer, U. (2013). Generation of scaffold free 3-d cartilage-like microtissues from human Chondrocytes. In A. Daskalaki (Ed.), *Medical advancements in aging and regenerative technologies: Clinical tools and applications* (pp. 169–194). Hershey, PA: Medical Information Science Reference.

Martins, P., Silva, S., Oliveira, C., Ferreira, C., Silva, A., & Teixeira, A. (2012). Polygonal mesh comparison applied to the study of European Portuguese sounds. [IJCICG]. *International Journal of Creative Interfaces and Computer Graphics*, *3*(1), 28–44. doi:10.4018/jcicg.2012010103

Masrom, M., Ling, E. L., & Din, S. (2013). e-Participation behavioral in e-government in Malaysia. In Z. Mahmood (Ed.), e-Government implementation and practice in developing countries (pp. 83-97). Hershey, PA: Information Science Reference. doi: doi:10.4018/978-1-4666-4090-0.ch004

Massaro, A., Spano, F., Caratelli, D., Yarovoy, A., Cingolani, R., & Athanassiou, A. (2013). New approaches of nanocomposite materials for electromagnetic sensors and robotics. In A. Lay-Ekuakille (Ed.), *Advanced instrument engineering: Measurement, calibration, and design* (pp. 57–73). Hershey, PA: Engineering Science Reference. doi:10.4018/978-1-4666-4165-5.ch005

Mauk, M. G. (2013). Image processing for solar cell analysis, diagnostics and quality assurance inspection. In S. Anwar, H. Efstathiadis, & S. Qazi (Eds.), *Handbook of research on solar energy systems and technologies* (pp. 338–375). Hershey, PA: Engineering Science Reference. doi:10.4018/978-1-4666-3994-2.ch071

Mawhinney, H. (2013). Artifacts of expansive learning in designing a web-based performance assessment system: Institutional effects of the emergent evaluative state of educational leadership preparation in the United States. In M. Khosrow-Pour (Ed.), *Cases on assessment and evaluation in education* (pp. 92–147). Hershey, PA: Information Science Reference.

Maxfield, M. B., & Romano, D. (2013). Videos with mobile technologies: Pre-service teacher observations during the first day of school. In J. Keengwe (Ed.), *Pedagogical applications and social effects of mobile technology integration* (pp. 133–155). Hershey, PA: Information Science Reference. doi:10.4018/978-1-4666-2985-1.ch008

McCrossan, B. A., & Casey, F. A. (2013). The role of telemedicine in paediatric cardiology. In V. Gulla, A. Mori, F. Gabbrielli, & P. Lanzafame (Eds.), *Telehealth networks for hospital services: New methodologies* (pp. 44–88). Hershey, PA: Medical Information Science Reference. doi:10.4018/978-1-4666-2979-0.ch004

McLaughlin, J., Fang, S., Jacobson, S. W., Hoyme, H. E., Robinson, L., & Foroud, T. (2013). Interactive feature visualization and detection for 3d face classification. In Y. Wang (Ed.), *Cognitive informatics for revealing human cognition: Knowledge manipulations in natural intelligence* (pp. 98–110). Hershey, PA: Information Science Reference.

Melo, A., Bezerra, P., Abelém, A. J., Neto, A., & Cerqueira, E. (2013). PriorityQoE: A tool for improving the QoE in video streaming. In D. Kanellopoulos (Ed.), *Intelligent multimedia technologies for networking applications: Techniques and tools* (pp. 270–290). Hershey, PA: Information Science Reference.

Men, H., & Pochiraju, K. (2013). Algorithms for 3D map segment registration. In I. Management Association (Ed.), Geographic information systems: Concepts, methodologies, tools, and applications (pp. 502-528). Hershey, PA: Information Science Reference. doi: doi:10.4018/978-1-4666-2038-4.ch031

Meng, T., Shyu, M., & Lin, L. (2013). Multimodal information integration and fusion for histology image classification. In S. Chen, & M. Shyu (Eds.), *Multimedia data engineering applications and processing* (pp. 35–50). Hershey, PA: Information Science Reference. doi:10.4018/978-1-4666-2940-0.ch003

Mihradi, S., Ferryanto, Dirgantara, T., & Mahyuddin, A. I. (2013). Tracking of markers for 2D and 3D gait analysis using home video cameras. [IJEHMC]. *International Journal of E-Health and Medical Communications*, 4(3), 36–52. doi:10.4018/jehmc.2013070103

Milovanovic, M., Minovic, M., Štavljanin, V., & Starcevic, D. (2013). Multimedia systems development. In P. Ordóñez de Pablos, H. Nigro, R. Tennyson, S. Gonzalez Cisaro, & W. Karwowski (Eds.), *Advancing information management through semantic web concepts and ontologies* (pp. 86–104). Hershey, PA: Information Science Reference.

Minovic, M., Štavljanin, V., & Milovanovic, M. (2012). Educational games and IT professionals: Perspectives from the field. [IJHCITP]. *International Journal of Human Capital and Information Technology Professionals*, 3(4), 25–38. doi:10.4018/jhcitp.2012100103

Mirbagheri, A., Baniasad, M. A., Farahmand, F., Behza-dipour, S., & Ahmadian, A. (2013). Medical robotics: State-of-the-art applications and research challenges. [IJHISI]. *International Journal of Healthcare Information Systems and Informatics*, 8(2), 1–14. doi:10.4018/jhisi.2013040101

Mobasheri, A. (2013). Regeneration of articular carti-lage: Opportunities, challenges, and perspectives. In A. Daskalaki (Ed.), *Medical advancements in aging and regenerative technologies: Clinical tools and applica-tions* (pp. 137–168). Hershey, PA: Medical Information Science Reference.

Modegi, T. (2013). Spatial and temporal position informa-tion delivery to mobile terminals using audio watermarking techniques. In K. Kondo (Ed.), *Multimedia information hiding technologies and methodologies for controlling data* (pp. 182–207). Hershey, PA: Information Science Reference.

Modiri, A., & Kiasaleh, K. (2013). Particle swarm op-timization algorithm in electromagnetics- Case studies: Reconfigurable radiators and cancer detection. In G. Fornarelli, & L. Mescia (Eds.), *Swarm intelligence for electric and electronic engineering* (pp. 72–99). Hershey, PA: Engineering Science Reference.

Monaghan, F., Handschuh, S., & O'Sullivan, D. (2013). ACRONYM: Context metrics for linking people to user-generated media content. In A. Sheth (Ed.), *Semantic web: Ontology and knowledge base enabled tools, services, and applications* (pp. 201–234). Hershey, PA: Information Science Reference.

Morell-Gimenez, V., Orts-Escolano, S., García-Rodríguez, J., Cazorla, M., & Viejo, D. (2013). A review of registra-tion methods on mobile robots. In J. Garcia-Rodriguez, & M. Cazorla Quevedo (Eds.), *Robotic vision: Technolo-gies for machine learning and vision applications* (pp. 140–153). Hershey, PA: Information Science Reference. doi:10.4018/978-1-4666-3994-2.ch029

Morsi, Y., Li, Z., & Wang, S. (2013). Heart valve diseases in the elderly: Current treatments and future directions. In A. Daskalaki (Ed.), *Medical advancements in aging and regenerative technologies: Clinical tools and applica-tions* (pp. 240–260). Hershey, PA: Medical Information Science Reference.

Mu, Z., Jing, L., Xiaohong, Z., Lei, T., Xiao-na, F., & Shan, C. (2013). Study on low-carbon economy model and method of chinese tourism industry. In Z. Luo (Ed.), *Technological solutions for modern logistics and supply chain management* (pp. 284–317). Hershey, PA: Business Science Reference. doi:10.4018/978-1-4666-2773-4.ch018

Mura, G. (2012). The MultiPlasticity of new media. In G. Ghinea, F. Andres, & S. Gulliver (Eds.), *Multiple sensorial media advances and applications: New developments in MulSeMedia* (pp. 258–271). Hershey, PA: Information Science Reference.

Murakami, S., Kim, H., Tan, J. K., Ishikawa, S., & Aoki, T. (2013). The development of a quantitative method for the detection of periarticular osteoporosis using density features within ROIs from computed radiography images of the hand. In J. Wu (Ed.), *Technological advancements in biomedicine for healthcare applications* (pp. 55–67). Hershey, PA: Medical Information Science Reference.

Muralidhar, G. S., Bovik, A. C., & Markey, M. K. (2012). Computer-aided detection and diagnosis for 3d x-ray based breast imaging. In K. Suzuki (Ed.), *Machine learning in computer-aided diagnosis: Medical imaging intelligence and analysis* (pp. 66–85). Hershey, PA: Medical Informa-tion Science Reference. doi:10.4018/978-1-4666-0059-1.ch003

Murray, N., Qiao, Y., Lee, B., Fallon, E., & Karunakar, A. (2013). Future multimedia system: SIP or the advanced multimedia system. In K. Curran (Ed.), *Pervasive and ubiquitous technology innovations for ambient intelligence environments* (pp. 18–30). Hershey, PA: Information Science Reference.

Nadin, M. (2012). Reassessing the foundations of semi-otics: Preliminaries. [IJSSS]. *International Journal of Signs and Semiotic Systems*, 2(1), 1–31. doi:10.4018/ijsss.2012010101

Naoui, O. E., Belalem, G., & Mahmoudi, S. (2013). A reflexion on implementation version for active appear-ance model. [IJCVIP]. *International Journal of Computer Vision and Image Processing*, 3(3), 16–30. doi:10.4018/ijcvip.2013070102

Nap, H. H., & Diaz-Orueta, U. (2013). Rehabilitation gaming. In S. Arnab, I. Dunwell, & K. Debattista (Eds.), *Serious games for healthcare: Applications and implications* (pp. 50–75). Hershey, PA: Medical Information Science Reference.

Nava-Muñoz, S., & Morán, A. L. (2013). A review of notifications systems in elder care environments: Challenges and opportunities. In M. Cruz-Cunha, I. Miranda, & P. Gonçalves (Eds.), *Handbook of research on ICTs for human-centered healthcare and social care services* (pp. 407–429). Hershey, PA: Medical Information Science Reference. doi:10.4018/978-1-4666-3986-7.ch022

Niimi, M., & Noda, H. (2013). Introduction to image steganography and steganalysis. In K. Kondo (Ed.), *Multimedia information hiding technologies and methodologies for controlling data* (pp. 209–237). Hershey, PA: Information Science Reference.

Nikolaos, P. (2013). Towards a beneficial formalization of cyber entities' interactions during the e-learning process in the virtual world of "second life". In P. Renna (Ed.), *Production and manufacturing system management: Coordination approaches and multi-site planning* (pp. 278–314). Hershey, PA: Engineering Science Reference.

Nikolaos, P. (2013). Towards a theoretical "cybernetic" framework: Discovering the pedagogical value of the virtual world "second life". In K. Buragga, & N. Zaman (Eds.), *Software development techniques for constructive information systems design* (pp. 128–181). Hershey, PA: Information Science Reference. doi:10.4018/978-1-4666-3679-8.ch008

Nikolaos, P., & Ioannis, K. (2013). A qualitative research approach for the investigation and evaluation of adult users' participation factors through collaborative e-learning activities in the virtual world of "second life". In V. Wang (Ed.), *Handbook of research on teaching and learning in K-20 education* (pp. 475–500). Hershey, PA: Information Science Reference. doi:10.4018/978-1-4666-4249-2.ch028

Nikolaos, P., & Ioannis, K. (2013). e-Learning quality through second life: Exploiting, investigating, and evaluating the efficiency parameters of collaborative activities in higher education. In V. Bryan, & V. Wang (Eds.), Technology use and research approaches for community education and professional development (pp. 250-273). Hershey, PA: Information Science Reference. doi: doi:10.4018/978-1-4666-2955-4.ch015

Ning, Z. (2013). Computer system attacks. In A. Miri (Ed.), *Advanced security and privacy for RFID technologies* (pp. 45–68). Hershey, PA: Information Science Reference. doi:10.4018/978-1-4666-3685-9.ch005

Nitta, T. (2012). Learning transformations with complex-valued neurocomputing. [IJOCI]. *International Journal of Organizational and Collective Intelligence*, *3*(2), 81–116. doi:10.4018/joci.2012040103

Nobre, F. S., & Walker, D. S. (2013). A dynamic ability-based view of the organization. In M. Jennex (Ed.), *Dynamic models for knowledge-driven organizations* (pp. 163–179). Hershey, PA: Business Science Reference.

Nogueira, A. F., & Silva, C. (2013). ChronoFindMe: Social networks' location-based services. In M. Cruz-Cunha, J. Varajão, & A. Trigo (Eds.), *Sociotechnical enterprise information systems design and integration* (pp. 171–187). Hershey, PA: Business Science Reference. doi:10.4018/978-1-4666-3664-4.ch011

O'Neill, E. J. (2013). Integrated cross-cultural virtual classroom exchange program: How adaptable public schools are in Korea and the USA? In A. Edmundson (Ed.), *Cases on cultural implications and considerations in online learning* (pp. 338–374). Hershey, PA: Information Science Reference.

Oigara, J. N. (2013). Integrating technology in teacher education programs. In J. Keengwe (Ed.), *Research perspectives and best practices in educational technology integration* (pp. 28–43). Hershey, PA: Information Science Reference.

Oigara, J. N., & Keengwe, J. (2013). Pre-service teachers and technology integration with smart boards. In L. Tomei (Ed.), *Learning tools and teaching approaches through ICT advancements* (pp. 1–9). Hershey, PA: Information Science Reference.

Orfescu, C. (2012). NanoArt: Nanotechnology and art. In A. Ursyn (Ed.), *Biologically-inspired computing for the arts: Scientific data through graphics* (pp. 125–137). Hershey, PA: Information Science Reference. doi:10.4018/978-1-4666-0942-6.ch008

Ostashewski, N., & Reid, D. (2013). The iPad in the classroom: Three implementation cases highlighting pedagogical activities, integration issues, and teacher professional development strategies. In J. Keengwe (Ed.), *Pedagogical applications and social effects of mobile technology integration* (pp. 25–41). Hershey, PA: Information Science Reference. doi:10.4018/978-1-4666-2985-1.ch002

Pal, S., Bhowmick, P., Biswas, A., & Bhattacharya, B. B. (2012). Understanding digital documents using gestalt properties of isothetic components. In C. Wei, Y. Li, & C. Gwo (Eds.), *Multimedia storage and retrieval innovations for digital library systems* (pp. 183–207). Hershey, PA: Information Science Reference. doi:10.4018/978-1-4666-0900-6.ch010

Pala, K., & Gangashetty, S. V. (2013). Virtual environments can mediate continuous learning. In D. Griol Barres, Z. Callejas Carrión, & R. Delgado (Eds.), *Technologies for inclusive education: Beyond traditional integration approaches* (pp. 90–121). Hershey, PA: Information Science Reference.

Paparountas, T., Nikolaidou-Katsaridou, M. N., Rustici, G., & Aidinis, V. (2012). Data mining and meta-analysis on DNA microarray data. [IJSBBT]. *International Journal of Systems Biology and Biomedical Technologies*, *1*(3), 1–39. doi:10.4018/ijsbbt.2012070101

Pastorino, M., & Randazzo, A. (2013). Nondestructive analysis of dielectric bodies by means of an ant colony optimization method. In G. Fornarelli, & L. Mescia (Eds.), *Swarm intelligence for electric and electronic engineering* (pp. 308–325). Hershey, PA: Engineering Science Reference.

Patterson, D. C., Reiniger, R. L., & Robertson, A. (2013). Cross-reality math visualization: The SubQuan system dream realizations in immersive environments, augmented realities, and virtual worlds. In S. D'Agustino (Ed.), *Immersive environments, augmented realities, and virtual worlds: Assessing future trends in education* (pp. 57–77). Hershey, PA: Information Science Reference.

Paukkunen, M., Linnavuo, M., Haukilehto, H., & Sepponen, R. (2012). A system for detection of three-dimensional precordial vibrations. [IJMTIE]. *International Journal of Measurement Technologies and Instrumentation Engineering*, *2*(1), 52–66. doi:10.4018/ijmtie.2012010104

Paul, A. (2013). Designing biomedical stents for vascular therapy: Current perspectives and future promises. In A. Daskalaki (Ed.), *Medical advancements in aging and regenerative technologies: Clinical tools and applications* (pp. 216–239). Hershey, PA: Medical Information Science Reference.

Paul, S., & Singh, S. P. (2013). ICT in libraries: Prospects and challenges. In T. Ashraf, & P. Gulati (Eds.), *Design, development, and management of resources for digital library services* (pp. 342–358). Hershey, PA: Information Science Reference.

Peden, B. F., & Tiry, A. M. (2013). Using web surveys for psychology experiments: A case study in new media technology for research. In N. Sappleton (Ed.), *Advancing research methods with new technologies* (pp. 70–99). Hershey, PA: Information Science Reference.

Peevers, G., Douglas, G., & Jack, M. A. (2013). Multimedia technology in the financial services sector: Customer satisfaction with alternatives to face-to-face interaction in mortgage sales. In A. Mesquita (Ed.), *User perception and influencing factors of technology in everyday life* (pp. 92–106). Hershey, PA: Information Science Reference.

Peksinski, J., Stefanowski, M., & Mikolajczak, G. (2013). Estimating the level of noise in digital images. In D. Kanellopoulos (Ed.), *Intelligent multimedia technologies for networking applications: Techniques and tools* (pp. 409–433). Hershey, PA: Information Science Reference.

Pellas, N. (2013). An innovative "cybernetic" organization improvement plan through participatory action research in persistent "open source" virtual worlds. In T. Issa, P. Isaías, & P. Kommers (Eds.), *Information systems and technology for organizations in a networked society* (pp. 107–129). Hershey, PA: Business Science Reference. doi:10.4018/978-1-4666-4062-7.ch007

Peng, F., Liu, J., & Long, M. (2013). Identification of natural images and computer generated graphics based on hybrid features. In C. Li (Ed.), *Emerging digital forensics applications for crime detection, prevention, and security* (pp. 18–34). Hershey, PA: Information Science Reference. doi:10.4018/978-1-4666-4006-1.ch002

Peng, F., Zhu, X., & Long, M. (2013). An effective selective encryption scheme for H.264 video based on chaotic Qi system. [IJDCF]. *International Journal of Digital Crime and Forensics, 5*(2), 35–49. doi:10.4018/jdcf.2013040103

Perez-Sala, X., Igual, L., Escalera, S., & Angulo, C. (2013). Uniform sampling of rotations for discrete and continuous learning of 2d shape models. In J. Garcia-Rodriguez, & M. Cazorla Quevedo (Eds.), *Robotic vision: Technologies for machine learning and vision applications* (pp. 23–42). Hershey, PA: Information Science Reference.

Pet, S. R., McVerry, J. G., & O'Byrne, W. I. (2013). Multimodal response and writing as poetry experience. In R. Ferdig, & K. Pytash (Eds.), *Exploring multimodal composition and digital writing* (pp. 201–225). Hershey, PA: Information Science Reference. doi:10.4018/978-1-4666-4345-1.ch013

Petre, R., & Zaharia, T. (2013). 3D model-based semantic categorization of still image 2D objects. In S. Chen, & M. Shyu (Eds.), *Multimedia data engineering applications and processing* (pp. 151–169). Hershey, PA: Information Science Reference. doi:10.4018/978-1-4666-2940-0.ch008

Petty, S., & Benedicenti, L. (2013). Interpretive strategies for analyzing digital texts. In T. Issa, P. Isaías, & P. Kommers (Eds.), *Information systems and technology for organizations in a networked society* (pp. 53–66). Hershey, PA: Business Science Reference. doi:10.4018/978-1-4666-4062-7.ch004

Peyton, L., & Hu, J. (2013). Identity management and audit trail support for privacy protection in e-health networks. In I. Management Association (Ed.), User-driven healthcare: Concepts, methodologies, tools, and applications (pp. 1112-1125). Hershey, PA: Medical Information Science Reference. doi: doi:10.4018/978-1-4666-2770-3.ch056

Pham, T. D. (2013). The hidden Markov brains. In A. Daskalaki (Ed.), *Medical advancements in aging and regenerative technologies: Clinical tools and applications* (pp. 195–214). Hershey, PA: Medical Information Science Reference.

Pillai, K., & Ozansoy, C. (2013). Web-based digital habitat ecosystems for sustainable built environments. In P. Ordóñez de Pablos (Ed.), *Green technologies and business practices: An IT approach* (pp. 185–199). Hershey, PA: Information Science Reference. doi:10.4018/978-1-4666-4852-4.ch031

Popescu, E., & Badica, C. (2013). Creating a personalized artificial intelligence course: WELSA case study. In J. Wang (Ed.), *Information systems and modern society: Social change and global development* (pp. 31–48). Hershey, PA: Information Science Reference. doi:10.4018/978-1-4666-2922-6.ch003

Postolache, O., Girão, P., & Postolache, G. (2013). Seismocardiogram and ballistocardiogram sensing. In A. Lay-Ekuakille (Ed.), *Advanced instrument engineering: Measurement, calibration, and design* (pp. 223–246). Hershey, PA: Engineering Science Reference. doi:10.4018/978-1-4666-4165-5.ch017

Puig, D. (2013). Collaborative exploration based on simultaneous localization and mapping. In J. Garcia-Rodriguez, & M. Cazorla Quevedo (Eds.), *Robotic vision: Technologies for machine learning and vision applications* (pp. 303–332). Hershey, PA: Information Science Reference. doi:10.4018/978-1-4666-4607-0.ch077

Qi, Y. (2012). Rendering of 3D meshes by feature-guided convolution. [IJAPUC]. *International Journal of Advanced Pervasive and Ubiquitous Computing, 4*(3), 81–90. doi:10.4018/japuc.2012070105

Quinaz, F., Fazendeiro, P., Castelo-Branco, M., & Araújo, P. (2013). Soft methods for automatic drug infusion in medical care environment. In M. Cruz-Cunha, I. Miranda, & P. Gonçalves (Eds.), *Handbook of research on ICTs and management systems for improving efficiency in healthcare and social care* (pp. 830–854). Hershey, PA: Medical Information Science Reference. doi:10.4018/978-1-4666-3990-4.ch043

Ramírez, E., & Coto, E. (2012). Implant deformation on digital preoperative planning of lower extremities fractures. [IJCICG]. *International Journal of Creative Interfaces and Computer Graphics, 3*(1), 1–15. doi:10.4018/jcicg.2012010101

Redondo, E., Navarro, I., Sánchez, A., & Fonseca, D. (2013). Implementation of augmented reality in "3.0 Learning" methodology: Case studies with students of architecture degree. In B. Pătruţ, M. Pătruţ, & C. Cmeciu (Eds.), *Social media and the new academic environment: Pedagogical challenges* (pp. 391–413). Hershey, PA: Information Science Reference. doi:10.4018/978-1-4666-2851-9.ch019

Reimann, D. (2012). Shaping interactive media with the sewing machine: Smart textile as an artistic context to engage girls in technology and engineering education. In I. Management Association (Ed.), Computer engineering: Concepts, methodologies, tools and applications (pp. 1342-1351). Hershey, PA: Engineering Science Reference. doi: doi:10.4018/978-1-61350-456-7.ch517

Reio, T. G., & Hill-Grey, K. (2013). Millennial adult learners in the 21st century: Implications for adult and community educators. In V. Wang (Ed.), *Handbook of research on technologies for improving the 21st century workforce: Tools for lifelong learning* (pp. 425–439). Hershey, PA: Information Science Publishing.

Reynolds, G. L. (2013). Technology use and research approaches for community education and professional development: Diffusion of innovation from medical schools to phlebotomy certificate programs. In V. Bryan, & V. Wang (Eds.), *Technology use and research approaches for community education and professional development* (pp. 194–208). Hershey, PA: Information Science Reference.

Richardson, J. C., Sadaf, A., & Ertmer, P. A. (2013). Relationship between types of question prompts and critical thinking in online discussions. In Z. Akyol, & D. Garrison (Eds.), *Educational communities of inquiry: Theoretical framework, research and practice* (pp. 197–222). Hershey, PA: Information Science Reference.

Robertson, L., & Thomson, D. (2013). Multiple solitudes: Digital curriculum access in the pan-Canadian context. In S. Siqueira (Ed.), *Governance, communication, and innovation in a knowledge intensive society* (pp. 168–179). Hershey, PA: Information Science Reference. doi:10.4018/978-1-4666-4157-0.ch014

Rodrigues, J., Lam, R., & du Buf, H. (2012). Cortical 3D face and object recognition using 2D projections. [IJCICG]. *International Journal of Creative Interfaces and Computer Graphics, 3*(1), 45–62. doi:10.4018/jcicg.2012010104

Romeo, L., Brennan, M., Peters, T. R., & Mitchell, D. (2013). A model for effective delivery of online instruction. In P. Pumilia-Gnarini, E. Favaron, E. Pacetti, J. Bishop, & L. Guerra (Eds.), *Handbook of research on didactic strategies and technologies for education: Incorporating advancements* (pp. 571–580). Hershey, PA: Information Science Reference.

Rosenbaum, R. (2013). Large imagery on small screens: Novel technology for device adaptation in mobile services. In I. Lee (Ed.), *Mobile services industries, technologies, and applications in the global economy* (pp. 134–152). Hershey, PA: Information Science Reference. doi:10.4018/978-1-4666-3994-2.ch067

Roussinos, D., & Jimoyiannis, A. (2013). Blended collaborative learning through a wiki-based project: A case study on students' perceptions. In A. Cartelli (Ed.), *Fostering 21st century digital literacy and technical competency* (pp. 130–145). Hershey, PA: Information Science Reference.

Roxo, D., Silva, J. S., Santos, J. B., Martins, P., Castela, E., & Martins, R. (2013). Cardiac chamber contour extraction: Performance evaluation of an algorithm and physicians. In R. Martinho, R. Rijo, M. Cruz-Cunha, & J. Varajão (Eds.), *Information systems and technologies for enhancing health and social care* (pp. 270–292). Hershey, PA: Medical Information Science Reference. doi:10.4018/978-1-4666-3667-5.ch018

Rubin, S., Kountchev, R., Milanova, M., & Kountcheva, R. (2012). Multispectral image compression, intelligent analysis, and hierarchical search in image databases. [IJMDEM]. *International Journal of Multimedia Data Engineering and Management, 3*(4), 1–30. doi:10.4018/jmdem.2012100101

Rusnak, L. M. (2012). Visual arts online educational trends. In S. Kelsey, & K. St. Amant (Eds.), *Computer-mediated communication: Issues and approaches in education* (pp. 89–98). Hershey, PA: Information Science Reference.

Russell, S. (2013). Knowledge management and project management in 3D: A virtual world extension. In I. Association (Ed.), *Enterprise resource planning: Concepts, methodologies, tools, and applications* (pp. 1452–1467). Hershey, PA: Business Science Reference. doi:10.4018/978-1-4666-4153-2.ch077

Rwabutaza, A., Yang, M., & Bourbakis, N. (2012). A comparative survey on cryptology-based methodologies. [IJISP]. *International Journal of Information Security and Privacy*, *6*(3), 1–37. doi:10.4018/jisp.2012070101

Saijo, Y. (2013). Biomedical application of multimodal ultrasound microscope. In J. Wu (Ed.), *Technological advancements in biomedicine for healthcare applications* (pp. 27–35). Hershey, PA: Medical Information Science Reference.

Sala, N. (2013). Fractals, computer science and beyond. In F. Orsucci, & N. Sala (Eds.), *Complexity science, living systems, and reflexing interfaces: New models and perspectives* (pp. 268–291). Hershey, PA: Information Science Reference.

Sambukova, T. V. (2013). Machine learning in studying the organism's functional state of clinically healthy individuals depending on their immune reactivity. In X. Naidenova, & D. Ignatov (Eds.), *Diagnostic test approaches to machine learning and commonsense reasoning systems* (pp. 221–248). Hershey, PA: Information Science Reference.

San Chee, Y., Gwee, S., & Tan, E. M. (2013). Learning to become citizens by enacting governorship in the statecraft curriculum: An evaluation of learning outcomes. In R. Ferdig (Ed.), *Design, utilization, and analysis of simulations and game-based educational worlds* (pp. 68–94). Hershey, PA: Information Science Reference. doi:10.4018/978-1-4666-4018-4.ch005

Sanford, K., & Merkel, L. (2012). Emergent/see: Viewing adolescents' video game creation through an emergent framework. In I. Management Association (Ed.), Computer engineering: Concepts, methodologies, tools and applications (pp. 924-939). Hershey, PA: Engineering Science Reference. doi: doi:10.4018/978-1-61350-456-7.ch409

Sannakki, S. S., Rajpurohit, V. S., Nargund, V. B., Kumar, A. R., & Yallur, P. S. (2013). Computational intelligence for pathological issues in precision agriculture. In S. Bhattacharyya, & P. Dutta (Eds.), *Handbook of research on computational intelligence for engineering, science, and business* (pp. 672–698). Hershey, PA: Information Science Reference. doi:10.4018/978-1-4666-3994-2.ch043

Santo, L., & Quadrini, F. (2013). Mold production by selective laser sintering of resin coated sands. [IJSEIMS]. *International Journal of Surface Engineering and Interdisciplinary Materials Science*, *1*(2), 1–13. doi:10.4018/ijseims.2013070101

Sasi, S. (2013). Security applications using computer vision. In J. Garcia-Rodriguez, & M. Cazorla Quevedo (Eds.), *Robotic vision: Technologies for machine learning and vision applications* (pp. 60–77). Hershey, PA: Information Science Reference.

Sato, T., & Minato, K. (2013). Differences in analysis methods of the human uncinate fasciculus using diffusion tensor MRI. In J. Wu (Ed.), *Biomedical engineering and cognitive neuroscience for healthcare: Interdisciplinary applications* (pp. 162–170). Hershey, PA: Medical Information Science Reference.

Schendel, E., Garrison, J., Johnson, P., & Van Orsdel, L. (2013). Making noise in the library: Designing a student learning environment to support a liberal education. In R. Carpenter (Ed.), *Cases on higher education spaces: Innovation, collaboration, and technology* (pp. 290–312). Hershey, PA: Information Science Reference.

Schlegel, K., Stegmaier, F., Bayerl, S., Kosch, H., & Döller, M. (2012). Exploring different optimization techniques for an external multimedia meta-search engine. [IJMDEM]. *International Journal of Multimedia Data Engineering and Management*, *3*(4), 31–51. doi:10.4018/jmdem.2012100102

Schnabel, M. A. (2012). Learning parametric designing. In N. Gu, & X. Wang (Eds.), *Computational design methods and technologies: Applications in CAD, CAM and CAE education* (pp. 56–70). Hershey, PA: Information Science Reference. doi:10.4018/978-1-61350-180-1.ch004

Schroeder, D. C., & Lee, C. W. (2013). Integrating digital technologies for spatial reasoning: Using Google SketchUp to model the real world. In D. Polly (Ed.), *Common core mathematics standards and implementing digital technologies* (pp. 110–127). Hershey, PA: Information Science Reference. doi:10.4018/978-1-4666-4086-3.ch008

Seddiqui, H., & Aono, M. (2012). Ontology instance matching based MPEG-7 resource integration. In S. Chen, & M. Shyu (Eds.), *Methods and innovations for multimedia database content management* (pp. 143–159). Hershey, PA: Information Science Reference. doi:10.4018/978-1-4666-1791-9.ch009

Sen, Z. (2013). New trends in fuzzy clustering. In V. Bhatnagar (Ed.), *Data mining in dynamic social networks and fuzzy systems* (pp. 248–288). Hershey, PA: Information Science Reference. doi:10.4018/978-1-4666-4213-3.ch012

Setiawan, A. W., Suksmono, A. B., Mengko, T. R., & Santoso, O. S. (2013). Performance evaluation of color retinal image quality assessment in asymmetric channel VQ coding. [IJEHMC]. *International Journal of E-Health and Medical Communications*, 4(3), 1–19. doi:10.4018/jehmc.2013070101

Shabayek, A. E., Morel, O., & Fofi, D. (2013). Visual behavior based bio-inspired polarization techniques in computer vision and robotics. In M. Pomplun, & J. Suzuki (Eds.), *Developing and applying biologically-inspired vision systems: Interdisciplinary concepts* (pp. 243–272). Hershey, PA: Information Science Reference. doi:10.4018/978-1-4666-3994-2.ch072

Shanthi, S., & Bhaskaran, V. M. (2013). A novel approach for detecting and classifying breast cancer in mammogram images. [IJIIT]. *International Journal of Intelligent Information Technologies*, 9(1), 21–39. doi:10.4018/jiit.2013010102

Sheeba, F., Nagar, A. K., Thamburaj, R., & Mammen, J. J. (2012). Segmentation of peripheral blood smear images using tissue-like P systems. [IJNCR]. *International Journal of Natural Computing Research*, 3(1), 16–27. doi:10.4018/jncr.2012010102

Shelton, B. E., Satwicz, T., & Caswell, T. (2013). Historical perspectives on games and education from the learning sciences. In P. Felicia (Ed.), *Developments in current game-based learning design and deployment* (pp. 339–364). Hershey, PA: Information Science Reference.

Shimogonya, Y., Ishikawa, T., Yamaguchi, T., Kumamaru, H., & Itoh, K. (2013). Computational study of the hemodynamics of cerebral aneurysm initiation. In J. Wu (Ed.), *Technological advancements in biomedicine for healthcare applications* (pp. 267–277). Hershey, PA: Medical Information Science Reference.

Shishakly, R. (2013). Computer assisted school administration in the United Arab Emirates. In F. Albadri (Ed.), *Information systems applications in the Arab education sector* (pp. 39–51). Hershey, PA: Information Science Reference.

Silva, W. B., & Rodrigues, M. A. (2013). Interactive rendering of indoor and urban environments on handheld devices by combining visibility algorithms with spatial data structures. In W. Hu, & S. Mousavinezhad (Eds.), *Mobile and handheld computing solutions for organizations and end-users* (pp. 341–358). Hershey, PA: Information Science Reference.

Siricharoen, W. V. (2013). Infographics: An approach of innovative communication tool for e-entrepreneurship marketing. [IJEEI]. *International Journal of E-Entrepreneurship and Innovation*, 4(2), 54–71. doi:10.4018/ijeei.2013040104

Smeda, N., Dakich, E., & Sharda, N. (2012). Digital storytelling with web 2.0 tools for collaborative learning. In A. Okada, T. Connolly, & P. Scott (Eds.), *Collaborative learning 2.0: Open educational resources* (pp. 145–163). Hershey, PA: Information Science Reference. doi:10.4018/978-1-4666-0300-4.ch008

Söffker, D., Fu, X., Hasselberg, A., & Langer, M. (2012). Modeling of complex human-process interaction as framework for assistance and supervisory control of technical processes. [IJITWE]. *International Journal of Information Technology and Web Engineering, 7*(1), 46–66. doi:10.4018/jitwe.2012010104

Sokouti, B., & Sokouti, M. (2013). Enhancing security at email end point: A feasible task for fingerprint identification system. In A. Elçi, J. Pieprzyk, A. Chefranov, M. Orgun, H. Wang, & R. Shankaran (Eds.), *Theory and practice of cryptography solutions for secure information systems* (pp. 361–404). Hershey, PA: Information Science Reference. doi:10.4018/978-1-4666-4030-6.ch015

Soon, L. (2013). e-Learning and m-learning: Challenges and barriers in distance education group assignment collaboration. In D. Parsons (Ed.), Innovations in mobile educational technologies and applications (pp. 284-300). Hershey, PA: Information Science Reference. doi: doi:10.4018/978-1-4666-2139-8.ch018

Soydemir, M., & Unay, D. (2013). Context-aware medical image retrieval for improved dementia diagnosis. In D. Kanellopoulos (Ed.), *Intelligent multimedia technologies for networking applications: Techniques and tools* (pp. 434–448). Hershey, PA: Information Science Reference.

Speaker, R. B., Levitt, G., & Grubaugh, S. (2013). Professional development in a virtual world. In J. Keengwe, & L. Kyei-Blankson (Eds.), *Virtual mentoring for teachers: Online professional development practices* (pp. 122–148). Hershey, PA: Information Science Reference.

Spencer, K., Lennex, L., & Bodenlos, E. (2013). 3D technology in P12 education: Cameras, editing, and apps. In K. Nettleton, & L. Lennex (Eds.), Cases on 3D technology application and integration in education (pp. 207-230). Hershey, PA: Information Science Reference. doi: doi:10.4018/978-1-4666-2815-1.ch009

Sriraam, N., Vijayalakshmi, S., & Suresh, S. (2012). Automated screening of fetal heart chambers from 2-d ultrasound cine-loop sequences. [IJBCE]. *International Journal of Biomedical and Clinical Engineering, 1*(2), 24–33. doi:10.4018/ijbce.2012070103

Sriraam, N., Vijayalakshmi, S., & Suresh, S. (2012). Computer-aided fetal cardiac scanning using 2d ultrasound: Perspectives of fetal heart biometry. [IJBCE]. *International Journal of Biomedical and Clinical Engineering, 1*(1), 1–13. doi:10.4018/ijbce.2012010101

Srivastava, R. (2013). PDE-based image processing: Image restoration. In I. Management Association (Ed.), Geographic information systems: Concepts, methodologies, tools, and applications (pp. 569-607). Hershey, PA: Information Science Reference. doi: doi:10.4018/978-1-4666-2038-4.ch035

Stock, M. (2012). Flow simulation with vortex elements. In A. Ursyn (Ed.), *Biologically-inspired computing for the arts: Scientific data through graphics* (pp. 18–30). Hershey, PA: Information Science Reference. doi:10.4018/978-1-4666-0942-6.ch002

Strader, T. J. (2013). Digital convergence and horizontal integration strategies. In I. Management Association (Ed.), IT policy and ethics: Concepts, methodologies, tools, and applications (pp. 1579-1607). Hershey, PA: Information Science Reference. doi: doi:10.4018/978-1-4666-2919-6.ch070

Swaminathan, R., Schleicher, R., Burkard, S., Agurto, R., & Koleczko, S. (2013). Happy measure: Augmented reality for mobile virtual furnishing. [IJMHCI]. *International Journal of Mobile Human Computer Interaction, 5*(1), 16–44. doi:10.4018/jmhci.2013010102

Swanger, T., Whitlock, K., Scime, A., & Post, B. P. (2013). ANGEL mining. In I. Association (Ed.), *Data mining: Concepts, methodologies, tools, and applications* (pp. 837–858). Hershey, PA: Information Science Reference.

Sylaiou, S., White, M., & Liarokapis, F. (2013). Digital heritage systems: The ARCO evaluation. In M. Garcia-Ruiz (Ed.), *Cases on usability engineering: Design and development of digital products* (pp. 321–354). Hershey, PA: Information Science Reference. doi:10.4018/978-1-4666-4046-7.ch014

Tamilselvi, P. R., & Thangaraj, P. (2012). A modified watershed segmentation method to segment renal calculi in ultrasound kidney images. [IJIIT]. *International Journal of Intelligent Information Technologies, 8*(1), 46–61. doi:10.4018/jiit.2012010104

Tamiya, T., Kawanishi, M., Miyake, K., Kawai, N., & Guo, S. (2013). Neurosurgical operations using navigation microscope integration system. In J. Wu (Ed.), *Technological advancements in biomedicine for healthcare applications* (pp. 128–138). Hershey, PA: Medical Information Science Reference.

Tan, K. L., Lim, C. K., & Talib, A. Z. (2013). Mobile virtual heritage exploration with heritage hunt with a case study of George Town, Penang, Malaysia. In S. Nasir (Ed.), *Modern entrepreneurship and e-business innovations* (pp. 115–126). Hershey, PA: Business Science Reference.

Tan, W. H. (2013). Game coaching system design and development: A retrospective case study of FPS trainer. [IJGBL]. *International Journal of Game-Based Learning*, 3(2), 77–90. doi:10.4018/ijgbl.2013040105

Tezcan, M. (2013). Social e-entrepreneurship, employment, and e-learning. In T. Torres-Coronas, & M. Vidal-Blasco (Eds.), *Social e-enterprise: Value creation through ICT* (pp. 133–147). Hershey, PA: Information Science Reference.

Thackray, S. D., Bourantas, C. V., Loh, P. H., Tsakanikas, V. D., & Fotiadis, D. I. (2013). Optical coherence tomography image interpretation and image processing methodologies. In I. Association (Ed.), *Image processing: Concepts, methodologies, tools, and applications* (pp. 513–528). Hershey, PA: Information Science Reference. doi:10.4018/978-1-4666-3994-2.ch026

Theng, Y., Luo, Y., & Sau-Mei, G. T. (2012). QiVMDL - Towards a socially constructed virtual museum and digital library for the preservation of cultural heritage: A case of the Chinese "Qipao". In C. Wei, Y. Li, & C. Gwo (Eds.), *Multimedia storage and retrieval innovations for digital library systems* (pp. 311–328). Hershey, PA: Information Science Reference. doi:10.4018/978-1-4666-0900-6.ch016

Theodosopoulou, M., & Papalois, V. (2013). Proverbial storytelling and lifelong learning in the home-school dialogue. In P. Pumilia-Gnarini, E. Favaron, E. Pacetti, J. Bishop, & L. Guerra (Eds.), *Handbook of research on didactic strategies and technologies for education: Incorporating advancements* (pp. 317–326). Hershey, PA: Information Science Reference. doi:10.4018/978-1-4666-4502-8.ch047

Torii, I., Okada, Y., Onogi, M., & Ishii, N. (2013). Inexpensive, simple and quick photorealistic 3DCG modeling. In I. Association (Ed.), *Image processing: Concepts, methodologies, tools, and applications* (pp. 550–561). Hershey, PA: Information Science Reference. doi:10.4018/978-1-4666-3994-2.ch028

Torrens, F., & Castellano, G. (2012). Bundlet model for single-wall carbon nanotubes, nanocones and nanohorns. [IJCCE]. *International Journal of Chemoinformatics and Chemical Engineering*, 2(1), 48–98. doi:10.4018/IJCCE.2012010105

Toumpaniaris, P., Lazakidou, A., & Koutsouris, D. (2013). Methods for the evaluation of right ventricular volume using ultrasound on a catheter, in intensive care unit. [IJSBBT]. *International Journal of Systems Biology and Biomedical Technologies*, 2(1), 35–50. doi:10.4018/ijsbbt.2013010104

Tournier, J., Donde, V., Li, Z., & Naef, M. (2012). Potential of general purpose graphic processing unit for energy management system. [IJDST]. *International Journal of Distributed Systems and Technologies*, 3(2), 72–82. doi:10.4018/jdst.2012040105

Truckenbrod, J. (2012). Physicalizing the image, physicalizing the digital. [IJACDT]. *International Journal of Art, Culture and Design Technologies*, 2(1), 1–9. doi:10.4018/ijacdt.2012010101

Tsolis, D., & Sioutas, S. (2013). Digital rights management in peer to peer cultural networks. In I. Association (Ed.), *Digital rights management: Concepts, methodologies, tools, and applications* (pp. 981–1002). Hershey, PA: Information Science Reference.

Turel, V., & McKenna, P. (2013). Design of language learning software. In B. Zou, M. Xing, Y. Wang, M. Sun, & C. Xiang (Eds.), *Computer-assisted foreign language teaching and learning: Technological advances* (pp. 188–209). Hershey, PA: Information Science Reference.

Tutwiler, M. S., & Grotzer, T. (2013). Why immersive, interactive simulation belongs in the pedagogical toolkit of "next generation" science: Facilitating student understanding of complex causal dynamics. In M. Khine, & I. Saleh (Eds.), *Approaches and strategies in next generation science learning* (pp. 127–146). Hershey, PA: Information Science Reference. doi:10.4018/978-1-4666-2809-0.ch007

Tzafestas, S. G., & Mantelos, A. (2013). Time delay and uncertainty compensation in bilateral telerobotic systems: State-of-art with case studies. In M. Habib, & J. Davim (Eds.), *Engineering creative design in robotics and mechatronics* (pp. 208–238). Hershey, PA: Engineering Science Reference. doi:10.4018/978-1-4666-4225-6.ch013

Unterweger, A. (2013). Compression artifacts in modern video coding and state-of-the-art means of compensation. In R. Farrugia, & C. Debono (Eds.), *Multimedia networking and coding* (pp. 28–49). Hershey, PA: Information Science Reference.

Upadhyay, S., Tiwari, S., & Singh, S. K. (2013). Intelligent video authentication: Algorithms and applications. In J. Tian, & L. Chen (Eds.), *Intelligent image and video interpretation: Algorithms and applications* (pp. 1–41). Hershey, PA: Information Science Reference.

Ursyn, A. (2012). Aesthetic expectations for information visualization. In B. Falchuk, & A. Fernandes-Marcos (Eds.), *Innovative design and creation of visual interfaces: Advancements and trends* (pp. 13–33). Hershey, PA: Information Science Reference. doi:10.4018/978-1-4666-0285-4.ch002

Ursyn, A. (2012). Looking at science through water. In A. Ursyn (Ed.), *Biologically-inspired computing for the arts: Scientific data through graphics* (pp. 161–206). Hershey, PA: Information Science Reference. doi:10.4018/978-1-4666-0942-6.ch011

Ursyn, A. (2012). Visual tweet: Nature inspired visual statements. In A. Ursyn (Ed.), *Biologically-inspired computing for the arts: Scientific data through graphics* (pp. 207–239). Hershey, PA: Information Science Reference. doi:10.4018/978-1-4666-0942-6.ch012

Ursyn, A., & Banissi, E. (2012). Selected ideas and methods in knowledge visualization. [IJCICG]. *International Journal of Creative Interfaces and Computer Graphics*, *3*(2), 1–7. doi:10.4018/jcicg.2012070101

Vardasca, R., Costa, A., Mendes, P. M., Novais, P., & Simoes, R. (2013). Information and technology implementation issues in AAL solutions. [IJEHMC]. *International Journal of E-Health and Medical Communications*, *4*(2), 1–17. doi:10.4018/jehmc.2013040101

Vento, M., & Foggia, P. (2013). Graph matching techniques for computer vision. In X. Bai, J. Cheng, & E. Hancock (Eds.), *Graph-based methods in computer vision: Developments and applications* (pp. 1–41). Hershey, PA: Information Science Reference.

Viet, V. Q., Negera, A. F., Thang, H. M., & Choi, D. (2013). Energy saving in forward fall detection using mobile accelerometer. [IJDST]. *International Journal of Distributed Systems and Technologies*, *4*(1), 78–94. doi:10.4018/jdst.2013010106

Vieyres, P., Sandoval, J., Josserand, L., Novales, C., Chiccoli, M., & Morette, N. et al. (2013). An anticipative control approach and interactive GUI to enhance the rendering of the distal robot interaction with its environment during robotized tele-echography: Interactive platform for robotized tele-echography. [IJMSTR]. *International Journal of Monitoring and Surveillance Technologies Research*, *1*(3), 1–19. doi:10.4018/ijmstr.2013070101

Walker, J. F. (2012). Getting closer to nature: Artists in the lab. In A. Ursyn (Ed.), *Biologically-inspired computing for the arts: Scientific data through graphics* (pp. 322–336). Hershey, PA: Information Science Reference. doi:10.4018/978-1-4666-0942-6.ch018

Walker, J. F. (2012). On not being able to draw a mousetrap. In B. Falchuk, & A. Fernandes-Marcos (Eds.), *Innovative design and creation of visual interfaces: Advancements and trends* (pp. 251–267). Hershey, PA: Information Science Reference. doi:10.4018/978-1-4666-0285-4.ch016

Wang, J. (2013). Design and virtual reality. In *Challenging ICT applications in architecture, engineering, and industrial design education* (pp. 177–197). Hershey, PA: Engineering Science Reference.

Wang, K., Lavoué, G., Denis, F., & Baskurt, A. (2013). Blind watermarking of three-dimensional meshes: Review, recent advances and future opportunities. In I. Association (Ed.), *Digital rights management: Concepts, methodologies, tools, and applications* (pp. 1559–1585). Hershey, PA: Information Science Reference.

Wang, S., & Zhan, H. (2012). Enhancing teaching and learning with digital storytelling. In L. Tomei (Ed.), *Advancing education with information communication technologies: Facilitating new trends* (pp. 179–191). Hershey, PA: Information Science Reference.

Wang, Y., Berwick, R. C., Haykin, S., Pedrycz, W., Kinsner, W., & Baciu, G. ... Gavrilova, M. L. (2013). Cognitive informatics and cognitive computing in year 10 and beyond. In Y. Wang (Ed.), Cognitive informatics for revealing human cognition: Knowledge manipulations in natural intelligence (pp. 140-157). Hershey, PA: Information Science Reference. doi: doi:10.4018/978-1-4666-2476-4.ch010

Wei, H., Zuo, Q., & Guan, X. (2013). Main retina information processing pathways modeling. In Y. Wang (Ed.), *Cognitive informatics for revealing human cognition: Knowledge manipulations in natural intelligence* (pp. 54–69). Hershey, PA: Information Science Reference.

Wei, L. H., Phooi, S. K., & Li-Minn, A. (2013). Audio and visual speech recognition recent trends. In J. Tian, & L. Chen (Eds.), *Intelligent image and video interpretation: Algorithms and applications* (pp. 42–86). Hershey, PA: Information Science Reference.

Weiss, A. (2013). Wrestling with mosasaurs: Results of the sternberg museum of natural history-forsyth library fossil digitization pilot project. In C. Cool, & K. Ng (Eds.), *Recent developments in the design, construction, and evaluation of digital libraries: Case studies* (pp. 103–124). Hershey, PA: Information Science Reference.

Werderich, D. E., & Manderino, M. (2013). The multimedia memoir: Leveraging multimodality to facilitate the teaching of narrative writing for preservice teachers. In R. Ferdig, & K. Pytash (Eds.), *Exploring multimodal composition and digital writing* (pp. 316–330). Hershey, PA: Information Science Reference. doi:10.4018/978-1-4666-4345-1.ch019

Wu, Y., & Koszalka, T. A. (2013). Instructional design of an advanced interactive discovery environment: Exploring team communication and technology use in virtual collaborative engineering problem solving. In I. Management Association (Ed.), Industrial engineering: Concepts, methodologies, tools, and applications (pp. 117-136). Hershey, PA: Engineering Science Reference. doi: doi:10.4018/978-1-4666-1945-6.ch009

Xiao, J. (2013). Motion segmentation and matting by graph cut. In X. Bai, J. Cheng, & E. Hancock (Eds.), *Graph-based methods in computer vision: Developments and applications* (pp. 95–117). Hershey, PA: Information Science Reference.

Xiao, Y., Yu, W., & Tian, J. (2013). Thresholding selection based on fuzzy entropy and bee colony algorithm for image segmentation. In J. Tian, & L. Chen (Eds.), *Intelligent image and video interpretation: Algorithms and applications* (pp. 165–183). Hershey, PA: Information Science Reference.

Xu, X., Chen, L., Zhang, X., Chen, D., Liu, X., & Fu, X. (2013). A survey of human activity interpretation in image and video sequence. In J. Tian, & L. Chen (Eds.), *Intelligent image and video interpretation: Algorithms and applications* (pp. 87–124). Hershey, PA: Information Science Reference.

Yanagida, Y., & Tomono, A. (2013). Basics for olfactory display. In T. Nakamoto (Ed.), *Human olfactory displays and interfaces: Odor sensing and presentation* (pp. 60–85). Hershey, PA: Information Science Reference.

Yang, J. (2012). Faculty adopters of podcasting: Satisfaction, university support and belief in podcasting. In L. Tomei (Ed.), *Advancing education with information communication technologies: Facilitating new trends* (pp. 356–370). Hershey, PA: Information Science Reference.

Yang, Y. (2013). A stereo matching system. [IJAPUC]. *International Journal of Advanced Pervasive and Ubiquitous Computing*, 5(2), 1–8. doi:10.4018/japuc.2013040101

Yokoi, H., Kato, R., Mori, T., Yamamura, O., & Kubota, M. (2013). Functional electrical stimulation based on interference-driven PWM signals for neuro-rehabilitation. In J. Wu (Ed.), *Technological advancements in biomedicine for healthcare applications* (pp. 180–192). Hershey, PA: Medical Information Science Reference.

Yu, Y., Yang, J., & Wu, J. (2013). Cognitive functions and neuronal mechanisms of tactile working memory. In J. Wu (Ed.), *Biomedical engineering and cognitive neuroscience for healthcare: Interdisciplinary applications* (pp. 89–98). Hershey, PA: Medical Information Science Reference.

Yu, Z., Liang, Y., Yang, Y., & Guo, B. (2013). Supporting social interaction in campus-scale environments by embracing mobile social networking. In G. Xu, & L. Li (Eds.), *Social media mining and social network analysis: Emerging research* (pp. 182–201). Hershey, PA: Information Science Reference.

Zemcik, P., Spanel, M., Krsek, P., & Richter, M. (2013). Methods of 3D object shape acquisition. In I. Association (Ed.), *Image processing: Concepts, methodologies, tools, and applications* (pp. 473–497). Hershey, PA: Information Science Reference. doi:10.4018/978-1-4666-3994-2.ch024

Zhang, L. (2013). A modified landweber iteration algorithm using updated sensitivity matrix for electrical impedance tomography. [IJAPUC]. *International Journal of Advanced Pervasive and Ubiquitous Computing*, *5*(1), 17–29. doi:10.4018/japuc.2013010103

Zhao, Z., Tian, X., Dong, Z., & Xu, K. (2012). Fabrication of nanoelectrodes by cutting carbon nanotubes assembled by di-electrophoresis based on atomic force microscope. [IJIMR]. *International Journal of Intelligent Mechatronics and Robotics*, *2*(3), 1–13. doi:10.4018/ijimr.2012070101

Zheng, D., Baciu, G., Hu, J., & Xu, H. (2012). Cognitive weave pattern prioritization in fabric design: An application-oriented approach. [IJCINI]. *International Journal of Cognitive Informatics and Natural Intelligence*, *6*(1), 72–99. doi:10.4018/jcini.2012010104

Zimeras, S. (2012). An efficient 3D segmentation method for spinal canal applied to CT volume sequence data. [IJRQEH]. *International Journal of Reliable and Quality E-Healthcare*, *1*(1), 33–42. doi:10.4018/ijrqeh.2012010104

Zimeras, S. (2012). Spatial resolution of shapes in gamma camera imaging using an exact formula for solid angle of view. [IJSBBT]. *International Journal of Systems Biology and Biomedical Technologies*, *1*(1), 35–51. doi:10.4018/ijsbbt.2012010104

Zinger, S., Do, L., de With, P. H., Petrovic, G., & Morvan, Y. (2013). Free-viewpoint 3DTV: View interpolation, coding, and streaming. In R. Farrugia, & C. Debono (Eds.), *Multimedia networking and coding* (pp. 235–253). Hershey, PA: Information Science Reference.

Zou, L., Wang, X., Shi, G., & Ma, Z. (2013). EEG feature extraction and pattern classification based on motor imagery in brain-computer interface. In Y. Wang (Ed.), *Advances in abstract intelligence and soft computing* (pp. 57–69). Hershey, PA: Information Science Reference.

Zygomalas, A., Giokas, K., & Koutsouris, D. (2012). Modular assembly micro-robots for natural orifice transluminal endoscopic surgery, the future of minimal invasive surgery. [IJRQEH]. *International Journal of Reliable and Quality E-Healthcare*, *1*(4), 43–55. doi:10.4018/ijrqeh.2012100104

Zygouris-Coe, V. I. (2013). A model for online instructor training, support, and professional development. In J. Keengwe, & L. Kyei-Blankson (Eds.), *Virtual mentoring for teachers: Online professional development practices* (pp. 97–121). Hershey, PA: Information Science Reference.

Compilation of References

Ackermann, U. (1992). *Essentials of human physiology*. St Louis, MO: Mosby Yearbook.

Aldous. (2004). Retrieved Nov 2012 from www.aldous.net/photo/project07

Aldridge, R., & Sykes, D. (2008). StereoData maker. *Journal of 3D Imaging. Stereoscopic Society, 180*, 25.

Arai, J., Kawai, H., & Okano, F. (2006). Microlens arrays for integral imaging system. *Applied Optics, 45*(36), 9066–9078. doi:10.1364/AO.45.009066 PMID:17151745

Arai, J., Okui, M., Yamashita, T., & Okano, F. (2006). Integral three-dimensional television using a 2000–scanning-line video system. *Applied Optics, 45*, 1704–1712. doi:10.1364/AO.45.001704 PMID:16572684

Arimoto, A., Ooshima, T., Tani, T., & Kaneko, Y. (1998). Wide viewing area glassless stereoscopic display using multiple projectors.[SPIE.]. *Proceedings of the Society for Photo-Instrumentation Engineers, 3295*, 186–191. doi:10.1117/12.307163

Baird, J. L. (1945). *Improvements in television* (UK Patent 573,008). London: UK Patent Office.

Bak, M. A. (2012). Democracy and discipline: Object lessons and the stereoscope in American education, 1870–1920. *Early Popular Visual Culture, 10*, 147–167. doi:10.1080/17460654.2012.664746

Balasubramonian, K. et al. (1982). On the merits of bicircular polarisation for stereo colour TV. *IEEE Transactions on Consumer Electronics, 28*.

Baloch, T. (2001). *Method and apparatus for displaying three-dimensional images* (United States Patent No. 6,201,565 B1). Washington, DC: US Patent Office.

Barabas, J., Jolly, S., Smalley, D. E., & Bove, V. M. (2011). Diffraction specific coherent panoramagrams of real scenes. In *Proceedings of the SPIE*, (vol. 7957). SPIE.

Barilleaux, R. P. (Ed.). (1984). Holography (re)defined: Innovation through tradition. New York: Museum of Holography.

Baths, R. (1982). Death of the author. In *A Barthes reader*. New York: Hill and Wang.

Benton, S. A. (2002). *Autostereoscopic display system* (United States Patent No. 6,351,280). Washington, DC: US Patent Office.

Benton, S. A. (1969). Hologram reconstruction with extended light sources. *Journal of the Optical Society of America, 59*, 1545.

Berkhout, R. (1989). Holography: Exploring a new art realm and shaping empty space with light. *Leonardo, 22*, 313–316. doi:10.2307/1575385

Biederman, I. (1987). Recognition-by-components: A theory of human image understanding. *Psychological Review, 94*(2), 115–147. doi:10.1037/0033-295X.94.2.115 PMID:3575582

Bjelkhagen, H. I., & Kostuk, R. K. (2008). Article. *Proceedings of the SPIE, 6912*.

Bjelkhagen, H. I. (1999). Lippmann photography: Reviving an early colour process. *History of Photography, 23*(3), 274–280.

Blanche, P.-A. et al. (2010). Holographic three-dimensional telepresence using large-area photorefractive polymer. *Nature, 468*(7320), 80–83. doi:10.1038/nature09521 PMID:21048763

Blundell, B. G., & Schwartz, A. J. (2000). *Volumetric three-dimensional display systems.* New York: Wiley-IEEE Press.

Boev, A., Raunio, K., Gotchev, A., & Egiazarian, K. (2008). GPU-based algorithms for optimized visualization and crosstalk mitigation on a multiview display. In *Proceedings of SPIE-IS&T Electronic Imaging* (vol. 6803). Retrieved 29 January 2012 from http://144.206.159.178/FT/CONF/16408309/16408328.pdf

Bonnet, M. (1942). *La photographie en relief. centre de documentation.* Maison de la Rhimie.

Bradshaw, G. (1953, July 25). We can have 3-D television. *Picture Post.*

Brewin, M., Forman, M., & Davies, N. (1995). Electronic capture and display of full parallax 3D images. *Proceedings of the Society for Photo-Instrumentation Engineers, 2409,* 118–124. doi:10.1117/12.205851

Brewster, D. (1856). *The stereoscope: Its history, theory and construction.* Paris: Triage Général de Stéréoscopie.

Brown, D. (2009). *Images across space.* London: Middlesex University Press.

Burckhardt, C. B. (1967). Optimum parameters and resolution limitation of integral photography. *Journal of the Optical Society of America, 58,* 71–76. doi:10.1364/JOSA.58.000071

Burckhardt, C. B., & Doherty, E. T. (1969). Beaded plate recording of integral photographs. *Applied Optics, 8,* 2329–2331. doi:10.1364/AO.8.002329 PMID:20076020

Caulfield, H. J. (2004). *The art and science of holography: A tribute to Emmett Leith and Yuri Denisyuk.* Bellingham, WA: SPIE Press.

Collender, R. (1986). 3-D television, movies and computer graphics without glasses. *IEEE Transactions on Consumer Electronics, 32*(1), 56–61. doi:10.1109/TCE.1986.290119

Cossairt, O., Napoli, J., Hill, S. L., Dorval, R. K., & Favalora, G. E. (2007). Occlusion-capable multiview volumetric three-dimensional display. *Applied Optics, 46*(8). doi:10.1364/AO.46.001244 PMID:17318244

Crary, J. (1990). *Techniques of the observer: On vision and modernity in the nineteenth century.* Cambridge, MA: MIT Press.

Crockett, R. (2008). *Ledametrix.* Retrieved Nov 2012 from www.ledametrix.com/lancshepherd/index.html

Crombie, A. C. (1990). *Science, optics, and music in medieval and early modern thought.* New York: Continuum International Publishing Group.

Cross, L. L., & Cross, C. (1992). HoloStories: Reminiscences and a prognostication on holography. *Leonardo, 25,* 421–424. doi:10.2307/1575748

Dalí, S., & Knoedler, M. (1972). *Holograms conceived by Dali.* New York: M. Knoedler.

David White Co. (1950). People who know picture-taking and picture-making prefer the stereo realist. *Popular Photography, 26*(1), 12.

Davies, N., McCormick, M., & Brewin, M. (1994). *Design and analysis of an image transfer system.* Unpublished.

Davies, N., McCormick, M., & Yang, L. (1988). Three-dimensional imaging systems: A new development. *Applied Optics, 27,* 4520–4528. doi:10.1364/AO.27.004520 PMID:20539602

de Montebello, R. L. (1977). Wide-angle integral photography – The integram system using microlens arrays. *Optical Engineering (Redondo Beach, Calif.), 33,* 3624–3633.

Denisyuk, Y. N. (1962). Photographic reconstruction of the optical properties of an object in its own scattered radiation field. *Soviet Physics, Doklady, 7,* 543–545.

Denisyuk, Y. N. (1963). On the reproduction of the optical properties of an object by the wave field of its scattered radiation. *Opt. Spectrosc. (USSR), 15,* 279–284.

Denisyuk, Y. N. (1965). On the reproduction of the optical properties of an object by the wave field of its scattered radiation II. *Opt. Spectrosc. (USSR), 18,* 152–157.

Dimenco. (2012). *Products – Displays 3D stopping power – 52 professional 3D display.* Retrieved 29 January 2012 from http://www.dimenco.eu/displays/

Dudnikov, Y. A., Rozhkov, B. K., & Antipova, E. N. (1980). Obtaining a portrait of a person by the integral photography method. *Sov. J. Opt. Tech., 47*(9).

Dudnikov, Y. A. (1970). Autostereoscopy & integral photography. *Optics Technology, 37*, 422–426.

Earl, E. W. (Ed.). (1979). *Points of view: The stereograph in America – A cultural history*. New York: Visual Studies Workshop Press.

Eco, U. (1986). *Travels in hyper reality – Essays*. San Diego, CA: Harcourt Brace Jovanovich.

Eco, U. (1995). *Faith in fakes: Travels in hyperreality*. London: Vintage.

Eichenlaub, J. B. (1990, May). Stereo display without glasses. *Advanced Imaging (Woodbury, N.Y.)*.

Eichenlaub, J. B. (1993). Developments in autostereoscopic technology at Dimension Technologies Inc. [SPIE.]. *Proceedings of SPIE Stereoscopic Displays and Applications, 1915*, 177–186.

Elheny, V. K. (n.d.). *The life and work of Edwin Land*. Retrieved 11 November 2011 from www.nap.edu/html/biomems/eland.html

Fan, F. C., Choi, S., & Jiang, C. C. (2010). Use of spatial spectrum of light to recover three dimensional holographic nature. *Applied Optics, 49*(14), 2676–2685. doi:10.1364/AO.49.002676

Favalora, G. E. (2009). Progress in volumetric three-dimensional displays and their applications. *Opt Soc Am*. Retrieved 29 Jan 2012 from http://www.greggandjenny.com/gregg/Favalora_OSA_FiO_2009.pdf

Florczak., et al. (2001). *Sheeting with composite image that floats* (US Patent 6,288,842). Washington, DC: US Patent Office.

Freund, Y., & Schapire, R. E. (1999). A short introduction to boosting. *Journal of Japanese Society for Artificial Intelligence, 14*(5), 771–780.

Gabor, D. (1948). A new microscopic principle. *Nature, 161*, 77–79. doi:10.1038/161777a0 PMID:18860291

Gabor, D. (1949). Microscopy by reconstructed wavefronts and communication theory. *Proceedings of the Physical Society. Section A, 194*, 454–487.

Gabor, D. (1949). Microscopy by reconstructed wavefronts. *Proceedings of the Royal Society, 197*(1051), 454–487. doi:10.1098/rspa.1949.0075

Good, J. P. (1920). The scope and outlook of visual education. *School Science and Mathematics, 20*, 82–87.

Goodman, J. W. (1968). *Introduction to Fourier optics*. San Francisco: McGraw-Hill.

Graham, C. H. (1965). *Visual space perception in vision and visual perception*. New York: Wiley.

Harman, P. (1996). Autostereoscopic display system. *Proceedings of the Society for Photo-Instrumentation Engineers, 2653*, 56–64. doi:10.1117/12.237458

Hattori, T. (2000a). *Sea phone 3D display 1-8*. Retrieved 28 January 2012 from http://home.att.net/~SeaPhone/3display.htm

Hattoti, T. (2000b). *Sea phone 3D display: 7-9*. Retrieved 28 January 2012 from http://home.att.net/~SeaPhone/3display.htm

Hattoti, T., Sakuma, S., Katayama, K., Omori, S., Hayashi, M., & Midori, Y. (1994). Stereoscopic liquid crystal display 1 (general description). *Proceedings of the Society for Photo-Instrumentation Engineers, 2177*, 146–147.

Haussler, R., Schwerdtner, A., & Leister, N. (2008). Large holographic displays as an alternative to stereoscopic displays. *Proceedings of the SPIE, 6803*.

Heidegger, M. (1969). *Identity and difference* (J. Stambaugh, Trans.). New York: Harper & Row.

Hembd, C., Stevens, R., & Hutley, M. (1997). Imaging properties of the Gabor superlens. *European Optical Society Topical Meetings Digest Series, 13*, 101–104.

Herapath, W. B. (1852). Quoted. *Philosophical Magazine, 3*, 161.

Hesselink, L., Orlov, S., & Bashaw, M. C. (2004). Holographic data storage systems. *I to Three Dimensional Displays, 92*, 1231–80.

Heyda, L. (n.d.). *Freewebs*. Retrieved Nov 2012 from www.freewebs.com/larryeda/

Hitachi. (2010). *Hitachi shows 10 glasses-free 3D display.* Retrieved 24 January 2012 from www.3d-display-info.com/hitachi-shows-10-glasses-free-3d-display

Holmes, O. W. (1859, June 1). The stereoscope and the stereograph. *Atlantic Monthly.*

Hoshino, H., Okano, F., Isono, H., & Yuyama, I. (1998). Analysis of resolution limitation of integral photography. *Journal of the Optical Society of America. A, Optics, Image Science, and Vision, 15,* 2059–2065. doi:10.1364/JOSAA.15.002059

Hubel, D. H., & Wiesel, T. N. (1979). *The brain: A scientific American book.* San Francisco: Freeman.

Huebschman, M. L., Munjuluri, B., & Garner, H. R. (2003). Dynamic holographic 3-D image projection. *Opt. Exp., 11,* 437–445. doi:10.1364/OE.11.000437 PMID:19461750

Huff, L., & Fusek, R. L. (1980). Color holographic stereograms. *Optical Engineering (Redondo Beach, Calif.), 19,* 691–695. doi:10.1117/12.7972589

Ichinose, S., Tetsutani, N., & Ishibashi, M. (1989). Full-color stereoscopic video pickup and display technique without special glasses. *Proceedings of the SID, 3014,* 319-323.

Igarashi, Y., Murata, H., & Ueda, M. (1978). 3-D display system using a computer generated integral photograph. *Journal of Applied Physics, 17,* 1683–1684. doi:10.1143/JJAP.17.1683

Ijzerman, W. (2005). Design of 2d/3d switchable displays. *Proc of the SID, 36*(1), 98-101.

Ives, F. E. (1903). *U.S. patent no 725567.* Washington, DC: US Patent Office.

Ives, H. E. (1933). Optical properties of Lippmann lenticulated sheet. *Journal of the Optical Society of America, 21,* 171–176. doi:10.1364/JOSA.21.000171

Jang, J.-S., & Javidi, B. (2002). Three-dimensional synthetic aperture integral imaging. *Optics Letters, 27,* 1144–1146. doi:10.1364/OL.27.001144 PMID:18026388

Javidi, B., Okano, F., & Son, J.-Y. (2009). *Three-dimensional imaging, visualization, and display.* Berlin: Springer–Verlag. doi:10.1007/978-0-387-79335-1

Jay, B. (1991). *Cyanide and spirits: An inside-out view of early photography.* Portland, OR: Nazraeli Press.

Jeong, T. H. (1967). Cylindrical holography and some proposed applications. *Journal of the Optical Society of America, 57,* 1396–1398. doi:10.1364/JOSA.57.001396

Johnston, S. F. (2001). *A history of light and colour measurement: Science in the shadows.* New York: Taylor and Francis. doi:10.1887/0750307544

Johnston, S. F. (2006). *Holographic visions: A history of new science.* Oxford, UK: Oxford University Press. doi:10.1093/acprof:oso/9780198571223.001.0001

Johnston, S. F. (2015). *Holograms: A cultural history.* Oxford, UK: Oxford University Press.

Jones, A., McDowall, I., Yamada, H., Bolas, M., & Debevec, P. (2007). Rendering for an interactive 360° light field display. *Siggraph 2007 Emerging Technologies.* Retrieved 29 January 2012 from http://gl.ict.usc.edu/Research/3DDisplay/

Jorke, H., & Fritz, M. (2012). *Infitec – A new stereoscopic visualization tool by wavelength multiplexing.* Retrieved 25 January 2012 from http://jumbovision.com.au/files/Infitec_White_Paper.pdf

Julesz, B. (1960). Binocular depth perception of computer-generated images. *The Bell System Technical Journal, 39*(5), 1125–1163. doi:10.1002/j.1538-7305.1960.tb03954.x

Kajiki, Y. (1997). Hologram-like video images by 45-view stereoscopic display. *Proceedings of the Society for Photo-Instrumentation Engineers, 3012,* 154–166. doi:10.1117/12.274452

Kanolt, C. W. (1918). *United States patent 1260682.* Washington, DC: US Patent Office.

Keystone View Company. (1942). The stereoscope goes to war. *The Educational Screen: The Magazine Devoted Exclusively to the Visual Idea in Education, 16,* 275.

Kim, Y., Hong, K., & Lee, B. (2010). Recent researches based on integral imaging display method. *3D Research, 1,* 17–27

Klooswijk, A. I. J. (1978). Natural and photographic stereo acuity. *Stereoscopy, 5.*

Klug, M. A., Halle, M. W., & Hubel, P. M. (1992). Full color ultragrams. *Proceedings of the Society for Photo-Instrumentation Engineers, 1667,* 110–119. doi:10.1117/12.59625

Korth, F. G. (1944). Take it close up. *Popular Mechanics, 81*(1), 118–121.

Krah, C. (2010). *Three-dimensional display system* (US Patent 7,843,449). Washington, DC: US Patent Office. Retrieved 4 February 2012 from www.freepatentsonline/7843339.pdf

Krauss, R. (1985). Corpus delicti. *October, 33,* 31-72.

Lambooij, M., IJsselsteijn, W., Fortuin, M., & Heynderickx, I. (2009). Visual discomfort and visual fatigue of stereoscopic displays: a review. *The Journal of Imaging Science and Technology, 53*(3). doi:10.2352/J.ImagingSci.Technol.2009.53.3.030201

Lee, B., Min, S.-W., & Javidi, B. (2002). Theoretical analysis for three-dimensional integral imaging systems with double devices. *Applied Optics, 41,* 4856–4865. doi:10.1364/AO.41.004856 PMID:12197653

Leith, E. N., & Upatnieks, J. (1962). Reconstructed wavefronts and communication theory. *Journal of the Optical Society of America, 52*(10), 1123–1130. doi:10.1364/JOSA.52.001123

Leith, E. N., & Upatnieks, J. (1963). Wavefront reconstruction with continuous-tone objects. *Journal of the Optical Society of America, 53*(12), 1377–1381. doi:10.1364/JOSA.53.001377

Leith, E. N., & Upatnieks, J. (1964). Wavefront reconstruction with diffused illumination and three-dimensional objects. *Journal of the Optical Society of America, 54*(11), 1295–1301. doi:10.1364/JOSA.54.001295

Lippmann, G. (1891). La photographie des couleurs. *Comptes Rendus de L'Academie des Sciences, 112,* 274–3.

Lippmann, M. G. (1891). La photographie des couleurs. *Comptes Rendus Hebdomadaires des Séances de l'Académie des Sciences, 112,* 274–275.

Lippmann, M. G. (1894). Sur la théorie de la photographie des couleurs simples et composées par la méthode interférentielle. *Journal of Physics, 3*(3), 97–107.

Lippmann, M. G. (1902). La photographie des couleurs[deuxième note]. *Comptes Rendus Hebdomadaires des Séances de l'Académie des Sciences, 114,* 961–962.

Lippmann, M. G. (1908). Epreuves reversibles donnant la sensation du relief. *Journal of Physics, 4,* 821–825.

Lippmann, M. G. (1908). Épreuves réversibles: Photographies integrals. *Comptes Rendus Hebdomadaires des Séances de l'Académie des Sciences, 146,* 446–451.

Lipton, L. (1988). Method and system employing a push-pull liquid crystal modulator. US patent 4792850.

Lohmann, A. W., & Paris, D. (1967). Binary Fraunhofer holograms generated by computer. *Applied Optics, 6*(10), 1739–1748. doi:10.1364/AO.6.001739 PMID:20062296

Loy, G. (2003). *Computer vision to see people: A basis for enhanced human computer interaction.* (PhD thesis). Australian National University, Canberra, Australia.

May, B., & Vidal, E. (2009). *A village lost and found.* Bristol, UK: Canopus Publishing.

McCarthy, S. (2010). *Glossary for video and perceptual quality of stereoscopic video.* Retrieved 24th January 2012 from http://www.3dathome.org/files/ST1-01-01_Glossary.pdf

McCormick, M. (1995). Integral 3D image for broadcast. In *Proceedings of the Second International Display Workshop.* ITE.

McCormick, M., Davies, N., & Chowanietz, E. G. (1992). Restricted parallax images for 3D TV. *IEE Colloq Stereoscopic Television, 173,* 3/1-3/4.

Microsoft. (2010). *The wedge - Seeing smart displays through a new lens.* Retrieved 4 February 2012 from http://www.microsoft.com/appliedsciences/content/projects/wedge.aspx

Milman, M. (1983). *Trompe-L'oeil painting: The illusions of reality.* Paris: Skira.

Mitchell, D. J. (2010). Reflecting nature: Chemistry and comprehensibility in Gabriel Lippmann's 'physical' method of photographing colours. *Notes and Records of the Royal Society, 64,* 319–337. doi:10.1098/rsnr.2010.0072

Moller, C., & Travis, A. (2004). Flat panel time multiplexed autostereoscopic display using an optical wedge waveguide. In *Proceedings of the 11th Int Display Workshops,* (pp. 1443-1446). Niigata, Japan: Academic Press.

Mora, B., Maciejewski, R., & Chen, M. (2008). Visualization and computer graphics on isotropically emissive volumetric displays. *IEEE Computer Society*. Retrieved 28 Jan 2012 from https://engineering.purdue.edu/purpl/level2/papers/Mora_LF.pdf

Morgan, W. D., & Lester, H. M. (1954). *Stereo realist manual*. New York: Morgan and Lester.

Navarro, H., Martínez-Cuenca, R., Saavedra, G., Martínez-Corral, M., & Javidi, B. (2010). 3D integral imaging display by smart pseudoscopic-to-orthoscopic conversion (SPOC). *Optics Express, 18*, 573–583. doi:10.1364/OE.18.025573 PMID:21164903

Needham, J. (1986). *Science and civilization in China: Physics and physical technology*. Taipei, Taiwan: Caves Books Ltd.

Norling, J. A. J. (1953). The stereoscopic art. *SMPTE, 60*(3), 286–308.

Nye, D. E. (1994). *American technological sublime*. Cambridge, MA: MIT Press.

Okano, F., Hoshino, H., Arai, J., & Yuyama, I. (1997). Real-time pickup method for a three-dimensional image based on integral photography. *Applied Optics, 36*, 1598–1603. doi:10.1364/AO.36.001598 PMID:18250841

Okoshi, T. (1976). *Three-dimensional imaging techniques*. Oxford, UK: Academic Press.

Okui, M., Arai, J., & Okano, F. (2007). *New integral imaging technique uses projector*. Retrieved 28 January 2012 from http://spie.org/x15277.xml?ArticleID=x15277

Okui, M., Arai, J., Nojiri, Y., & Okano, F. (2006). Optical screen for direct projection of integral imaging. *Applied Optics, 45*, 9132–9139. doi:10.1364/AO.45.009132 PMID:17151752

Otsuka, R., Hoshino, T., & Horry, Y. (2006). Transpost: 360 deg-viewable three-dimensional display system. *Proceedings of the IEEE, 94*(3). doi:10.1109/JPROC.2006.870700

Pastoor, S. (1992). Human factors of 3DTV: An overview of current research at Heinrich-Hertz-Institut Berlin. *IEE Colloquium 'Stereoscopic Television', 173*, 1/3.

Perlin, K. (1999). *A sisplayer and a method for displaying* (International Publication No WO 99/38334).

Perrin, J. (Ed.). (1997). *Jules Richard et la magie du relief*. Paris, France: Prodieux.

Phillips, D. (1996). *Art for industry's sake: Halftone technology, mass photography, and the social transformation of American print culture 1880-1920*. (PhD thesis). Yale University, New Haven, CT.

Photographer, Uknown. (1937). Cover: Simultaneous calculator. *Technology Review, 39*(3), 1.

Photographer, Unknown. (1940). Unusual photos made with camera mirrors. *Popular Mechanics, 74*(3), 406.

Photographer, Unknown. (1947).Cover. *Discovery: The Magazine of Scientific Progress, 8*(11), 1.

Pihlainen, K. (2012). Cultural history and the entertainment age. *Cultural History, 1*, 168–179. doi:10.3366/cult.2012.0019

Pinker, S. (1997). *How the mind works*. London: Penguin Books.

Pokescope. (2005). Retrieved Nov 2012 from www.pokescope.com

Purcher, J. (2011). *Apple wins a surprise 3D display and imaging patent stunner*. Retrieved 4 February 2012 from http://www.patentlyapple.com/patently-apple/2011/09/whoa-apple-wins-a-3d-display-imaging-system-patent-stunner.html

Richardson, M. (2006). *The prime illusion: Modern holography in the new age of digital media*. London: Holographic Studio.

Rowlands, J. J., & Killian, J.R., Jr. (1937). Seeing solid: The third dimension at work and play. *Technology Review, 39* (5), 191-5, 182.

Sandlin, D. J., Margolis, T., Dawe, G., Leigh, J., & DeFanti, T. A. (2001). Varrier autostereographic display: Stereoscopic displays and virtual reality systems VIII. *Proceedings of the Society for Photo-Instrumentation Engineers, 4297*, 204–211. doi:10.1117/12.430818

Sawyer's Inc. (1949). Sell your products with view-master. *Business Screen, 10*, 8.

Saxby, G. (1988). *Practical holography*. Bristol, UK: Institute of Physics Press. doi:10.1887/0750309121

Schwartz, A. (1985). Head tracking stereoscopic display. In *Proceedings of IEEE International Display Research Conference*, (pp. 141-144). IEEE.

Schwerdtner, A., & Heidrich, H. (1998). The Dresden 3D display (D4D). *Proceedings of the Society for Photo-Instrumentation Engineers, 3295*, 203–210. doi:10.1117/12.307165

Sexton, I., & Crawford, D. (1989). Parallax barrier 3DTV. *Proceedings of the SPIE Three-Dimensional Visualization and Display Technologies, 1083*, 84–94. doi:10.1117/12.952875

Shang, X., Fan, F. C., Jiang, C. C., Choi, S., Dou, W., Yu, C., & Xu, D. (2009). Demonstration of a large size real time full color three dimensional display. *Optics Letters, 34*(24), 3803–3805. doi:10.1364/OL.34.003803 PMID:20016619

Sherman, R. A. (1953). Benefits to vision through stereoscopic films. *Journal of the Society of Motion Picture and Television Engineers, 61*, 295–308.

Slinger, C. W. (2004). Recent development in computer generated holography: Towards a practical electroholography system for interactive 3D visualization. *Proceedings of the Society for Photo-Instrumentation Engineers, 5290*, 27–41. doi:10.1117/12.526690

Sobel, D. (1996). *Longitude: The true story of a lone genius who solved the greatest scientific problem of his time*. London: Fourth Estate.

Son, J. Y., & Shestak, S. A. KIM, S.-S., & Choi, Y. J. (2001). A desktop autostereoscopic display with head tracking capability. *Proceedings of the SPIE, 4297*, 160-164.

Soneira, R. M. (2012). *3D TV display technology shootout*. Retrieved 27 January 2012 from http://www.displaymate.com/3D_TV_ShootOut_1.htm

St Hilaire, M., Lucente, P., & Benton, S. A. (1992). Synthetic aperture holography: A novel approach. *Journal of the Optical Society of America. A, Optics and Image Science, 9*, 1969–1978. doi:10.1364/JOSAA.9.001969

St Hilaire, P. (1994). Modulation transfer function and optimum sampling of holographic stereograms. *Applied Optics, 33*(5), 768–774. doi:10.1364/AO.33.000768 PMID:20862073

Stetson, K. A., & Powell, R. L. (1966). Hologram interferometry. *Journal of the Optical Society of America, 56*, 1161–1166. doi:10.1364/JOSA.56.001161

Street, G. S. B. (1998). *Autostereoscopic image display adjustable for observer location and distance* (United States Patent 5,712,732). Washington, DC: US Patent Office.

Sukhanov, V. I., & Denisyuk Yu, N. (1970). On the relationship between spatial frequency spectra of a three-dimensional object and its three-dimensional hologram. *Optics and Spectroscopy, 28*(1), 63–66.

Sullivan, A. (2004). DepthCube solid-state 3D volumetric display. *Proceedings of the SPIE, 5291*.

Suto, M., & Sykes, D. (2006). *Stereo*. Retrieved Nov 2012 from www.stereo.jpn.org

Suto, M., & Sykes, D. (2007). *Stereo*. Retrieved Nov 2012 from www.stereo.jpn.org

Swindell, W. (1975). *Polarized light – Benchmark papers in optics/1*. Stroudsburg, PA: Dowden, Hutchinson & Ross.

Tanaka, K., & Aoki, S. (2006). A method for the real-time construction of a full parallax light field. *Proceedings of the Society for Photo-Instrumentation Engineers, 6055*.

Tay, S. (2008). An updatable holographic three-dimensional display. *Nature, 451*, 694–698. doi:10.1038/nature06596 PMID:18256667

Tetsutani, N., Ichinose, S., & Ishibashi, M. (1989). 3D-TV projection display system with head tracking. *Japan Display, 89*, 56–59.

Tetsutani, N., Omura, K., & Kishino, F. (1994). A study on a stereoscopic display system employing eye-position tracking for multi-viewers. *Proceedings of the Society for Photo-Instrumentation Engineers, 2177*, 135–142. doi:10.1117/12.173868

Themelis, G. A. (1999). *How to use and maintain your stereo realist*. Cleveland, OH: Author's Publication.

Tilton, H. B. (1987). *The 3-D oscilloscope - A practical manual and guide.* Upper Saddle River, NJ: Prentice Hall Inc.

Traub, A. C. (1967). Stereoscopic display using rapid varifocal Mmrror oscillations. *Applied Optics, 6*(6), 1085–1087. doi:10.1364/AO.6.001085 PMID:20062129

Travis, A., Emerton, N., Large, T., Bathiche, S., & Rihn, B. (2010). Backlight for view-sequential autostereo 3D. *SID 2010 Digest,* 215–217.

Trayner, D., & Orr, E. (1996). Autostereoscopic display using holographic optical elements. *Proceedings of the Society for Photo-Instrumentation Engineers, 2653,* 65–74. doi:10.1117/12.237459

Treisman, A. M., & Gelade, G. (1980). A feature-integration theory of attention. *Cognitive Psychology, 12*(1), 97–136. doi:10.1016/0010-0285(80)90005-5 PMID:7351125

Tsai, C. H., Lee, K., Hseuh, W. J., & Lee, C. K. (2001). Flat panel autostereoscopic display. *Proceedings of the Society for Photo-Instrumentation Engineers, 4297,* 165–174. doi:10.1117/12.430815

Tupitsyn, M. (1996). *The Soviet photograph, 1924-1937.* New Haven, CT: Yale University Press.

Twist, S. H. (1912, August 17). The picture of the future: A reverie. *New York Clipper,* p. 7.

un She, S. S. & Wang, Q. (2005). Optimal design of a chromatic quarter-wave plate using twisted nematic crystal cells. *Optical and Quantum Electronics, 37*(7), 625-34.

Valyus, N. A. (1966). *Stereoscopy.* London: Focal Press.

Van Ekeren, J. (2011). *Ederen3d.* Retrieved Nov 2012 from www.ekeren3d.com

Viola, P., & Jones, M. (2001). Rapid object detection using a boosted cascade of simple features. In *Proceedings of IEEE Conference on Computer Vision and Pattern Recognition.* IEEE.

Warhol, A. (1977). *The philosophy of Andy Warhol: From A to B and back again.* Toronto, Canada: Harvest.

Wheatstone, C. (1838). Contributions to the physiology of vision: Part 1: On some remarkable, and hitherto unobserved, phenomena of binocular vision. *Phil. Trans. Royal Society, 1,* 371–394. doi:10.1098/rstl.1838.0019

Woodgate, G. J., Ezra, D., Harrold, J., Holliman, N. S., Jones, G. R., & Moseley, R. R. (1997). Observer tracking autostereoscopic 3D display systems. *Proceedings of the Society for Photo-Instrumentation Engineers, 3012,* 187–198. doi:10.1117/12.274457

Wu, P., Dunn, D. S., Smithson, R. L., & Rhyner, S. J. (2009). Development of integral images. In *Proceedings of the Frontiers in Optics (FiO), 3-D Capturing, Visualization and Displays.* Stockholm, Sweden: Rolling Optics AB.

Zebra Imaging. (2008). Retrieved 24 January 2012 from http:// www.zebraimaging.com

Zone, R. (2007). *Stereoscopic cinema: The origins of 3-D films, 1838–1952.* Lexington, KY: Univ. of Kentucky. doi:10.5810/kentucky/9780813124612.003.0010

Zworykin, V. K. (1938). *Television system* (US Patent no 2,107,464). Washington, DC: US Patent Office.

About the Contributors

Martin Richardson gained the world's first PhD in Display Holograms from The Royal College of Art in London. In 1999, he was awarded the Millennium Fellowship by the UK Millennium Government Commission and in 2009 became an Associate Member to the Royal Photographic Society when he was awarded the RPS SAXBY Award for his contributions to 3-D imaging. He is currently Professor of Modern Holography at De Montfort University, where he leads the Imaging and Displays Research Group in the faculty of Technology conducting innovative research into three-dimensional imaging and holographic wavefront applications for industry.

* * *

Hans I. Bjelkhagen, Hansholo Consulting Ltd. and Professor Emeritus of Interferential Imaging Sciences, Glyndŵr University, Centre for Modern Optics (CMO), located in North Wales, UK, was awarded his Doctoral Degree in 1978 by the Royal Institute of Technology in Stockholm, Sweden. Over the last 15 years, Bjelkhagen has received much international recognition for his work in the field of colour holography and holographic recording materials. He has specialised in recording Denisyuk-type colour holograms. He has researched and improved Lippmann photography over a period of many years. In 1983, Bjelkhagen joined CERN in Geneva, Switzerland, where he was involved in the development of bubble chamber holography. A year later, he participated in an international team project, recording holograms in the 15-foot bubble chamber at Fermilab in Batavia, IL, USA. Between 1985 and 1991, he was employed at Northwestern University, Illinois, USA, working on medical applications of holography. In 1997, Bjelkhagen was invited by Professor Nick Phillips to join him at CMO at De Montfort University, Leicester, England. In 2004, CMO moved to the then newly established OpTIC in Wales. In addition to scientific applications, Bjelkhagen is a well-known holographer who has recorded many holograms for 3D display purposes. From his early years in the field, he has been involved in large-format, high-quality display holography, using both pulsed and cw lasers. He has recorded many unique art objects, such as, for example, the Swedish "Coronation Crown of Erik XIV" (the crown dates back to 1561), or the Chinese "Flying Horse from Kansu" (from 100 A.D.). Bjelkhagen has worked with a number of famous artists, for example, Carl Fredrik Reuterswärd, creating holograms exhibited in many art museums and galleries around the world. Bjelkhagen has also used pulsed holography to record a number of holographic portraits. In 1989, he recorded a portrait of the inventor of single-beam reflection holography, Yuri Denisyuk. The most famous person recorded by Bjelkhagen was President Ronald Reagan. His portrait was recorded on 24 May 1991. This was the first and, so far, the only holographic portrait recorded of an American President. A copy of this holographic portrait is held in The National

Portrait Gallery of the Smithsonian Institution in Washington, DC. Bjelkhagen has published over 100 papers in refereed journals and conference proceedings, and holds 14 international patents. His most important academic contributions are a book on Silver-Halide Recording Materials for Holography and Their Processing published by Springer and Ultra-Realistic Imaging, Advanced Techniques in Analogue and Digital Colour Holography by CRC Press. He is a member of the Optical Society of America (OSA) and a fellow of the International Society for Optical Engineering (SPIE). He is the Chairman of SPIE's Photonics West Practical Holography Conference and the SPIE's Holography Technical Group. He is an Accredited Senior Imaging Scientist and Fellow of The Royal Photographic Society (RPS) as well as Chairman of the RPS 3D Imaging and Holography Group. In 2001, he received the RPS Saxby Award for his work in holography and in 2011 the Denisyuk Medal, from the D.S. Rozhdestvensky Optical Society, Russia.

John Emmett joined the Engineering Department of Thames Television in London after gaining a Ph.D. at Durham University, and starting the UK audio equipment manufacturers EMO Systems Ltd. Whilst at Thames Television, he worked on subjects as diverse as film archive formats and psychoacoustics and along the way gained six international patents as sole inventor, and jointly with Lee Lighting, a Technical Oscar for developing the flicker-free Lighting Ballast. He is currently Technical director and Chief Executive of Broadcast Project Research, an independent studio-based research group.

Frank C. Fan was born in 1964. He got his Bachelor Degree of Engineering in 1985 from University of Electronic Science and Technology in Chengdu, China. Since 1986, he has engaged himself in Display Holography and Information Optics; he got his Master degree of Engineering in 1988 from Beijing University of Technology in Beijing, China, and Doctor Degree of Science in 1991 from Sichuan University in Chengdu, China. He set up AFC Technology Co., Ltd. in 1999 in Shenzhen to produce holograms for Packaging and Security. He sponsored the 8th International Symposium on Display Holography in Shenzhen. He also invented the "real-time holographic 3D display."

Sean F. Johnston, Professor of Science, Technology, and Society at the University of Glasgow, obtained degrees in Physics from Simon Fraser University in Canada and his PhD in History and Philosophy of Science from the University of Leeds, UK. His career began as a physicist working at national laboratories (Canada's principal particle accelerator facility, its national optics laboratory, and national satellite image analysis centre) and as an R&D manager in Canadian and British engineering firms. One of his instruments was designed for the Space Shuttle, another comprised a measurement network for year-round observations of the aurora borealis in the Canadian arctic, and yet others measured greenhouse gases. These varied working environments, clients, and sponsors provided experience directly relevant to his subsequent research in the history and sociology of science and technology. His research interests focus on how new technical professions and scientific knowledge develop hand in hand. An ongoing research interest has been the historical sociology of holography, and his most recent work has focused on the emergence of nuclear specialists, the cultural history of holography, and twentieth-century amateur science.

Geoff Ogram graduated in Industrial Metallurgy at the University of Birmingham. After gaining his PhD, he stayed on for a short time as a research fellow, but later joined the GKN Research Laboratory in Wolverhampton in 1963, working on the thermo-mechanical treatment of steel. A further change took him into teaching physical metallurgy for 30 years at Wednesbury Technical College, which later became Sandwell College. Following early retirement in 1995, his long interest in stereoscopy led to his writing and self-publishing in 2001 a book on 3-D photography, which was favourably received both in the UK and abroad. He joined the Stereoscopic Society in 1997, contributing to its activities with enthusiasm, which led to his spending three years as Chairman followed by two years as Honorary President.

Graham Saxby obtained his first degree from London University in combined Maths/Science in 1945. He subsequently had a career as a photographer in the RAF, culminating in seven years as OC Photographic Science at the RAF School of Photography. On his retirement from the RAF in 1974, he joined what is now Wolverhampton University as a senior lecturer, where he taught modern optics and conducted research into practical display holography. He has been the recipient of a number of prestigious awards for his teaching and writing, including Fellowship of the Institute of Physics and Honorary Fellowship of the Royal Photographic Society. He retired formally in 1993 and now lives in Wolverhampton.

Phil Surman is a Senior Visiting Research Fellow in the Imaging and Displays Research Group at De Montfort University in the UK. He received his PhD in the subject "Head Tracking Two-Image 3D Television Displays" from De Montfort University in 2003. He has carried out research into autostereoscopic television displays for many years and holds the patents on the multi-user autostereoscopic television displays developed in the European Union-funded ATTEST, MUTED, and HELIUM3D projects. The two displays are both head tracked where the tracker controls an active LCD backlight that forms multiple exit pupil pairs that follow the viewers' eye positions.

Roger Taylor has had photography at the heart of his professional career for over 50 years, either as a practitioner, educator, or curator. With numerous publications, monographs, and exhibition catalogues to his name, he is widely regarded as one of Britain's foremost photographic historians.

Index